Disaster Management and Environmental Sustainability

Scrivener Publishing
100 Cummings Center, Suite 541J
Beverly, MA 01915-6106

Publishers at Scrivener
Martin Scrivener (martin@scrivenerpublishing.com)
Phillip Carmical (pcarmical@scrivenerpublishing.com)

Disaster Management and Environmental Sustainability

Edited by

Sanjay Kumar

Suraj Kumar Singh

Shruti Kanga

Gowhar Meraj

Majid Farooq

and

M. S. Nathawat

Scrivener Publishing

WILEY

This edition first published 2024 by John Wiley & Sons, Inc., 111 River Street, Hoboken, NJ 07030, USA and Scrivener Publishing LLC, 100 Cummings Center, Suite 541J, Beverly, MA 01915, USA

For more information about Scrivener publications please visit www.scrivenerpublishing.com.

Wiley Global Headquarters
111 River Street, Hoboken, NJ 07030, USA

For details of our global editorial offices, customer services, and more information about Wiley products visit us at www.wiley.com.

Library of Congress Cataloging-in-Publication Data

ISBN 978-1-394-16635-0

Front cover images supplied by Pixabay.com and Adobe Firefly
Cover design by Russell Richardson

Set in size of 11pt and Minion Pro by Manila Typesetting Company, Makati, Philippines

Printed in the USA

10 9 8 7 6 5 4 3 2 1

Contents

Editorial

In an era marked by environmental crises and frequent natural disasters, the urgency to understand and address these issues is palpable. The book *Disaster Management and Environmental Sustainability* enters this crucial dialogue, bringing to light a plethora of interconnected themes, insights, and solutions for tackling these complex challenges, with a specific focus on the diverse context of India.

The authors embark on a meticulous exploration of the key themes surrounding disaster management and environmental sustainability. The book effectively maps the terrain of these topics through fifteen comprehensive chapters. The journey starts with an examination of the geographically vulnerable areas of Madhya Pradesh, laying the groundwork for understanding the significant impact of disasters on communities.

"Disaster Management and Environmental Sustainability" does not shy away from delving into controversial topics. It explores the paradox of development, and its double-edged impact on environmental degradation and human progress. The authors weave together an intricate narrative, exploring the multifaceted nature of sustainability, including the impacts on safety, law enforcement, agriculture, and economic development. It is a testament to the book's strength that it is able to navigate these interwoven themes, providing valuable insights for both the uninitiated and the experienced reader.

As we progress into a digital era, the book also navigates the role of modern technologies such as geographic information systems (GIS) and remote sensing in managing disasters and their potential influence on environmental sustainability. In today's interconnected world, the examination of these digital tools feels both relevant and timely.

However, it is not just about the 'what' but also the 'how'. The book's approach to these complex subjects is evidence-based, rigorous, and comprehensive. Each chapter offers a deep dive into the topic at hand, underpinned by meticulous research and objective analysis. This fact-based approach helps the reader to navigate the often confusing landscape of environmental sustainability and disaster management, providing clarity amidst the complexity.

This book shall serve both a guide and an inspiration for its readers. It calls on us all to become agents of change, making informed decisions for the benefit of our communities and future generations. It is a call that resonates, reminding us of the interconnected nature of our world and our shared responsibility towards it.

From understanding the dangers of illegal sand mining to exploring the potential of solar power as a sustainable energy alternative, from delineating the historical transformation of Berhampur to the detrimental effects of human activities on the environment, and from the geographical aspects of urban crime to the significance of agricultural development, this book covers a broad spectrum. It also discusses the implications of the recent pandemic on migrant workers, a reality that still resonates across the globe.

"Disaster Management and Environmental Sustainability" is not just a book; it is a blueprint for understanding our relationship with the environment and how we can navigate the challenges that arise from this relationship. It invites us to not only understand the present but to take action for a sustainable future. This book is a valuable resource for anyone interested in environmental sustainability and disaster management, and it is an essential read for those who dare to make a difference.

Editors
Sanjay Kumar
Suraj Kumar Singh
Shruti Kanga
Gowhar Meraj
Majid Farooq
M. S. Nathawat

Preface

In this critical juncture of human history, the importance of an understanding and commitment to environmental sustainability and disaster management cannot be overemphasized. This book, titled *Disaster Management and Environmental Sustainability*, is a significant endeavor towards developing a more in-depth understanding of these complex issues, focusing primarily on the diverse context of India. The chapters herein are designed to serve as a resource for academics, practitioners, policymakers, students, and anyone interested in grasping the intricacies of our natural world and its governance. The book's fifteen chapters explore a range of related themes, each contributing to a holistic understanding of the environmental and disaster management landscape. The earlier chapters provide a foundation in understanding the geography of disaster-prone areas in Bihar and the subsequent impact on its citizens. We delve into the role of urbanization in shaping our environment and analyze both its boons and banes. The book brings to light the paradoxical relationship between development and environmental conservation. While development is seen as a symbol of human progress, it can also lead to environmental degradation, underlining the need for sustainable practices. We also underscore the need for legal measures and regulations to ensure environmental preservation, detailing the evolution of environmental laws in India. In subsequent chapters, the narrative shifts towards the issues of urban crime, agricultural growth, and forest resources, demonstrating the multifaceted nature of environmental sustainability. These chapters show how every aspect of our society is intertwined with the environment, from safety and law enforcement to agriculture and economic development. The later chapters incorporate the use of modern technologies such as geographic information systems (GIS) and remote sensing in managing disasters like urban flooding. We also examine the role of the internet in our lives, a pertinent issue given the digital revolution and the post-pandemic scenario. This chapter will help readers understand how these technological advancements can both aid and pose challenges to environmental sustainability. Each chapter has been meticulously researched and is built on evidence-based facts, aiming to enlighten and engage readers in these crucial topics. It is my hope that this book will not only serve as a comprehensive guide to disaster management and environmental sustainability but also inspire readers to become proactive agents of change in their communities. This book is a testament to the belief that despite the complexity and magnitude of the challenges we face, collective understanding, sustained effort, and persistent innovation can indeed foster a harmonious coexistence between humanity and the environment. It is a call to all of us to tread the path of sustainability, making informed decisions that are beneficial not just for us, but for generations to come.

Chapter 1 delves into the alarming consequences of illegal sand mining on river ecosystems, with an in-depth focus on the Chambal River. The researchers leverage advanced geospatial techniques to map the changes in river morphology due to unlawful extraction activities. The findings illuminate the pressing need for sustainable sand mining practices to safeguard our rivers and their delicate ecosystems. **Chapter 2** shifts our attention to the energy sector, focusing on the enormous potential of solar power as a green energy alternative. The research identifies suitable areas for solar power plant installation in four districts of Himachal Pradesh, India, with the help of geospatial technology. It spotlights the role of solar energy in relieving the increasing pressure on exhaustible resources, like coal and petroleum, imposed by population growth and industrial expansion. **Chapter 3** is the third chapter in our journey. It deals with the essential role of potable water in households and takes a close look at the consumption patterns in Idah LGA, Kogi State. It presents an analysis of water sources, household water demand, and influential factors, underpinned by a scientific method involving questionnaires, interviews, and observations. The findings provide a basis for effective water supply planning.

Chapter 4 presents a broad overview of disasters, emphasizing their omnipresence and destructive nature. It underscores the necessity of comprehensive knowledge about disasters to manage their risks and impacts effectively. The chapter elucidates the concept of disasters, frequency of occurrences, and how to build resilience against these calamities, drawing from the author's vast expertise in the field. **Chapter 5** focuses on the detrimental effects of human activities on the environment. Detailing various types of pollution, it underlines their consequences on human health. Stressing the need for environmental education, the chapter calls for simple techniques to measure pollution and appropriate remedies to mitigate their effects. **Chapter 6** discusses the interplay between environmental degradation and human progress. Using an allegory from mythology, it stresses the necessity of balancing development and environmental preservation. The chapter emphasizes the urgency to address environmental issues while ensuring economic growth and poverty eradication.

Chapter 7 addresses the effects of the pandemic and the ensuing lockdown on migrants in India. Using Rohtak, a significant migrant destination, as a case study, it provides insights into the experiences of migrant laborers during these trying times. This chapter paints a vivid picture of the human toll of the pandemic, extending beyond health issues to include broader socio-economic implications. **Chapter 8** delves into the geography of crime, a social issue whose variable nature provides geographers with a dynamic field to study. Despite its relevance, this discipline has struggled to capture the attention of mainstream geographers. Taking the city of Patna as a case study, we explore how urban transformations and economic, political, and social changes influence the city's crime rates, with a specific focus on crimes against women. The intricate and layered nature of urban crime is discussed, alongside the impacts of fear perception and gender differences in such contexts. **Chapter 9** discusses the significance of cropping intensity as an indicator of agricultural development. The potential of multiple cropping as a solution to maximize agricultural production and augment farmer incomes in India is examined, taking into account irrigation impacts on cropping intensity. Analyzing secondary data from 1991–92 to 2017–18 in Haryana, we delve into how higher degrees of cropping intensity can signify a higher level of agricultural development. **Chapter 10** underlines the importance of forests as crucial ecological resources, detailing the direct and indirect uses they serve. With Bihar's forest cover at a mere 7.75%, we discuss the reasons behind forest cover disappearance and the prospects of forest resources in the state. Measures to improve and sustain forest cover are suggested, all based on an empirical study conducted via secondary data, literature surveys, and self-observation.

Chapter 11 emphasizes the profound challenges of environmental degradation in the modern century, a result of population growth and the overexploitation of natural resources. It introduces sustainable development as a crucial solution, stressing the need for a rational and equitable balance between economic growth and environmental conservation for the sake of future generations. **Chapter 12** provides a historical overview of environmental laws in India, beginning with the single law of the colonial era—the Indian Forest Act (1927)—and tracing the evolution to the current comprehensive suite of laws and regulatory bodies that protect and improve the environment. **Chapter 13** explores how remote sensing and GIS have revolutionized the approach to flood management, mapping flood risks and defining flood zones. The chapter evaluates various techniques, methodologies, and models used globally for urban flood management, underscoring the role of technology in keeping hazards under control. **Chapter 14** dissects the unique challenges of urbanization in Bihar, where a rapidly growing urban population grapples with environmental degradation and social disparities. The focus lies on the cities of Patna, Gaya, Bhagalpur, Muzaffarpur, and Darbhanga, as we explore how unchecked migration, illegal settlements, socio-cultural disparities, and infrastructure shortcomings contribute to environmental deterioration and threaten the quality of life.

Chapter 15 presents an in-depth analysis of the Network—the Internet—and its profound effects on our social life. As the world increasingly shifts online, we examine how this digital realm shapes our daily interactions, from shopping to learning, and the ramifications of this transformation on our societal structures, especially in the wake of the COVID-19 pandemic. **Chapter 16** unveils the grave reality of natural and man-made disasters affecting both developed and developing nations. It emphasizes the dire need for tools and methodologies to mitigate the disastrous impacts on human life, economy, infrastructure, and the environment. Highlighting the importance of technological advancements, the chapter delves into the role of remote sensing and GIS in disaster management, drawing from comprehensive reviews of existing research. In **Chapter 17,** the reader is taken through the historical transformation of Berhampur from an army cantonment chosen by the British to a bustling town. It underlines the interplay between politics, trade, and demography, leading to increased demand for quality water for various uses. The chapter provides insights into the complex dynamics of urban growth and the consequential strain on resources.

In **Chapter 18**, the researchers apply the genetic algorithm (GA) technique to optimize reservoir operation rule curves. The application was implemented in the KLRS Pulichintala project in India, resulting in operational insights and effectiveness projections under varying climate scenarios. **Chapter 19** examines the essential role of rivers in human well-being and the threats they face from sub-optimal investment decisions. It presents a conceptual framework explaining the connections between rivers and their societal, economic, and strategic benefits, emphasizing the importance of sustainable river management. Lastly, **Chapter 20** explores the issue of water shortage, focusing on groundwater availability for agriculture in the Pisangan Watershed. The researchers combine remote sensing data and GIS to identify potential groundwater zones, underscoring the necessity of modern techniques in water resource management.

Acknowledgments

The completion of this book, *Disaster Management and Environmental Sustainability*, is the culmination of numerous contributions from various individuals and institutions, without whom this monumental task could not have been realized. We would first and foremost like to extend our deepest appreciation to the reviewers who invested their time and expertise to refine the quality of our work. Their feedback and constructive criticism played an indispensable role in sharpening our arguments and illuminating the nuances of the complex topics we tackled. Their meticulous reviews not only ensured the accuracy of the presented information but also significantly enhanced its readability and accessibility. We also owe a significant debt of gratitude to our families, who provided us with the much-needed support, both emotionally and practically, throughout this demanding journey. Their patience, understanding, and unwavering belief in our efforts provided the constant encouragement that propelled us through the challenges and late nights. Their role in the creation of this book is as crucial as ours.

Our institutional heads deserve a special note of thanks. Their leadership, guidance, and vision were critical in providing the conducive environment necessary for undertaking a project of such a vast scale. They supported us unconditionally, lending us resources and fostering an atmosphere of intellectual curiosity that was critical for our work.

To our graduate students, we express our heartfelt gratitude. Their tireless efforts, spirited discussions, and intellectual vigor greatly enriched our research. Their fresh perspectives often led us to reconsider established notions and explore new directions. Their enthusiasm and dedication served as a constant reminder of why we embarked on this journey in the first place.

We would also like to extend our thanks to the support staff at our institutions. Their administrative, technical, and logistical assistance was invaluable in facilitating our research and ensuring our focus remained undivided on the intellectual tasks at hand. Lastly, we want to express our gratitude to our readers for embarking on this journey with us. We sincerely hope this book brings value to your understanding of environmental sustainability and disaster management.

Thank you, everyone, for making "Disaster Management and Environmental Sustainability" a reality. The journey has been as rewarding as the outcome, and we are glad to have undertaken it with all of you by our side.

Monday, 26 June 2023

Sanjay Kumar
Suraj Kumar Singh
Shruti Kanga
Gowhar Meraj
Majid Farooq
M. S. Nathawat
Editors

1

Assessment of Changes in River Morphology Due to Illegal Sand Mining by Geospatial Techniques

Saurabh Singh[1]*, Deepak Kumar Prajapat[2], Vikas Pandey[2], Rajat Agrawal[1] and Jerripotu Gopala Rao[1]

[1]*Centre for Climate Change and Water Research, Suresh Gyan Vihar University, Rajasthan, Jaipur, India*
[2]*Department of Civil Engineering, Suresh Gyan Vihar University, Rajasthan, Jaipur, India*

Abstract
Illegal sand mining poses a significant threat to river ecosystems, causing detrimental impacts on river morphology and ecological balance. This chapter presents a comprehensive assessment of the changes in river morphology resulting from illegal sand mining on the Chambal River using advanced geospatial techniques. The study focused on the Aandva Purani site, heavily affected by illegal sand mining, and employed the end point rate (EPR) methodology to evaluate erosion and accretion processes. The analysis involved the examination of satellite data from Landsat 5 OLI (1990), Landsat 5 OLI (2000), Landsat 7 ETM+ (2010), and Landsat 8 OLI (2020) to track changes over time. Geometric correction and digitization techniques were applied to accurately delineate the river shoreline and identify illegal sand mining sites. The results revealed a significant proportion of erosional transects (52.17%) and accretional transects (47.83%) within the study area. The average erosion rate was calculated as –0.16 m/y, indicating the adverse effects of sand extraction on the river's morphology. Conversely, the average accretion rate was estimated as 0.19 m/y, highlighting the deposition of sediments resulting from altered flow patterns caused by illegal sand mining activities. Visual representations of the changes in river morphology between different time intervals were provided, emphasizing the temporal dynamics and trends observed. The findings underscore the urgency of implementing sustainable sand mining practices and regulations to preserve the Chambal River's integrity and protect its delicate ecosystems. This research serves as a valuable resource for policymakers, environmental agencies, and local communities in formulating strategies and policies to mitigate the adverse impacts of illegal sand mining on river systems and promote sustainable management practices.

Keywords: Illegal sand mining, geospatial techniques, end point rate (EPR), erosion and accretion, sustainable sand mining practices, river morphology, average erosion rate, average accretion rate

**Corresponding author*: saurabh.singh@mygyanvihar.com

Sanjay Kumar, Suraj Kumar Singh, Shruti Kanga, Gowhar Meraj, Majid Farooq and M. S. Nathawat (eds.)
Disaster Management and Environmental Sustainability, (1–12)

1.1 Introduction

According to the sand mining framework, India has an annual sand requirement of 70 million tons, predominantly driven by the booming construction industry. With each passing year, the demand for sand continues to rise steadily at a significant growth rate of 7% (Singh *et al.*, 2022a; Singh *et al.*, 2023b). This surge in demand can be attributed to the escalating construction activities across the country. As the construction sector expands, sand mining plays a vital role in meeting the soaring demand for this essential resource. Sand and gravel are crucial components used in various construction processes, including the production of concrete, mortar, and other building materials. The reliance on sand for construction purposes has transformed it into a highly sought-after commodity. Due to the tremendous demand–supply gap, sand has become a valuable resource, both economically and environmentally. The scarcity of quality sand reserves, coupled with the need for sustainable extraction Practices, has further contributed to the increasing value of sand. The escalating prices and the competition for limited sand resources have led to intensified efforts to regulate and manage the sand mining industry (Upadhyay *et al.*, 2023; Debnath *et al.*, 2022a). Efforts are being made to promote alternative materials and construction techniques to reduce the dependence on sand (Singh *et al.*, 2023b; Debnath *et al.*, 2022a). Additionally, governments and regulatory bodies are implementing stricter guidelines and policies to ensure responsible sand mining practices, safeguarding the environment, and mitigating the adverse impacts associated with excessive extraction. As India continues its rapid urbanization and infrastructure development, the demand for sand is expected to persistently rise. Balancing this growing demand with sustainable extraction practices and exploring alternatives will be crucial in ensuring the availability of this essential resource for construction activities while minimizing environmental consequences. Sand mining has adverse impacts on the lives and livelihoods of the people, especially those staying in close proximity to the river banks (Joy *et al.*, 2020; Debnath *et al.*, 2023b). One of the common problems faced due to sand extraction is river bank erosion. During heavy rains, water enters into the fields and other areas, destroying the crops. It also pollutes the local drinking water resources in the village. Sand mining has affected the ground water tables significantly, which has caused shortage of drinking water (Singh *et al.*, 2023; Debnath *et al.*, 2023c). When the rate of extraction of sand, gravel, and other materials surpasses the natural replenishment rate, it leads to a range of environmental issues. Sand mining, in particular, has been known to have detrimental effects on ecosystems, disrupting the natural cycle of sedimentation and impacting various habitats. The extraction of sand and gravel often involves the removal of topsoil, vegetation, and the alteration of landscapes (Farooq *et al.*, 2022b). This process not only leads to the loss of biodiversity but also disrupts the intricate balance within ecosystems. The morphological changes caused by mining activities can have long-lasting consequences on the surrounding environment. One of the primary impacts of sand mining is the alteration of riverbeds and coastal areas. Excessive extraction of sand from river channels and coastal regions can disrupt the natural flow of water, affecting sediment transport and leading to erosion and flooding in some cases. These changes in hydrological patterns can adversely impact aquatic ecosystems and the organisms that rely on them. Furthermore, sand mining can result in the destruction of habitats for various plant and animal species. Wetlands, dunes, and riverbanks often harbor unique ecosystems that are highly sensitive to disturbances. The removal of sand disrupts these habitats, displacing organisms and reducing biodiversity. Additionally, the excavation and transportation of sand can cause air and noise pollution. Dust emissions during the mining process can have adverse effects on air quality, potentially leading to respiratory problems for both humans and wildlife. The operation of heavy machinery and transportation vehicles can generate noise pollution, further disturbing local ecosystems and wildlife. To address these environmental concerns, it is crucial to implement sustainable sand mining practices (Farooq *et al.*, 2022c). This includes the implementation of regulations and monitoring mechanisms to ensure responsible extraction, restoration of mined areas, and the promotion of alternatives such as recycled materials or artificial substitutes for sand in certain applications. By striking a balance between the demand for sand and the preservation of ecosystems, it is possible to mitigate the negative impacts of sand mining. Collaboration among stakeholders, including government bodies, mining companies, environmental organizations, and local communities, is essential to develop and enforce sustainable practices that protect the environment while meeting the needs of construction and development activities (Rafiq *et al.*, 2022).

A case has been filed against illegal sand mining in the Chambal National Sanctuary (Figure 1.1) to address violations, hold responsible parties accountable, and protect endangered species. Illegal sand mining disrupts the river's morphology and threatens the sanctuary's delicate ecosystems.

Demand for sand is increasing non-linearly, particularly in Asia, and especially for concrete production. China used more sand for concrete production in 2011 and 2013 than what the United States used during the entire 20th century. The scale of the problem can be better assessed if one considers the fact that 1 ton of cement can require up

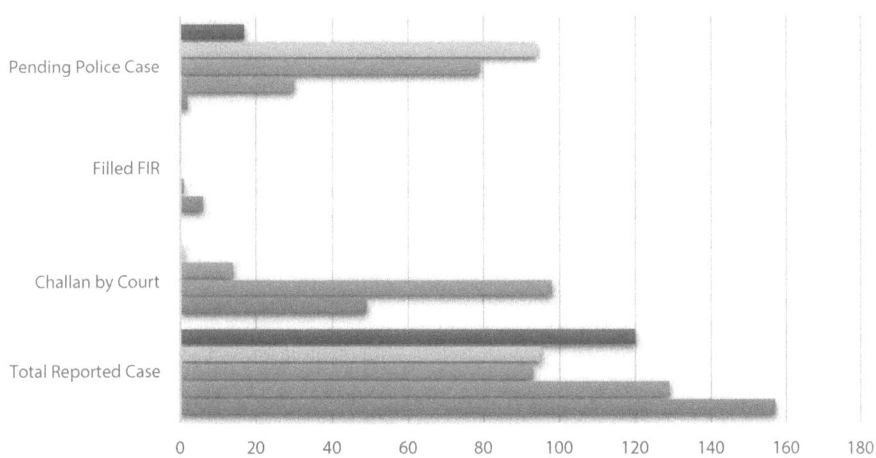

	Total Reported Case	Challan by Court	Filled FIR	Pending Police Case
2022-23	120	0	0	17
2021-22	95	1	0	94
2020-21	93	14	0	79
2019-20	129	98	1	30
2018-19	157	49	6	2

2022-23 2021-22 2020-21 2019-20 2018-19

Figure 1.1 Filed case against illegal sand mining in the Chambal National Sanctuary.

to 10 tons of sand to make concrete (Singh *et al.*, 2022b). In order to put an end to river and mining, The Ministry of Environment, Forest and Climate Change formulated the Sustainable Sand Management Guidelines 2016 (Singh *et al.*, 2023b), which focuses on the management of sand mining in the country. The study, The Monitoring of Riverbank Erosion and Shoreline Movement at Amu River Using Remote Sensing and GIS, found that the river movement was toward the Afghanistan side, *i.e.*, southward. The threat of losing Afghanistan land was estimated to be ca. 362.4 ha/y. The total eroded area was calculated to be 1984 ha, 1410 ha, and 1680 ha during the period of 2000–2005, 2005–2010, and 2010–2014, respectively. The shoreline was demarcated in each image and the area was categorized into three zones, corresponding to the erosion trend, *i.e.*, lower as moderate eroded zone, central as high eroded zone, and upper as lower eroded zone (Singh *et al.*, 2022). Figure 1.2 illustrates the direct and indirect impacts of the Chambal River, which have far-reaching consequences on both the natural environment and human communities. The Chambal River serves as a lifeline for the surrounding regions, providing water for irrigation, drinking, and other domestic purposes. Its fertile banks support agriculture and contribute to the livelihoods of local communities. However, the illegal sand mining activities taking place along the Chambal River have unleashed a series of detrimental effects. The extraction of sand disrupts the river's natural flow and alters its morphological characteristics. This leads to changes in sediment transport, erosion patterns, and overall river dynamics, negatively impacting the river's ecosystem.

The Chambal River, selected as the study area for assessing changes in river morphology due to illegal sand mining, is a significant river system located in central India. With a total length of approximately 960 km, it traverses through the states of Madhya Pradesh, Rajasthan, and Uttar Pradesh. As a tributary of the Yamuna River, it eventually joins the Ganges River, forming an essential part of the larger river network in the region.

Figure 1.2 Illustration of the direct and indirect impacts of illegal sand mining on the Chambal River.

1.2 Materials and Methods

The study incorporated satellite data from multiple time periods, including Landsat 5 OLI (1990), Landsat 5 OLI (2000), Landsat 7 ETM+ (2010), and Landsat 8 OLI (2020). These satellite images were used to assess the changes in the Chambal River over the course of several years. To begin the image processing, geometric correction was applied to the satellite images. This process involved rectifying any distortions or misalignments in the images to ensure accurate spatial representation. The next step involved the digitization of the baseline of the Chambal River using a topographic sheet from the survey of India. This provided a reference for the river's original position. Subsequently, the river shoreline was digitized using the satellite data from each respective year (1990, 2000, 2010, and 2020). This entailed delineating the water boundary along the river's edges in the satellite images. Figure 1.3 Study area map. (a) Map of India, illustrates the geographic context of the study area, (b) map showcases the state of Rajasthan in India, emphasizing the district divisions within the state, and (c) map focuses on the specific sites of illegal sand mining along the Chambal River.

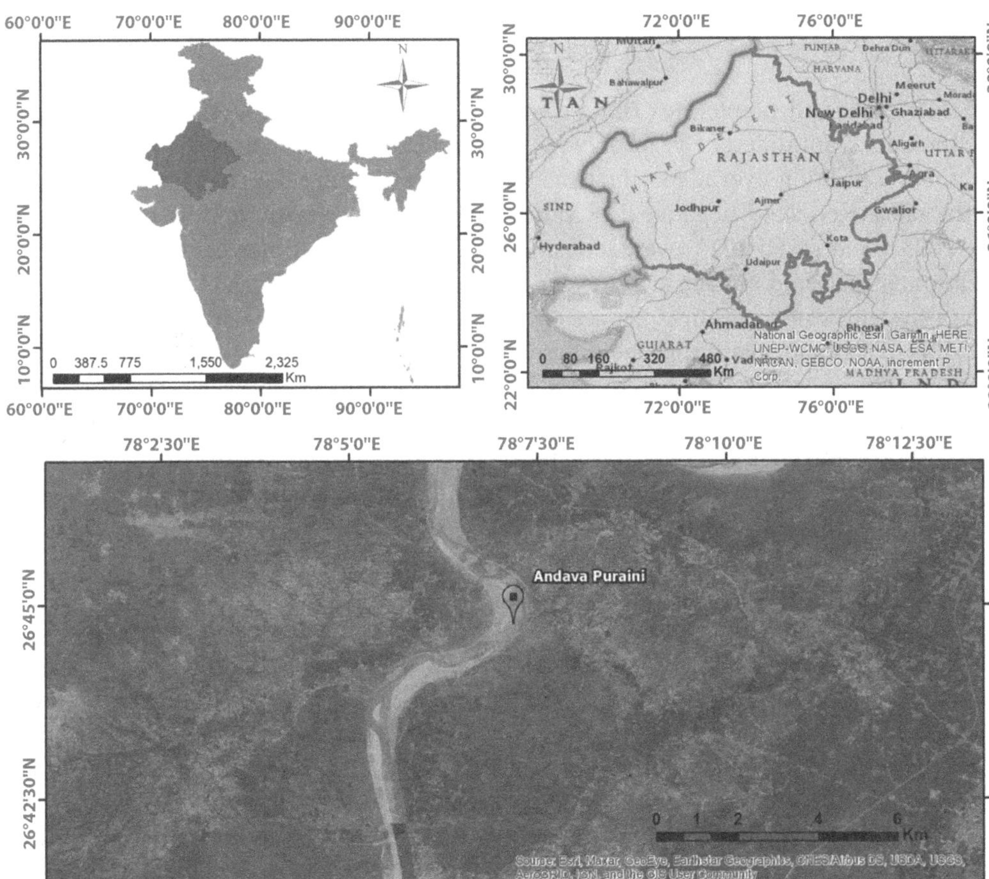

Figure 1.3 Study area map. (a) Map of India, illustrates the geographic context of the study area, (b) map showcases the state of Rajasthan in India, emphasizing the district divisions within the state, and (c) map focuses on the specific sites of illegal sand mining along the Chambal River.

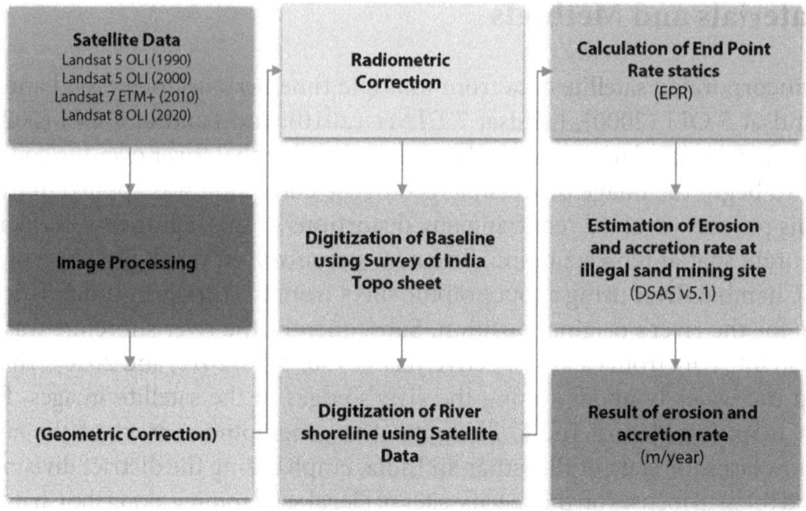

Figure 1.4 Methodology flow chart for assessing changes in river morphology due to illegal sand mining on the Chambal River using geospatial techniques.

To calculate the end point rate (EPR) statistics, a comparative analysis of the digitized shoreline data from different years was performed. The EPR measures the net change in the shoreline position over a specific time frame. By analyzing the changes in the shoreline, the rates of erosion (negative EPR values) and accretion (positive EPR values) (Singh *et al.*, 2023a; Sud *et al.*, 2023) were determined. The estimation of erosion and accretion rates at the illegal sand mining site was conducted using the DSAS v5.1 (Digital Shoreline Analysis System) software. This software enables quantitative analysis of shoreline changes. The digitized shoreline data from the different years were inputted into the software, which calculated the erosion and accretion rates specifically for the area affected by illegal sand mining. The results of the erosion and accretion rates were reported in meters per year (m/y), indicating the amount of erosion and accretion occurring over time.

1.3 Results and Discussion

By presenting the changes in river shoreline over different time intervals, Figure 1.5 allows for a comprehensive visual assessment of the Chambal River's erosion and accretion. The analysis focused on the Aandva Purani area and employed the EPR methodology to evaluate erosion and accretion processes. The results revealed a total of 23 transects analyzed, among these, 12 transects exhibited erosional characteristics, representing 52.17% of the total transects. The average erosion rate was calculated to be –0.16 m/y, indicating erosion on the illegal sand mining sites of Aandva Purani. Conversely, 11 transects showed signs of accretion, accounting for 47.83% of the transects, with an average accretion rate of 0.19 m/y, as show in in Table 1.1.

Figure 1.5 depicts the changes in river morphology over time, highlighting the alterations observed during different periods. Subfigure (a) specifically focuses on the changes that occurred between 1990 and 2000, providing insights into the river's morphology

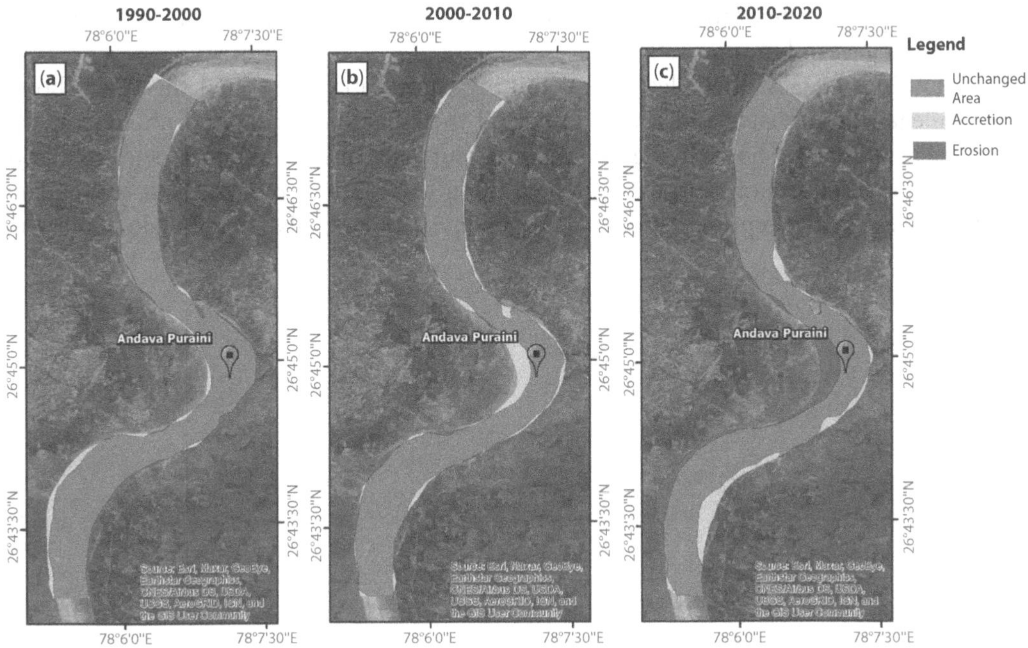

Figure 1.5 Changes in river morphology. (a) Showing the changes between 1990 and 2000, (b) showing the changes between 1990 and 2000, and (c) showing the changes between 2010 and 2020.

Table 1.1 End point rate of Aandva Purani.

Total number of transects 23
Number of erosional transects 12
Percent of all transects that are erosional 52.17%
Average of all erosional rates −0.16
Number of accretional transects 11
Percent of all transects that are accretional 47.83%
Average of all accretional rates 0.19

during that decade. Subfigure (b) presents the changes in river morphology between 2000 and 2010, shedding light on the transformations that took place during this period. By examining this subfigure, one can observe any shifts in channel alignment, changes in the river's shape, and alterations in sediment distribution. Comparing the river's morphology between these two time periods helps to identify any significant trends or patterns in river behavior and geomorphological changes. Lastly, subfigure (c) focuses on the changes in river morphology between 2010 and 2020, providing a more recent perspective on the river's evolution. This subfigure enables the examination of any recent developments, including changes in river course, erosion or deposition hotspots, and overall modifications in channel characteristics. Understanding the river's morphology during this period helps to assess the current state of the river and identify any ongoing trends or potential concerns regarding illegal sand mining activities.

The findings suggest that illegal sand mining significantly impacted the river shoreline at illegal mining sites and it also impacted the Chambal River's morphology at the Aandva Purani site, as shown in Figure 1.6. The occurrence of erosional transects highlights the detrimental effects of sand extraction, leading to sediment removal and subsequent erosion. These visual representations aid researchers, policymakers, and stakeholders in

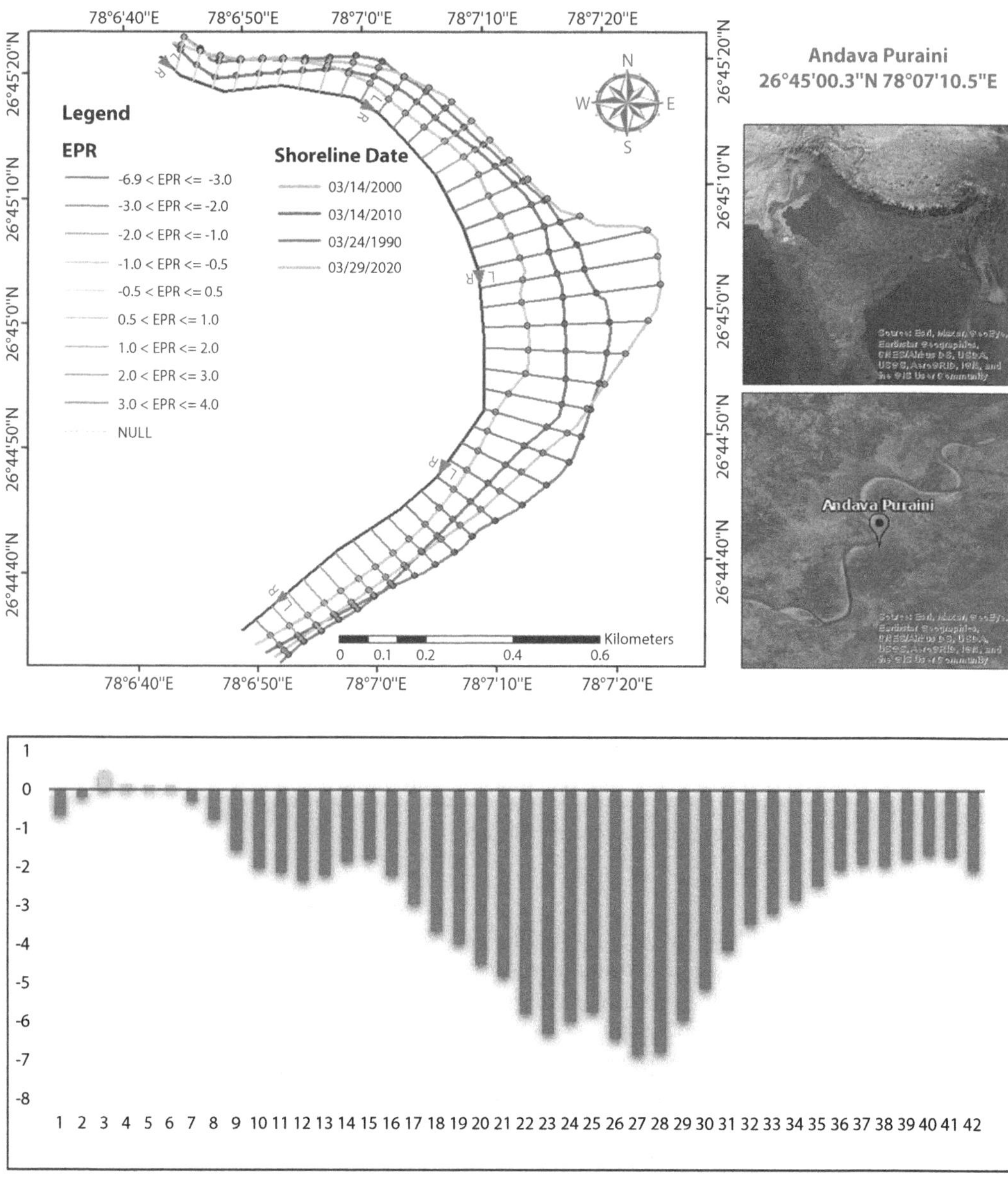

Figure 1.6 The erosion and accretion map by DSAS. (a) Showing end point rate of each transects and (b) showing rate of erosion and accretions of each transects in m/y.

understanding the temporal dynamics of the river, highlighting the need for sustainable management strategies to preserve the integrity and ecological balance of the Chambal River system. This phenomenon disrupts the river's natural equilibrium, resulting in channel instability and alterations to the riverbed. On the other hand, the presence of accretional transects indicates sediment deposition, potentially caused by altered river flow patterns resulting from illegal sand mining activities. Accretion can lead to the formation of sandbars, affecting the river's course and the overall dynamics of the ecosystem. The results emphasize the urgency of implementing effective measures to regulate and combat illegal sand mining within the Chambal River. It is crucial to safeguard the river's morphology and preserve its delicate ecosystem by enforcing sustainable sand mining practices. The findings of this study can serve as a valuable resource for policymakers, environmental agencies, and local communities in formulating appropriate strategies and policies to mitigate the adverse impacts of illegal sand mining on river systems.

1.4 Conclusion

In this study, we conducted a comprehensive assessment of the changes in river morphology resulting from illegal sand mining on the Chambal River using advanced geospatial techniques. Our findings provide compelling evidence of the direct and indirect impacts of this destructive activity on the river ecosystem. The analysis focused on the Aandva Purani site, which has been heavily affected by illegal sand mining. By employing the EPR methodology, we evaluated the erosion and accretion processes occurring in this area. The results revealed alarming trends, with a significant proportion of transects exhibiting erosional characteristics. Approximately 52.17% of the analyzed transects experienced erosion, indicating the detrimental effects of sand extraction on the river's morphology. Furthermore, our study identified 11 transects showing signs of accretion, representing 47.83% of the total transects. This suggests that sediment deposition is occurring, potentially due to altered flow patterns resulting from illegal sand mining activities. The average accretion rate of 0.19 m/y underscores the importance of understanding the dynamics of sedimentation in mitigating the impacts of sand mining.

This study emphasizes the severe impacts of illegal sand mining on the Chambal River shoreline. Erosion caused by sand extraction can disrupt the river's natural equilibrium, leading to channel instability and the destruction of the riverbed. Conversely, accretion can result in the formation of sandbars, altering the river's course and affecting the delicate balance of the ecosystem.

The implications of our findings are significant for policymakers, environmental agencies, and local communities. It is crucial to develop and enforce sustainable sand mining practices that prioritize the preservation of the Chambal River's morphology and ecological integrity. By implementing effective regulations and interventions, we can mitigate the adverse impacts of illegal sand mining and ensure the long-term sustainability of this invaluable river system.

Overall, our research highlights the urgent need to address the challenges posed by illegal sand mining on the Chambal River. By protecting the river's morphology and ecosystem, we can safeguard the diverse range of species, including endangered gharials, dolphins, turtles, and migratory birds, that rely on the river for their survival.

References

Singh, S., Singh, S. K., Kumar, R., Shrama, A., & Kanga, S. (2022). Finding Alternative to River Sand in Building Construction.

Singh, S., Meraj, G., Kumar, P., Singh, S. K., Kanga, S., & Johnson, B. A., (2023). Decoding Chambal River Shoreline Transformations: A Comprehensive Analysis Using Remote Sensing, GIS, and DSAS. *Water*, 15(9), 1793.

Joy, K. J., Neha Bhadbhade, Sarita Bhagat, Kiran Lohakare, and Abraham Rao, Nagmani Samuel. (2020). India River Week 2020 Is Sand Mining Killing Our Rivers ? Mining Riparian Health: *West Zone Regional Report*.

Singh, S., Singh, S. K., Prajapat, D. K., Pandey, V., Kanga, S., Kumar, P., & Meraj, G. (2023). Assessing the Impact of the 2004 Indian Ocean Tsunami on South Andaman's Coastal Shoreline: A Geospatial Analysis of Erosion and Accretion Patterns. *J. Mar. Sci. Eng.*, 11(6), 1134.

Rafiq, M., Meraj, G., Kesarkar, A. P., Farooq, M., Singh, S. K., Kanga, S., & Bhardwaj, V. (2022). Hazard Mitigation and Climate Change in the Himalayas–Policy and Decision Making. In *Disaster Management in the Complex Himalayan Terrains. Geography of the Physical Environment*, S. Kanga, G. Meraj, M. Farooq, S. K. Singh, & M. S. Nathawat (Eds.), Springer, Cham. https://doi.org/10.1007/978-3-030-89308-8_12.

Singh, A., Singh, S. K., Meraj, G., Kanga, S., Farooq, M., Kranjcic, N., Đurin, B., & Sudhanshu. (2022b). Designing Geographic Information System Based Property Tax Assessment in India. *Smart Cities*, 5, 364–381.

Singh, H., Meraj, G., Singh, S., Shrivastava, V., Sharma, V., Farooq, M., Kanga, S., & Kumar, P. (2022a). Status of Air Pollution during COVID-19-Induced Lockdown in Delhi, India. *Atmosphere*, 13(12), 2090.

Singh, S. K., Gupta, S. K., Meraj, G., Kumar, P., & Kanga, S. (2023b). Uncovering the hydrometeorological drivers responsible for forest fires utilizing geospatial techniques. *Theor. Appl. Climatol.*, https://doi.org/10.1007/s00704-023-04497-y

Singh, S. K., Meraj, G., Mondal, N., Bera, A., Verma, M. K., Tomar, J. S., & Kanga, S. (2022). Assessment of seasonal vegetation dynamics over parts of thar desert using geospatial techniques. *ANGRAU*, 105.

Singh, S., Singh, S. K., Prajapat, D. K., Pandey, V., Kanga, S., Kumar, P., & Meraj, G. (2023a). Assessing the Impact of the 2004 Indian Ocean Tsunami on South Andaman's Coastal Shoreline: A Geospatial Analysis of Erosion and Accretion Patterns. *J. Mar. Sci. Eng.*, 11(6), 1134.

Sud, A., Kanga, R., Singh, S. K., Meraj, G., Kanga, S., Kumar, P., Ramanathan, A. L., & Bhardwaj, V. (2023). Simulating Groundwater Potential Zones in Mountainous Indian Himalayas—A Case Study of Himachal Pradesh. *J. Geophys. Res.: Earth Surf.*, 10(3), 65.

Upadhyay, R. K., Tripathi, G., Đurin, B., Šamanović, S., Cetl, V., Kishore, N., Sharma, M., Singh, S. K., Kanga, S., Wasim, M., Rai, P. K., & Bhardwaj, V. (2023). Groundwater Potential Zone Mapping in the Ghaggar River Basin, North-West India, Using Integrated Remote Sensing and GIS Techniques. *Water*, 15(5), 961.

Debnath, J., Sahariah, D., Lahon, D., Nath, N., Chand, K., Meraj, G., Kumar, P., Singh, S. K., Kanga, S., & Singh, S. K. (2022a). Geospatial modeling to assess the past and future land use-land cover changes in the Brahmaputra Valley, NE India, for sustainable land resource management. *Environ. Sci. Pollut. Res.*, 1-24.

Debnath, J., Sahariah, D., Lahon, D., Nath, N., Chand, K., Meraj, G., Kumar, P., Singh, S. K., Kanga, S., & Farooq, M. (2023b). Assessing the impacts of current and future changes of the planforms of River Brahmaputra on its land use-land cover. *Geosci. Front.*, https://doi.org/10.1016/j.gsf.2023.101557.

Debnath, J., Sahariah, D., Saikia, A., Meraj, G., Nath, N., Lahon, D., Annayat, W., Kumar, P., Chand, K., Singh, S. K., & Kanga, S. (2023c). Shifting Sands: Assessing Bankline Shift Using an Automated Approach in the Jia Bharali River, India. *Land*, 12(3), 703.

Farooq, M., Meraj, G., Rishabh, Kanga, S., Nathawat, R., Singh, S. K., & Ranga, V. (2022a). Slum Categorization for Efficient Development Plan—A Case Study of Udhampur City, Jammu and Kashmir Using Remote Sensing and GIS. In *Geospatial Technology for Landscape and Environmental Management: Sustainable Assessment and Planning* (pp. 283-299). Cham: Springer International Publishing.

Farooq, M., Mushtaq, F., Meraj, G., Singh, S. K., Kanga, S., Gupta, A., Kumar, P., Singh, D., & Avtar, R. (2022b). Strategic Slum Upgrading and Redevelopment Action Plan for Jammu City. *Resources*, 11(12), 120.

Farooq, M., Rashid, H., Meraj, G., Kanga, S., & Singh, S. K. (2022c). Assessing the Microclimatic Environmental Indicators of Climate Change of a Temperate Valley in the Western Himalayan region. In *Climate Change, Disaster and Adaptations - Contextualizing human responses to ecological change*, A. R. Siddiqui and Avijit Sahay (Eds.), (pp. 47–61). https://link.springer.com/chapter/10.1007/978-3-030-91010-5_5

2

Feasibility of Solar Power Generation Potential in Una, Bilaspur, Solan, and Sirmaur Districts of Himachal Pradesh Using Geospatial Techniques

Lalit Jain, Krishan Chand*, Rohit Chauhan and Brijesh Saklani

Science, Technology & Environment, GoHP AGiSAC - HIMCOSTE, Kasumpti Shimla, India

Abstract
As the population is increasing day by day, industries are also increasing to meet the demands of the increased population, and obviously, the pressure on resources like coal and petroleum is increasing as well. Solar energy is a huge source of energy which can be an alternative and green source of energy. Potential areas for setting up solar power plants have been identified in this study with the help of geospatial technology. Typically, PV plant should be installed in high radiation areas, preferably flat land or slopes facing south with minimum undulations, and free from shading objects. According to expert knowledge, about 2.4 ha of land is required for setting up a 1-MW solar power plant in hilly areas. Certain geographical parameters, *i.e.*, slope, aspect, wasteland, nearest electricity substation, and road connectivity have been followed to identify suitable land banks for the set-up of solar power plants in four districts of Himachal Pradesh, namely, Una, Bilaspur, Solan, and Sirmaur. The ground truth observations were carried out at random after receiving the results of the analysis. An area of about 24,000 ha has been identified with land banks of more than 2 ha, thus, having an estimated capacity of 10,000 MW of solar power generation.

Keywords: GIS, solar power, ArcGIS, site suitability, photovoltaic, geospatial technology

2.1 Introduction

Energy is the essential to economic and social development of any state (Yang *et al.*, 2012). The growth of world human population and the industrialization of developing countries have put immense pressure on fossil fuel. The increase in global energy consumption is expected to continue, and resources are also depleting gradually (Shorabeh *et al.*, 2019; Muhammad *et al.*, 2017; Baquiran *et al.*, 2020). Energy demand has increased considerably because of urbanization, consumption pattern, and changes in life style. More and more energy consumption leads to increase of carbon concentration and average global temperature, which adversely affect the biosphere (Kashani *et al.*, 2013; Debnath *et al.*, 2023b).

**Corresponding author*: krishan.hpushimla@gmail.com

Sanjay Kumar, Suraj Kumar Singh, Shruti Kanga, Gowhar Meraj, Majid Farooq and M. S. Nathawat (eds.)
Disaster Management and Environmental Sustainability, (13–38) © 2024 Scrivener Publishing LLC

Availability of energy sources plays an important role in the development of any region. However, energy shortages in recent times and the threat of climate change are some of the challenges our society has to deal with, nowadays, humans have focused on the use of renewable sources of energy (Ramachandra *et al.*, 2012). The use of renewable energies is one of the most effective tools against climate change and to attain short- and long-term sustainable development goals (Quiros *et al.*, 2018; Haddad *et al.*, 2021; Suuronen *et al.*, 2017; Debnath *et al.*, 2022a). Solar energy is abundant, offers significant opportunities for climate change mitigation, and the energy potential of solar energy is enormous and is an important energy resource compared to the various renewable energy options; it is also clean, inexhaustible, and eco-friendly, and can be used to meet a variety of energy service requirements. Solar energy activities depend on the available potential of any given geographical location because it has been limited to certain areas due to higher insolation requirements (Pomares *et al.*, 2017; Tarai & Kale, 2016; Firozjaei *et al.*, 2019). Over the last few decades, advances in solar energy technology have made it possible to produce power by substituting carbonaceous fuels with high temperature thermal energy from the sun (Tlhalerwa & Mulalu, 2019; Ziuku, 2014). The operating efficiency of photovoltaic (PV) solar cells depends on local meteorological conditions, which vary significantly throughout the large territories (Simioni & Schaeffer, 2019). Therefore, prior to installation of solar power plants at any location for these activities, assessment of solar potential is very important due to costly technologies (Singh *et al.*, 2023b; Sud *et al.*, 2023; Upadhyay *et al.*, 2023). In energy planning problems, site identification for the installation of a solar PV power plant is a difficulty that also confuse decision makers or planners. In the case of solar photovoltaic system (SPVS), the highest solar resources are not always a viable factor, they are a range of factors that depend on site selection of the SPVS (Settou *et al.*, 2020; Chiemelu *et al.*, 2021). Methodology and results for a geographical information system (GIS) analysis are used to identify the locations of the most suitable lands for concentrating solar thermal power (Kanga *et al.*, 2022a).

A GIS integrates hardware, software, and data for capturing, managing, analyzing, and displaying all forms of geographically referenced information (Debnath *et al.*, 2023c; Singh *et al.*, 2023a). A GIS allows us to view, understand, question, interpret, and visualize data in many ways that reveal relationships, patterns, and trends in the form of maps, globes, reports, and charts (Singh *et al.*, 2022a). A GIS helps in spatial and temporal analysis of the resources and demand, and also aids as a decision support system while implementing location-specific renewable energy technologies (Ramachandra & Shruthi, 2005). Geographical information system (GIS) technology can be integrated into any enterprise information system framework. A GIS is a set of information transformation tools that derive new geographic datasets from existing datasets. These geo-processing functions take information from existing datasets, apply analytic functions, and write results into new derived datasets (Singh *et al.*, 2022b). Present study has focused on identifying suitable sites for the installation of solar power and to find out the solar energy potential in four industrial districts of Himachal Pradesh. Geographic information system (GIS) presents one of the best supports in the spatial analysis which can be combined with multi-criteria decision method and it became a successful approach to solve the energy planning problem such as site selection for solar plants (Kanga *et al.*, 2022b; Singh *et al.*, 2022; Sajan *et al.*, 2022).

Objective:

- The objective of this study includes the solar potential estimation for four districts of the H.P.

Study Area:

Four districts of Shiwalik Himalaya were selected to identify the solar power potential. District Una has a geographical area of 1540 skm^2 and shares its border with district Punjab. It lies in the southwest part of Himachal, with altitude ranging from 400 m to 1000 m. Swan is the main river of the district which is basically a seasonal. Secondly, Bilaspur district has an area of 1167 km^2 and has the famous Govind Sagar Dam on the river Satluj. This district has a hilly topography and also shares its border with Punjab state. Third district Solan is the industrial hub of the Himachal Pradesh. It is also known for its pharmaceutical industries and small-scale and medium industrial units, and play a role in the economy of the inhabitants of Solan. Fourthly, District Sirmaur is located in the outer Himalayas, which is commonly known as Shivalik range, and shares its border with Uttrakhand in the east, Haryana in the south, and Solan in the northwest. The river Giri is the biggest river in the district, which originates from Kotkhai/Jubbal Tehsil of Shimla district, and flows down in the southeast direction. Another important river which forms the eastern border of the Sirmaur district is the river Tons (Figure 2.1).

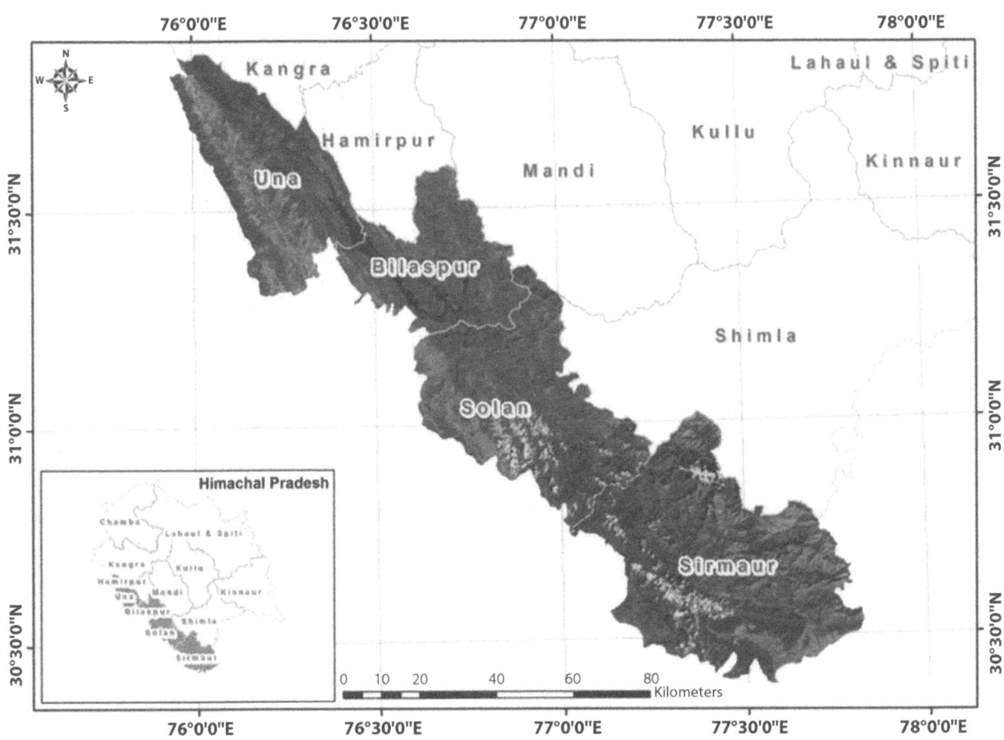

Figure 2.1 Study area.

2.2 Material and Methods

To identify the solar power potential sites and to prepare a feasibility inventory for the specific sites, the guidelines of the Himachal Pradesh Electricity Regulatory Commission were followed:

HPERC Guidelines:

- Typically, PV plant should be installed in high radiation areas, preferably flat land or slopes facing south with minimum undulations, free from shading objects.
- Presently, 1-MW plant needs at least 2-ha/5-acre/25-bigha/50-kanal land. We should look for areas having a capacity of about 50 MW but not less than around 10 MW. Therefore, land should be around 1000 bighas to 2000 bighas in a compact area, with minimum parcel not less than say 200 bighas, if land is available in clusters.
- Should be closer to load centers and closer to evacuation substations.
- Sites should be preferably shamlat land or government wasteland with least forest plants. These could be a mix of private lands, also like ghasni, changer, and barren land, which do not have better alternative use and farmers are willing to sell or lease for 25 to 35 years.
- Should be accessible, with no major costs of infrastructure like roads.

In the present study, to identify the sites for solar power generation, initially, wasteland was mapped which includes open land, grass land, shrubs, barren, and rocky land parcels. Beside this, spatial mapping of major roads, settlement clusters, and electricity power substations were done. Road connectivity is essential for the identified site so that one can easily reach the site. For the consumption of power, settlement clusters and industrial hubs are also required near the identified sites. ArcGIS 10.6.1 software was used to generate the GIS-based database for the present study (Figures 2.2 and 2.3).

Further, slope and its aspect is the most important part of the study. Aspect identifies the downslope direction of the maximum rate of change in value from each cell to its neighbor. It can be thought of as the slope direction. The values of each cell in the output raster indicate the compass direction that the surface faces at that location. It is measured clockwise in degrees from 0 (due north) to 360 (again due north), coming full circle. Flat areas having no downslope direction are given a value of –1.

Conceptually, the aspect tool fits a plane to the z-values of a 3 x 3 cell neighborhood around the processing or center cell. The direction the plane faces is the aspect for the processing cell.

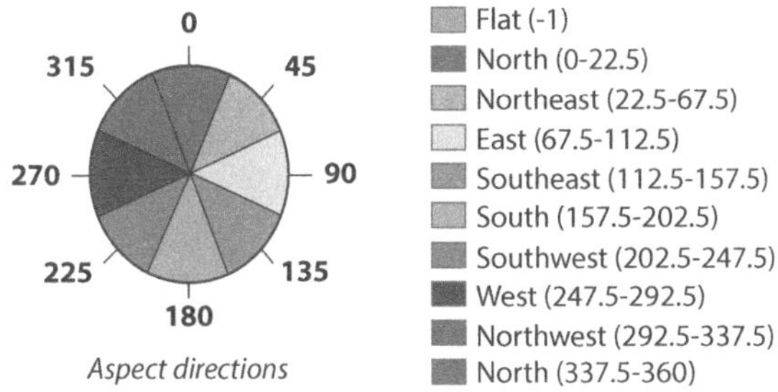

Figure 2.2 Slope aspect directions.

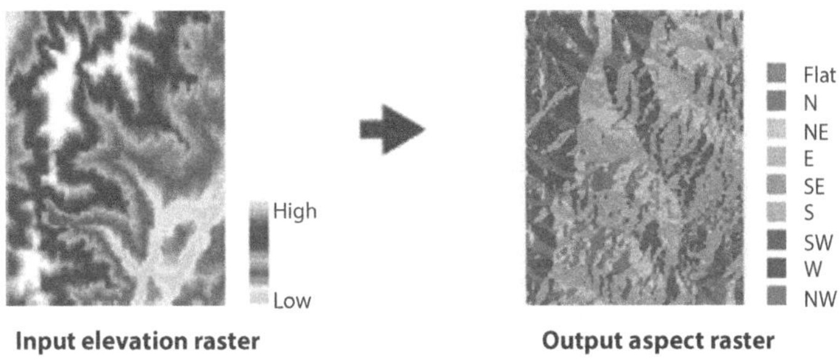

Figure 2.3 Slope aspect directions.

The following diagram shows an input elevation dataset and the output aspect raster.

Slope and its aspect of the study area was calculated from ASTER (Advanced Spaceborne Thermal Emission and Reflection Radiometer) GDEM data of 30-m resolution in GIS environment through using ArcGIS software.

The conventional wisdom (in the northern hemisphere) is that the best direction to face solar panels is south, since that is generally where t receive the most sunlight. Aspect can have a strong influence on temperature. This is because of the angle of the sun in the northern and southern hemispheres, which is less than 90° or directly overhead. In the northern hemisphere, the north side of slopes is often shaded, while the southern side receives more solar radiation for a given surface area insolation because the slope is tilted toward the sun and is not shaded directly by the earth itself. In the present study, flat, south, southeast, and southwest slope directions were preferred for suitable sites. In the present study, slope is categorized into 0°–10°, 10.1°–20°.......... 70.1°–80° (Figure 2.4).

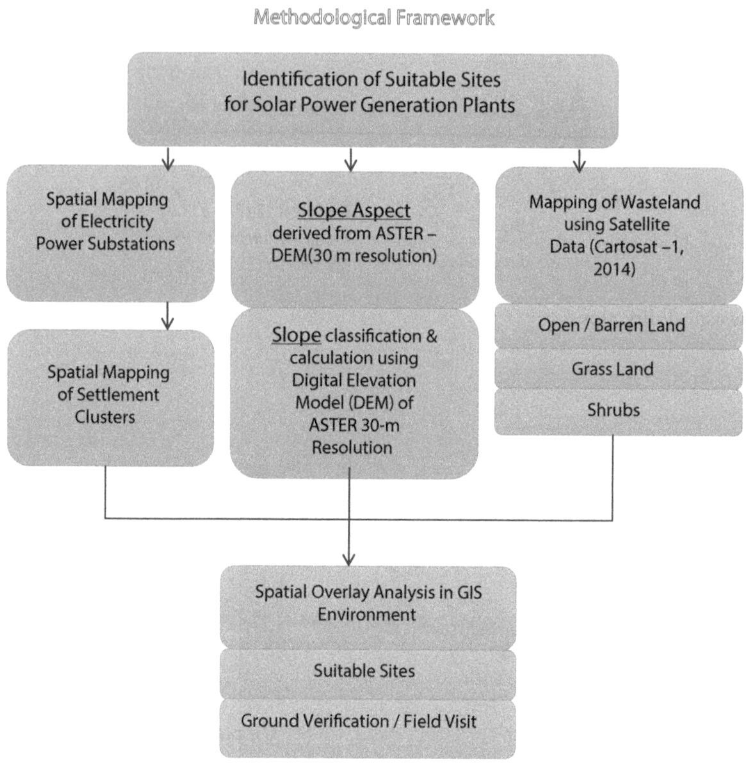

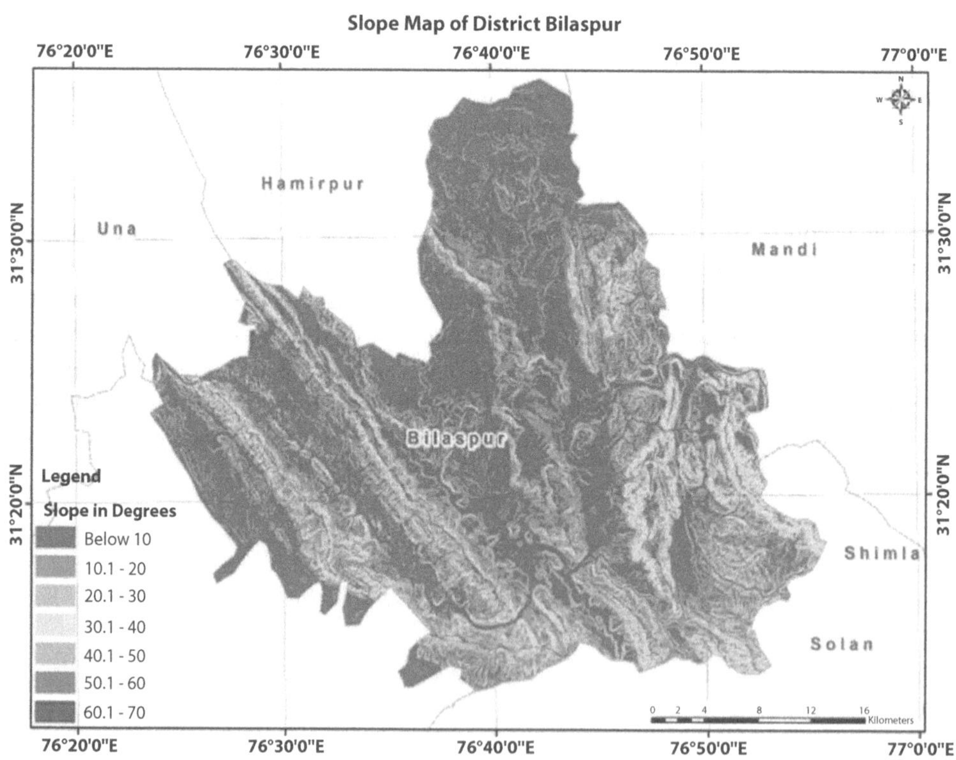

Figure 2.4 Slope map of Bilaspur District.

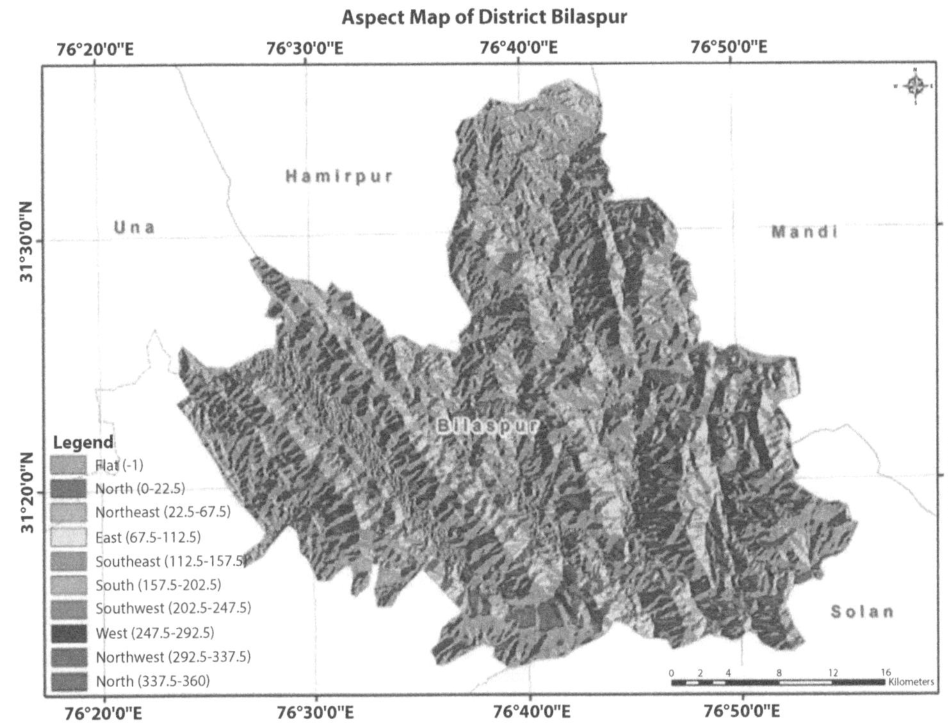

Figure 2.5 Aspect map of Bilaspur District.

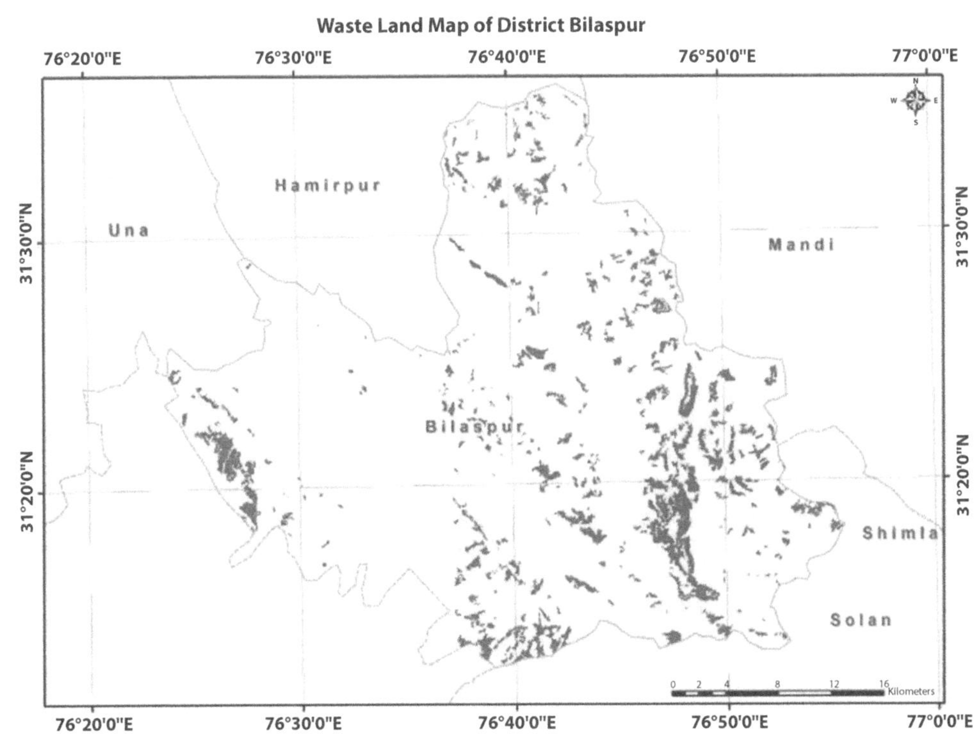

Figure 2.6 Wasteland map of Bilaspur District.

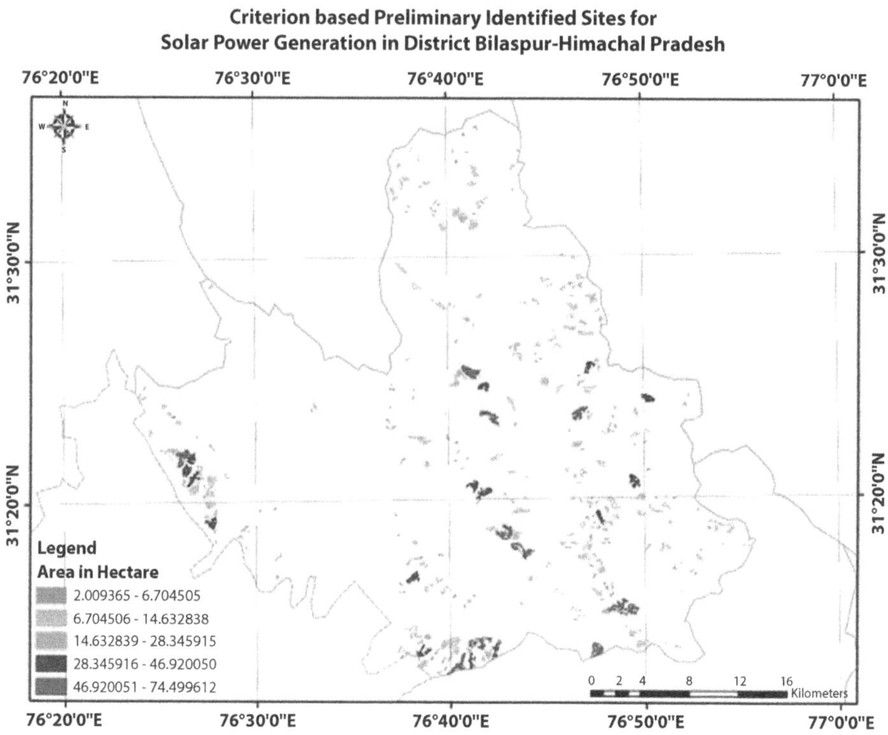

Figure 2.7 Preliminary solar power sites map of Bilaspur District.

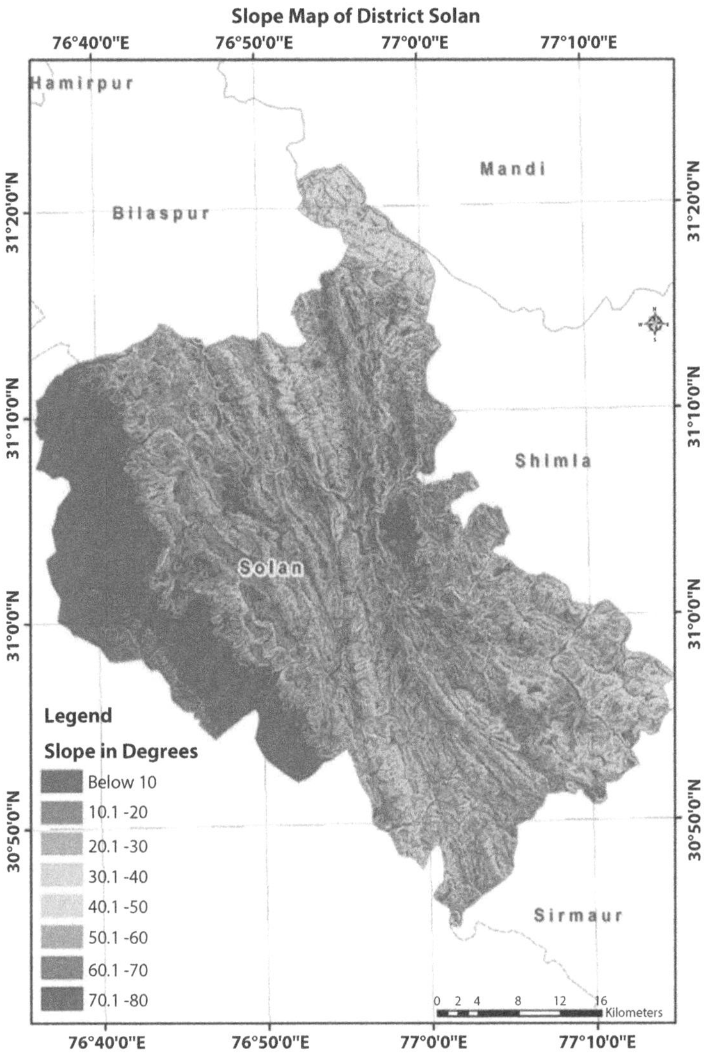

Figure 2.8 Slope map of Solan District.

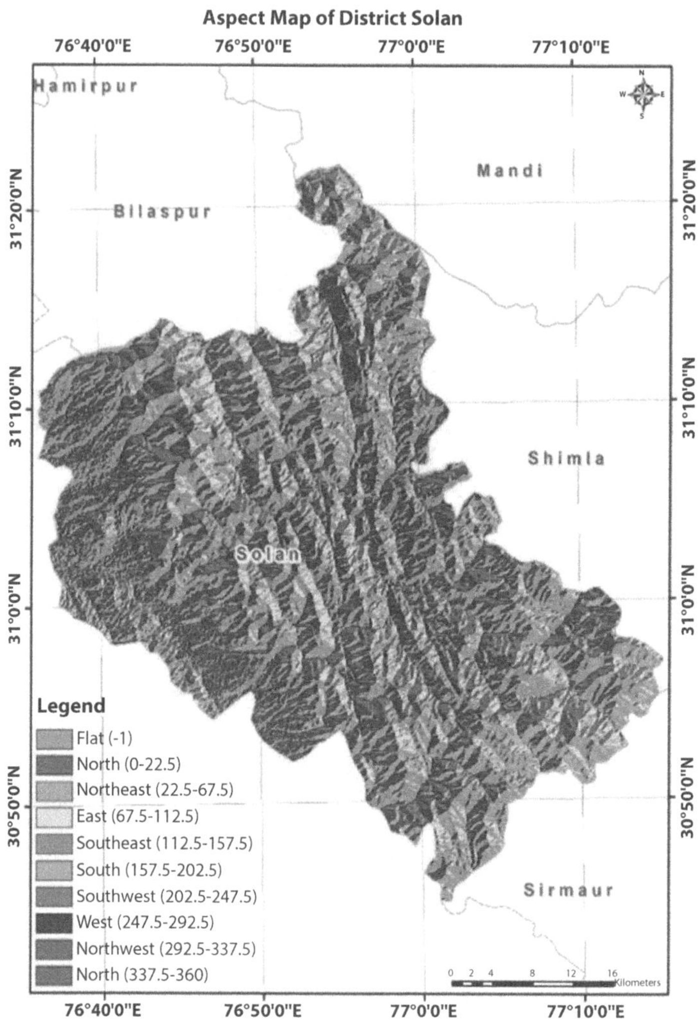

Figure 2.9 Aspect map of Solan District.

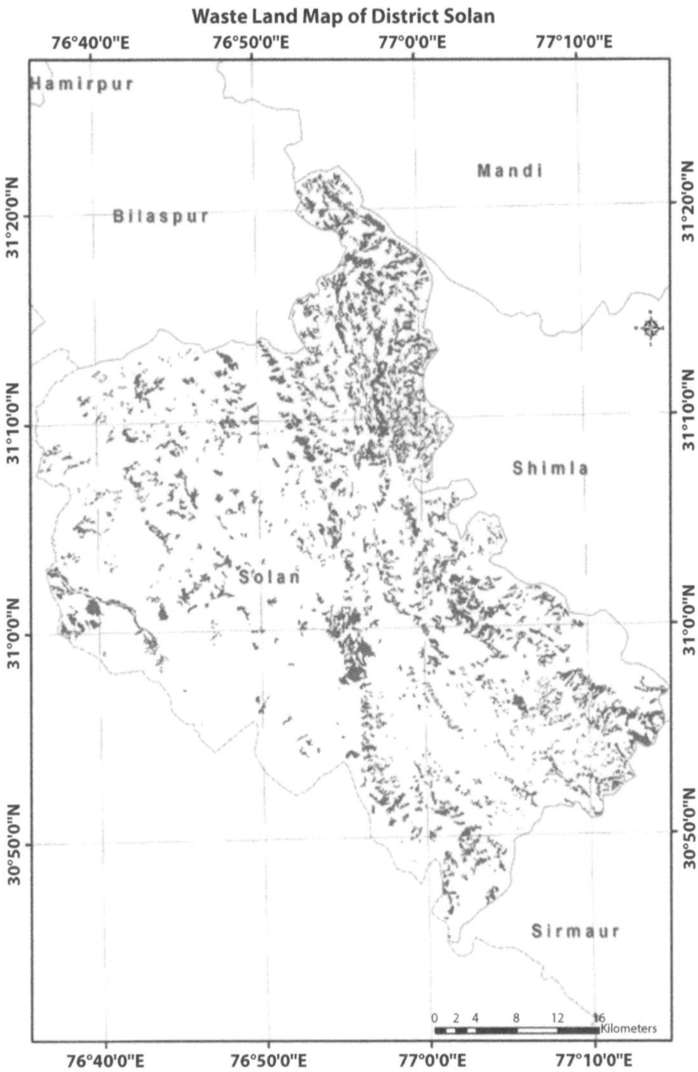

Figure 2.10 Wasteland map of Solan District.

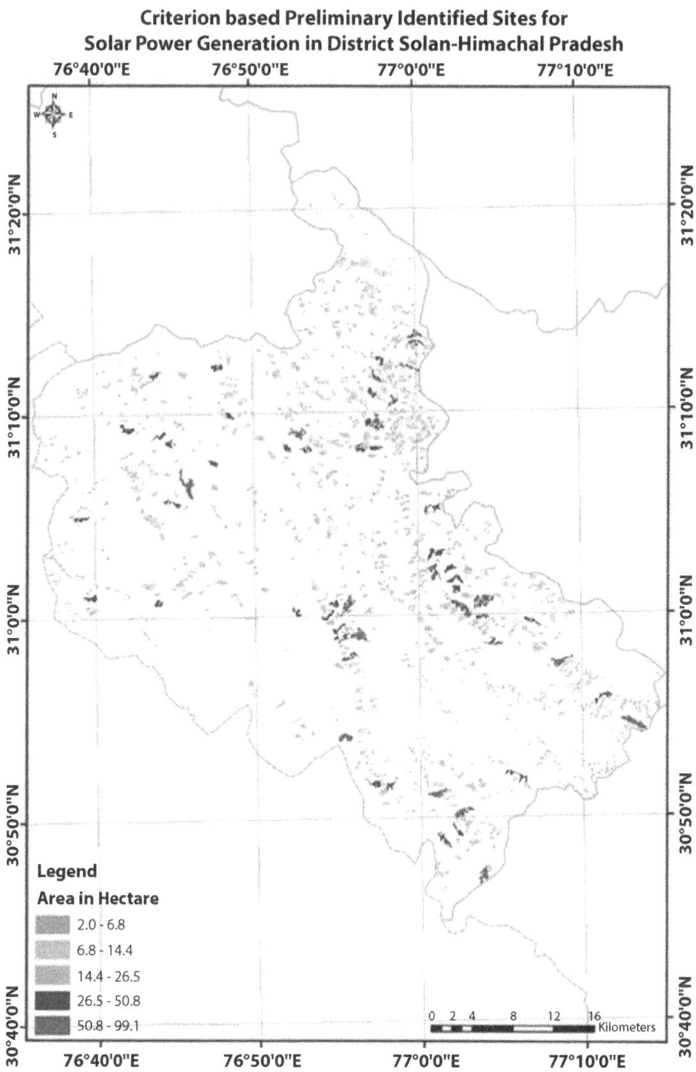

Figure 2.11 Preliminary solar power sites map of Solan District.

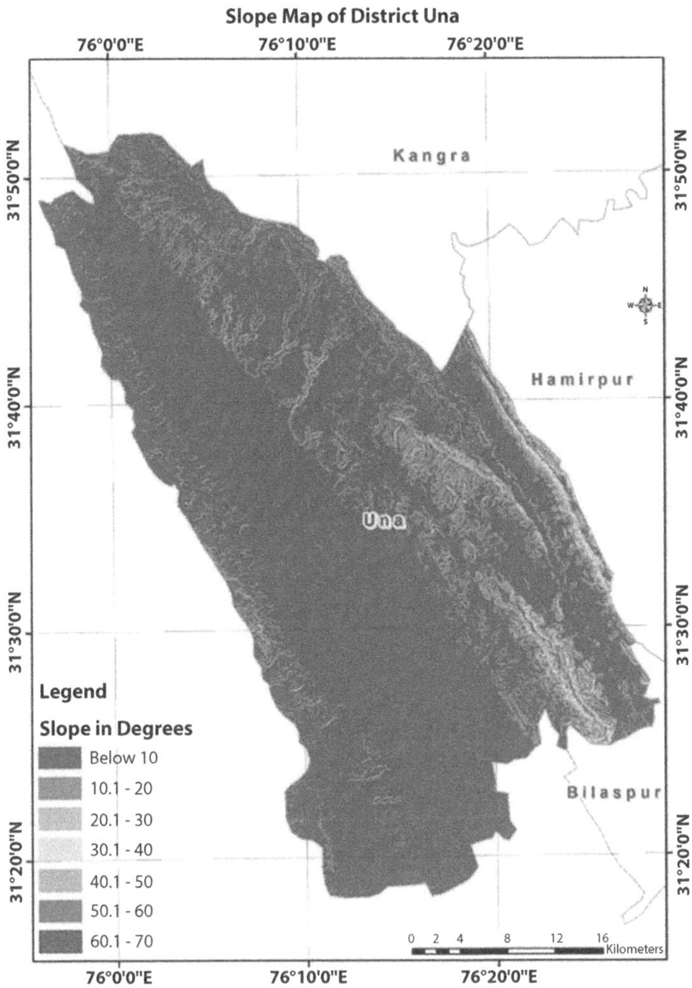

Figure 2.12 Slope map of Una District.

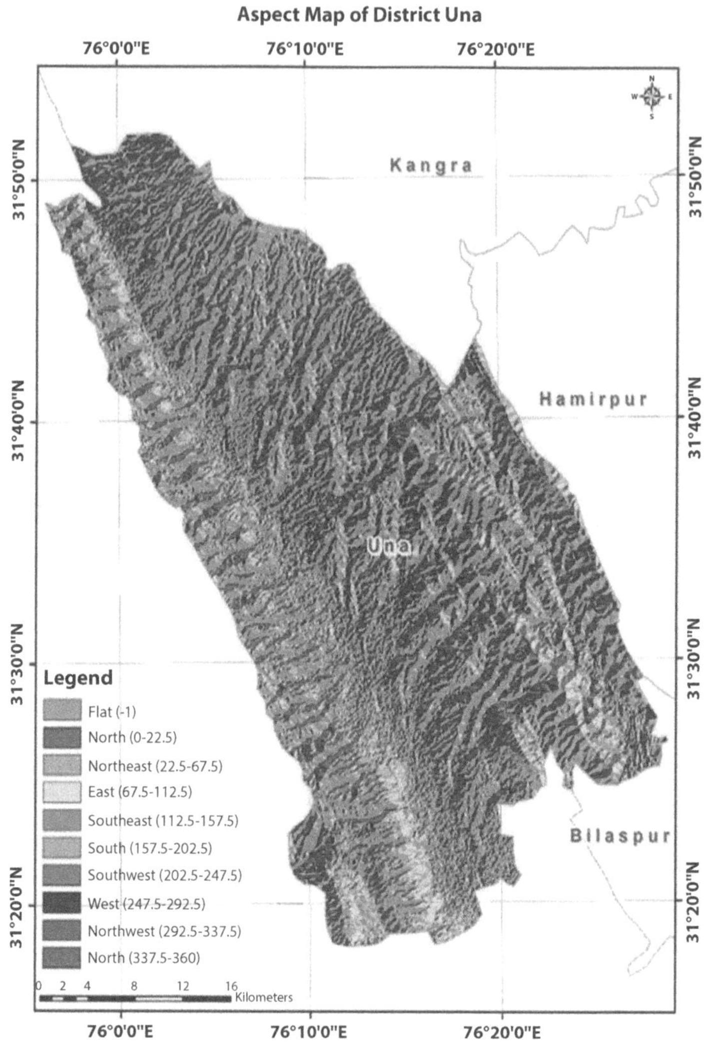

Figure 2.13 Aspect map of Una District.

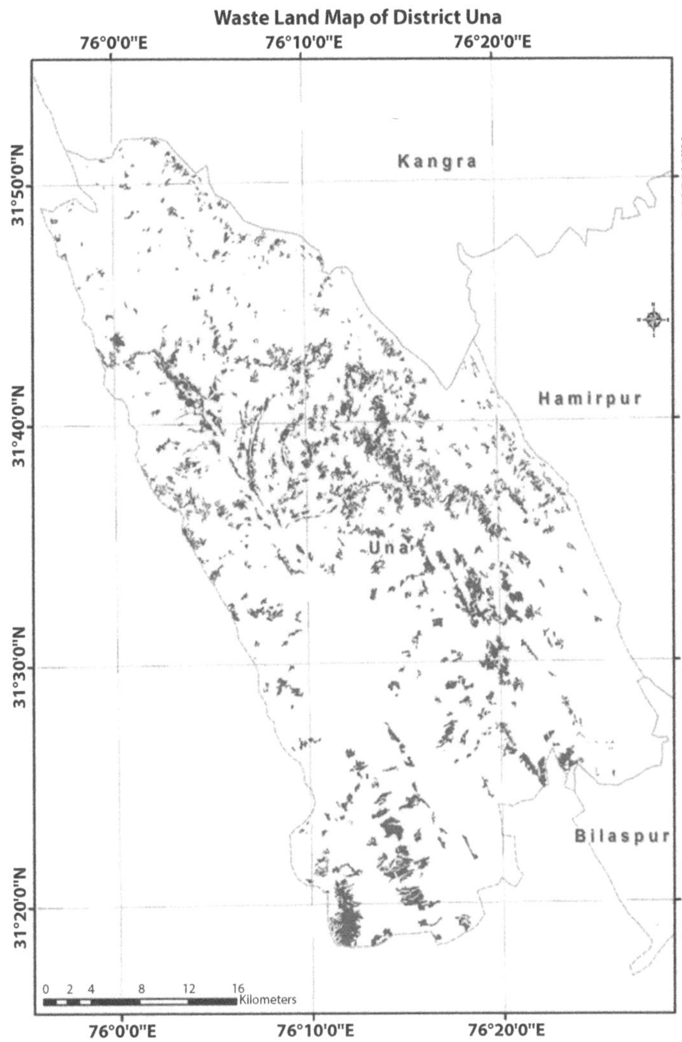

Figure 2.14 Wasteland map of Una District.

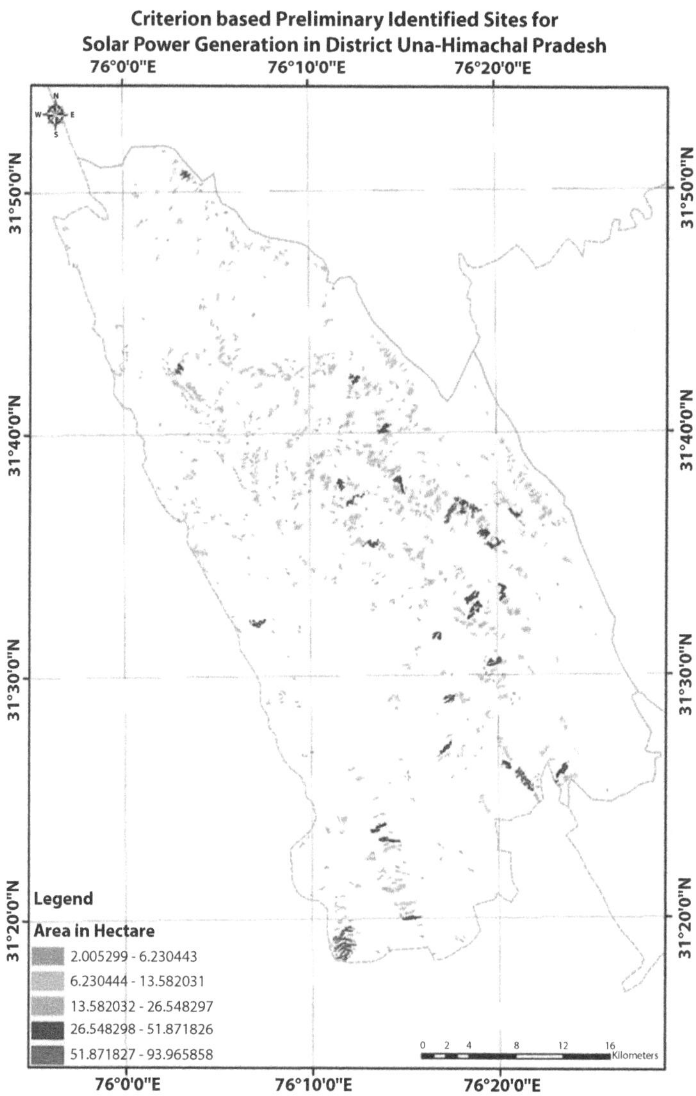

Figure 2.15 Preliminary solar power sites map of Una District.

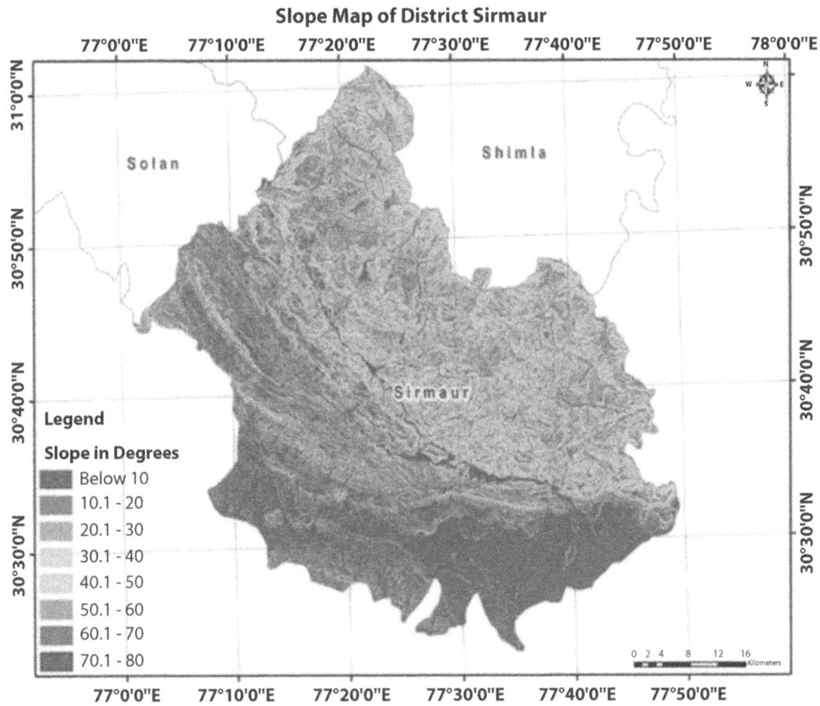

Figure 2.16 Slope map of Sirmaur District.

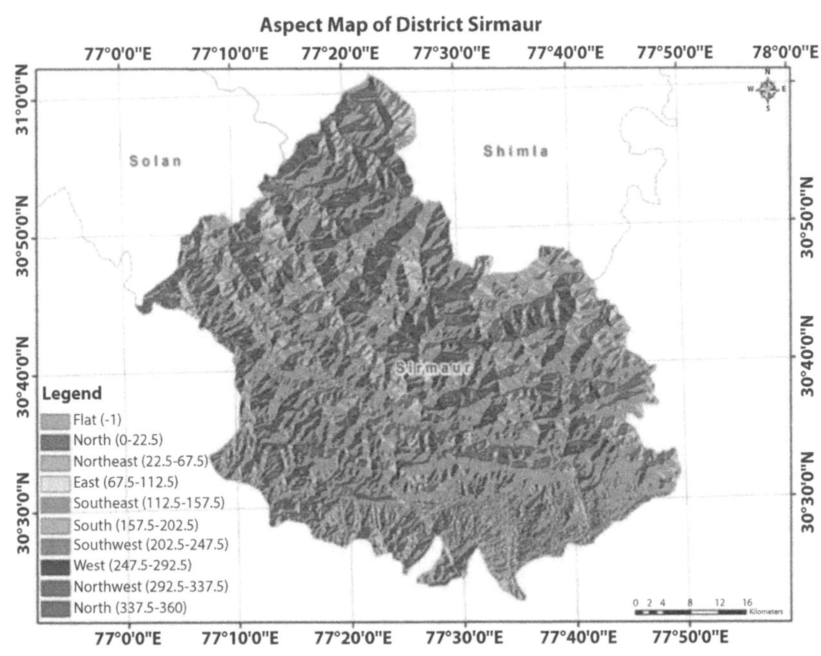

Figure 2.17 Aspect map of Sirmaur District.

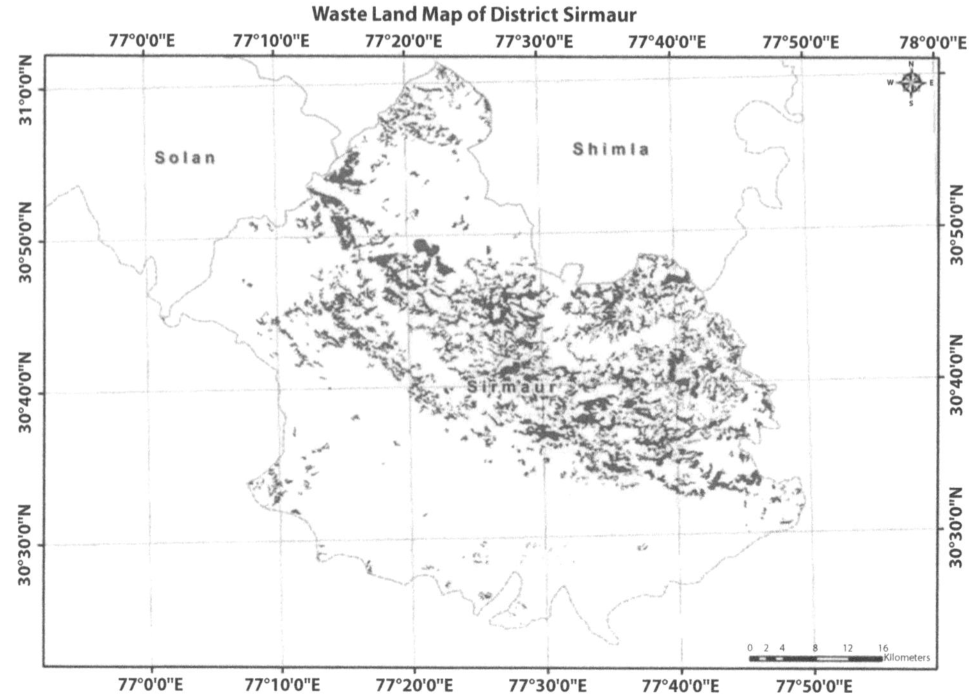

Figure 2.18 Wasteland map of Sirmaur District.

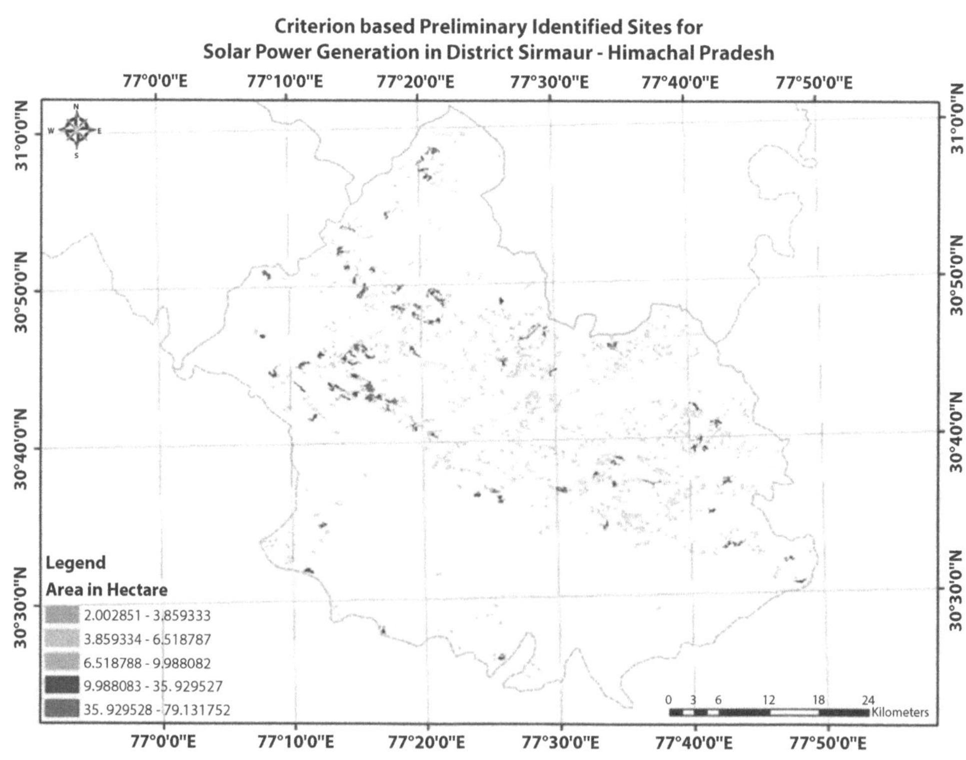

Figure 2.19 Preliminary solar power sites map of Sirmaur District.

2.3 Statistical Summary, Results, and Analysis

Wasteland maps for all the four districts were prepared through online satellite image interpretation and calculated that out of the total geographical area of the respective district, 10% area in district Una and 13% area in district Solan were identified; Bilaspur district accounts for 8%, and Sirmaur district accounts for 13% area.

All the four districts, Una, Solan, Bilaspur, and Sirmaur, have almost flat terrain and about 90% of the area has slope less than 30°. District Una is mostly flat terrain and 82% of the total geographical area is below 10° and about 15% area has 10°–20° slope and 3% is between 20° and 30° slopes. Around 90% of the total geographical area of the district Solan is below 30° of slope, one-third of the area has a slope below 10°. In District Bilaspur, half of the area out of the total geographical area has less than 10° and one-third of the area has a slope between 10° and 20°, and in district Sirmaur, around three-fourths of the area has a slope below 30°. The slope maps of districts Bilaspur, Solan, Una and Sirmaur are shown in Figure 2.4, 2.8, 2.12, and 2.16, respectively.

Slope aspect is the most important component to identify the potential sites for solar power generation. South-, southeast-, and southwest-facing slopes get maximum insolation and as per the map prepared in all four districts, around 40% of the area lies towards the south, southeast, and southwest slopes.

The aspect maps of districts Bilaspur, Solan, Una and Sirmaur are shown in Figure 2.5, 2.9, 2.13, and 2.17, respectively. Whileas The wasteland maps of districts Bilaspur, Solan, Una and Sirmaur are shown in Figure 2.6, 2.10, 2.14 and 2.18 respectively.

Table 2.1 Total wasteland identified in four districts.

District	Identified wasteland area (in hectare)	Identified wasteland area (in %)	Figure number
Una	14,798	9.61	2.14
Solan	24,557	12.68	2.10
Bilaspur	9438	8.09	2.6
Sirmaur	35,993	12.74	2.18

Table 2.2 Slope area below 30°.

District	Slope (area in hectare)				Figure number
	Below 10° (1)	10°-20° (2)	20°–30° (3)	Total (1+2+3)	
Una	12,6334 (82.04%)	22,569 (14.66%)	4306 (2.80%)	153,209 (99.49%)	2.12
Solan	57,987 (29.95%)	67,880 (35.06%)	50,970 (26.33%)	176,837 (91.34%)	2.8
Bilaspur	54,027 (46.3 %)	36,746 (31.25%)	19,438 (16.66%)	110,211 (94.44%)	2.4
Sirmaur	63,691 (22.55%)	77,885 (27.57%)	79,946 (28.30%)	221,522 (78.41%)	2.16

Table 2.3 Slope aspect area (flat, southeast, south, southwest).

District	Aspect (in hectare)					Figure number
	Flat(1)	Southeast(2)	South(3)	Southwest(4)	Total (1+2+3+4)	
Una	2523 (1.64%)	20,313 (13.19%)	20,671 (13.42%)	23,645 (15.35%)	67,152 (43.61%)	2.13
Solan	790 (0.41%)	17,942 (9.27%)	25,272 (13.05%)	32,863 (16.97%)	76,867 (39.70%)	2.9
Bilaspur	218 (0.19 1%)	11,129 (9.54%)	13,833 (11.85%)	19,229 (16.48%)	44,409 (38.05%)	2.5
Sirmaur	338 (0.12%)	34,314 (12.15%)	43,219 (15.30%)	40,313 (14.27%)	118,184 (41.84%)	2.17

Table 2.4 Statistics of identified sites in Una, Solan, Bilaspur, and Sirmaur Districts of Himachal Pradesh.

District	Suitable area (hectare)	Suitable area (in %)	Solar power potential (MW) (as per expert knowledge, 2.4 ha for 1 MW, since the terrain is hilly)	Figure number
Una	5666	3.68	2361	2.15
Solan	9094	4.70	3789	2.11
Bilaspur	3640	3.12	1517	2.7
Sirmaur	6097	2.16	2540	2.19
Total	**24,497**		**10,207**	

The area estimates of wasteland, slope, aspect, and estimated power plants in the four districts are shown in Table 2.1 to 2.4 respectively.

After following the methodological framework and by weighted overlay analysis, it was found out that in district Una, 5666-ha, in Solan, 9094-ha, in Bilaspur, 3640-ha, and in Sirmaur, 6097-ha area are suitable for the installation of solar plants. In the hilly region, about 2.4 ha of land is required to generate 1 MW of solar power, hence, 24,497 ha of suitable land in all the four districts, have the potential to generate 10,207 MW of solar light. Thus, a total of 679 suitable sites were identified in Una district, 987 in Solan district, 339 in Bilaspur district, and 844 in Sirmaur district, for setting up solar power plants.

A field survey was conducted to verify the sites on random basis at ground level in all four districts to verify name of site, aspect, maximum slope aspect, slope (in degrees), latitude, longitude, elevation (m), area (in hectares), distance from road, nearest substation, and distance in kilometers (Figure 2.2).

The solar power plants of districts Bilaspur, Solan, Una and Sirmaur are shown in Figures 2.7, 2.11, 2.15, and 2.19 respectively.

The districtwise number of suitable sites for installation of solar power plants is shown in Table 2.5.

Table 2.5 District-wise number of suitable sites for installation of solar power plants.

Suitable sites (area in hectare)	District Una		District Solan		District Bilaspur		District Sirmaur	
	No. of sites	Area (ha)	No. of sites	Area (ha)	No. of sites	Area (ha)	No. of sites	Area (ha)
2.4–10	533	2615	732	3609	243	1293	768	3692
10.1–20	94	1316	165	2297	55	771	-	-
20.1–30	28	685	47	1156	16	388	45	1103
30.1–40	12	398	20	674	6	213	16	539
40.1–50	7	306	9	395	9	395	9	389
Above 50	5	346	14	963	10	580	6	374
Total	**679**	**5666**	**987**	**9094**	**339**	**3640**	**844**	**6097**

Ground Truthing

Uperla Nagal
Nalagarh, Solan

Description

Site	**21**
Name of Site	Uperla Nangal
Village/Block	Amli/Nalagarh
Landmark	Near Nalagarh
Aspect	S/SE/SW/Flat
Maximum Aspect	SW
Slope	<30
Latitude	31° 4' 23.352" N
Longitude	76° 43' 2.551" E
Elevation (m)	350-370
Area (in Ha.)	38
Distance from Road	Link Road from Goyala Jamwal
Remarks	Shrub/Elavation 300-400m

Distance from Substations

Substation	Distance (Km)
Sub-Station Uperla Nagal	0.8
Ramshehr, Solan	8
Manjholi, Solan	9
Baglaher, Solan	9
Haripur, Solan	13
33/11 KV Swarghat, Bilaspur	17
33/11 KV Jabli, Bilaspur	19
Kunihar	24
33/11 KV Namhol, Bilaspur	24
33/11 KV Beri, Bilaspur	27

Figure 2.20 Field photograph of Solan District for ground truthing.

Takarla
Amb, Una

Description

Site	
Name of Site	**Takarla (Two Sites)**
Village/Block	Dhanri/Charu/Amb
Landmark	Radha Swami Satsang
Aspect	S,SE,SW, Flat
Maximum Aspect	Flat
Slope (Degree)	<10
Latitude	31° 36' 2.979" N
Longitude	76° 8' 29.807" E
Elevation (m)	400-430
Area (in Ha.)	94
Distance from Road	On Una Amb Road
Remarks	Transmission Line/Flat/

Distance from Substations

Substation	Distance (Km)
33/11 KV Baruhi (Bhaira), Una	4
Un-manned 33/11 KV Khad	9
Un-manned 33/11 KV JeetpurBaheri, Una	10
33/11 KV Basal, Una	11
33/11 KV Gagret, Una	11
33/11 KV Thanakalan, Una	16
33/11 KV Bangana, Una	18
33/11 KV Bharwain, Una	20
33/11 KV Haroli, Una	21
33/11 KV Dhaneta, Hamirpur	21

Figure 2.21 Field photograph of Una District for ground truthing.

2.4 Conclusion

In today's fast-paced time, where, in a way, technology is getting advanced day by day, our lifestyle is also changing. In modern times, excessive pressure on natural resources has forced us to promote renewable energy sources such as solar, wind, and hydropower. The state of Himachal Pradesh is playing a leading role in this direction. Himachal is playing a leading role in hydropower, apart from making its efforts in wind and solar energy. In this direction, this study aimed to identify potential solar power plants in four major industrial centers of Himachal Pradesh, where energy demand is high due to the presence of multiple industries. In this study, using geospatial technology and considering the parameters required for the construction of solar power plants, it was revealed that the solar power generation capacity of the four districts is about 10,000 MW. This is more than enough to meet future energy needs.

Limitations of the Study

The study was carried out through GIS and remote sensing technology and a few sites were visited physically for ground truthing. Cartosat - 1 data (2.5 m) and ASTER GDEM data of 30-m resolution were used. Thus, the accuracy on the ground may vary slightly. It was not possible to visit physically each identified site for the generation of solar power. Solar power potential was calculated for the entire area of the selected four districts and absolute solar power potential depends on road accessibility of the identified sites.

References

Baquiran, A., Pacis, D.,J., Javier, T.,B., Palchan, D., L., Dalit, R.,J., Abana, E., Manuel, J.,J., and Manuel, J.,. 2020. Mapping of Solar Power Potential Structures Using Geographic Information System. *Eng. Res. Bull.*, 8 (2020):5-9

Brahim, H., Pilar D.,C., Paula, F., Ahmed, D.,and Juan, P., P., 2021. Mapping concentrated solar power site suitability in Algeria. *Renew. Energy*, 168 (2021): 838-853.

Chiemelua, N.,E., Obinna, C.,D., Anejionub, R., I., and Ndukwuc, F., I. O., 2021. Assessing the potentials of largescale generation of solar energy in Eastern Nigeria with geospatial technologies. *Sci. Afr.*, 12 (2021):1-15.

Debnath, J., Sahariah, D., Lahon, D., Nath, N., Chand, K., Meraj, G., Kumar, P., Singh, S. K., Kanga, S., & Singh, S. K. (2022a). Geospatial modeling to assess the past and future land use-land cover changes in the Brahmaputra Valley, NE India, for sustainable land resource management. *Environ. Sci. Pollut. Res.*, 1-24.

Debnath, J., Sahariah, D., Lahon, D., Nath, N., Chand, K., Meraj, G., Kumar, P., Singh, S. K., Kanga, S., & Farooq, M. (2023b). Assessing the impacts of current and future changes of the planforms of River Brahmaputra on its land use-land cover. *Geosci. Front.*. https://doi.org/10.1016/j.gsf.2023.101557.

Debnath, J., Sahariah, D., Saikia, A., Meraj, G., Nath, N., Lahon, D., Annayat, W., Kumar, P., Chand, K., Singh, S. K., & Kanga, S. (2023c). Shifting Sands: Assessing Bankline Shift Using an Automated Approach in the Jia Bharali River, India. *Land*, 12(3), 703.

Firozjaei, M., K., Omid, N., Mijani,N., Saman N., S., Hamzeh, K., F., and Toomanian, A. 2019. An integrated GIS-based Ordered Weighted Averaging analysis for solar energy evaluation in Iran: Current conditions and future planning. *Renew. Energy*, 136, 1130-1146.

Golara, G., Younes, N., Hamed, A., Mousa, M., and Mahmoud, S., 2019. Theoretical and technical potential evaluation of solar power generation in Iran. *Renew. Energy*, 138 (2019) 1250-1261. *J. Maps* 14 (1): 44–51.

Joshi, J. 2013. A research study on developing solar potential map using GIS. *Int. J. Emerg. Trends Technol. Comput. Sci. (IJETTCS)*, 2 (3):220-224.

Kanga, S., Meraj, G., Farooq, M., Singh, S. K., & Nathawat, M. S. (Eds.). (2022a). Disaster Management in the Complex Himalayan Terrains. *Geography of the Physical Environment*. Springer, Cham.

Kanga, S., Singh, S. K., Meraj, G., Kumar, A., Parveen, R., Kranjˇ, N., & Đurin, B. (2022b). Assessment of the Impact of Urbanization on Geoenvironmental Settings Using Geospatial Techniques: A Study of Panchkula District, Haryana. *Geographies*, 2(1), 1-10. ISSN No. 2673-7086.

Kashani , A., H., Izadkhast, P., S., and Asnaghi, A., 2014. Mapping of solar energy potential and solar system capacity in Iran. *Int. J. Sustain. Energy*, 33, (4): 883–903.

Muhammad, F., Raza M.,W., Khan, S., Khan, F., 2017. Different Solar Potential Co-Ordinates of Pakistan. *Innov. Ener. Res.*, 6: 173.

Ouammi,A., Zejli,D., Dagdougui,H., Benchrifa, R., 2017. Artificial neural network analysis of Moroccan solar potential. *Renew. Sustain. Energy Rev.*, 16 (2012): 4876–4889.

Pomaresa,L.,M., Martínezb,D., Jesús, P., Daniel, P.,A.,and Dunia, B., 2017, Antonio Sanfilippo Analysis of the long-term solar potential for electricity generation in Qatar. *Renew. Sustain. Energy Rev.*, 73 (2017): 1231-1246.

Quiros, E., Pozo, M., and Ceballos, J., 2018. Solar potential of rooftops in Caceres City, Spain.

Ramachandra, T.,V., and Shruthi, B.,V., 2007. Spatial mapping of renewable energy potential. *Renew. Sustain. Energy Rev.*, 11:1460–80.

Ramachandra, T.,V., Krishnadas, G., and Jain, R., 2012. Solar potential in Himalayan Landscape. *Renew. Energy*, 12: 1–13.

Sajan, B., Mishra, V. N., Kanga, S., Meraj, G., Singh, S. K., & Kumar, P. (2022). Cellular automata-based artificial neural network model for spatial-temporal dynamics and land use/land cover change prediction. *Agronomy*, 12, 2772.

Settou, B., Settou, N., Gouareh, A., Negrou, B., Mokhtara, C., and Messaoudi, D., 2020. A highresolution geographic information systemanalytical hierarchy processbased method for solar PV power plant site selection: a case study Algeria. *Clean Technol. Environ. Policy*, 23 (1): 219-234.

Shorabeh S., N., Firozjaei, M., K., Nematollahi, O., Firozjaei, H., K., and Niaraki, M., J., 2019. A risk-based multi-criteria spatial decision analysis for solar power plant site selection in different climates: A case study in Iran. *Renew. Energy*, 13:958-973.

Simioni, T., and Schaeffer, R., 2019. Georeferenced operating-efficiency solar potential maps with local weather conditions – An application to Brazil. *Sol. Energy*, 184 (2019): 345–355.

Singh, A., Singh, S. K., Meraj, G., Kanga, S., Farooq, M., Kranjcic, N., Đurin, B., & Sudhanshu. (2022b). Designing Geographic Information System Based Property Tax Assessment in India. *Smart Cities*, 5, 364–381.

Singh, H., Meraj, G., Singh, S., Shrivastava, V., Sharma, V., Farooq, M., Kanga, S., & Kumar, P. (2022a). Status of Air Pollution during COVID-19-Induced Lockdown in Delhi, India. *Atmosphere*, 13(12), 2090.

Singh, S. K., Gupta, S. K., Meraj, G., Kumar, P., & Kanga, S. (2023b). Uncovering the hydro-meteorological drivers responsible for forest fires utilizing geospatial techniques. *Theor. Appl. Climatol.*, https://doi.org/10.1007/s00704-023-04497-y

Singh, S. K., Meraj, G., Mondal, N., Bera, A., Verma, M. K., Tomar, J. S., & Kanga, S. (2022). Assessment of seasonal vegetation dynamics over parts of thar desert using geospatial techniques. *ANGRAU*, 105.

Singh, S., Singh, S. K., Prajapat, D. K., Pandey, V., Kanga, S., Kumar, P., & Meraj, G. (2023a). Assessing the Impact of the 2004 Indian Ocean Tsunami on South Andaman's Coastal Shoreline: A Geospatial Analysis of Erosion and Accretion Patterns. *J. Mar. Sci. Eng.*, 11(6), 1134.

Suuronen, A., Lensu, A., Kuitunen, M., Andrade-Alvear, R., Celis, N. G., Miranda, M., Perez, M., and Kukkonen, J. (2017). Optimization of photovoltaic solar power plant locations in northern Chile. *Environ. Earth Sci.*, 76(24):824.

Tarai, R.,K., and Kale, P., Development of rasterized map using PVGIS for assessment of solar PV energy potential of Odisha. *Int. J. Renewable Energy Res.* 6(1): 61-73.

Tlhalerwa,K., and Mulalu, M., 2019. Assessment of the concentrated solar power potential in Botswana. *Renew. Sustain. Energy Rev.*, 109 (2019): 294–306.

Yang, L.,V, Xianfeng, Z., Yu, L., 2012. Solar Energy Potential Mapping at A Building Scale. *International Conference on Computer Vision in Remote Sensing.*

Ziuku, S., Seyitinib, L., Mapuris, B, Chikodzib, D., and Koen,V.,K., 2014. Potential of Poncentrated Solar Power (CSP) in Zimbabwe. *Energy Sustain. Dev.*, 23 (2017): 220-227.

Sud, A., Kanga, R., Singh, S. K., Meraj, G., Kanga, S., Kumar, P., Ramanathan, A. L., & Bhardwaj, V. (2023). Simulating Groundwater Potential Zones in Mountainous Indian Himalayas—A Case Study of Himachal Pradesh. *J. Geophys. Res.: Earth Surf.*, 10(3), 65.

Upadhyay, R. K., Tripathi, G., Đurin, B., Šamanović, S., Cetl, V., Kishore, N., Sharma, M., Singh, S. K., Kanga, S., Wasim, M., Rai, P. K., & Bhardwaj, V. (2023). Groundwater Potential Zone Mapping in the Ghaggar River Basin, North-West India, Using Integrated Remote Sensing and GIS Techniques. *Water*, 15(5), 961.

3

Assessment of the Drivers of Domestic Water Consumption Pattern in Growing Population of Idah LGA, Kogi State, Nigeria

Ogunkolu Ayodeji Bolade[1]*, Ogbole Moses[2], Stephen Favour Ojonuba[1] and Oyetunji Adewale Bashir[2]

[1]*Geography and Environmental Studies Department, Kogi State University, Anyigba, Kogi State, Nigeria*
[2]*Geography Department, Nigeria Defence Academy, Kaduna, Kaduna State, Nigeria*

Abstract

Household access to potable water is required for sanitation and general well-being. This study assessed the drivers of domestic water consumption pattern in Idah LGA, Kogi State. It focuses on the analysis of the sources of water, the magnitude of household water demand, consumption pattern of water, factors influencing consumption pattern of household water, and the relationship between domestic water supply and consumption pattern in Idah. The methodology involves the use of questionnaires, oral interview, and personal observation to gather necessary information. The sample size of four hundred (400) was selected using simple random technique. The result from the study was subjected to SPSS for both descriptive and inferential statistics. Descriptive statistics employed for this study include frequency, percentages mean, and standard deviation, while the inferential statistics adopted in the study is correlation analysis. The result from the study revealed that boreholes (48.6%) and rivers (47.5%) are the major sources of water to households. The study further revealed that consumption and supply of water are 105.1 L and 110.4 L, respectively per person per day. It was also revealed that the household activity that consumed more water is clothes washing. The study also revealed that household size (33.6%) is the major determinant of water consumption pattern in the study area. The conclusion from the result of correlation coefficient is that there exists a very high positive linear relationship between the household water consumed and water supplied. The study concluded that the planners need to consider these variables (factors) in water supply planning. The study therefore recommended that investigations should be conducted on the reliability of water sources in terms of its quality to ensure its conformity with the standard recommended to safeguard human health, also, there should be a concerted effort by both water users and supplier to control water wastage, and the Government should endeavor to come to the water board's aid with proper funding so as to purchase the material needed to ensure efficient and effective functioning of the water board in Idah.

***Keywords*:** Water use, household consumption pattern, water supply, drivers, population growth

**Corresponding author*: princeogunkolu4real@gmail.com

Sanjay Kumar, Suraj Kumar Singh, Shruti Kanga, Gowhar Meraj, Majid Farooq and M. S. Nathawat (eds.)
Disaster Management and Environmental Sustainability, (39–60) © 2024 Scrivener Publishing LLC

3.1 Introduction

Water supply, consumption, and spatial distribution are closely associated with economic growth and the development of society. Apart from air, water is indispensable to life, it is a foundation for human prosperity because adequate and high quality water supplies provide a basis for the growth and development of human social, economic, and culture of people (Young, 2006). The World Health Organization (WHO) defined domestic water as water used for all domestic purposes including drinking, bathing, and food preparation. Domestic water consumption is a significant component of the total water use and it varies according to living standards of the consumers in urban and rural areas (Mohammed and Sanaullah, 2017).

There are factors which affect the amounts of water withdrawn by individual households thus causing the variation of water consumption between households. These are the size of family, distance, income level, education, cultural heritage, character of water supply, cost of obtaining water as measured by energy or cash expenditure, climate, and terrain (Farooq *et al.*, 2022c).

In Africa today, water scarcity is a menace and it has been estimated that by 2030, 75 to 250 million people will be living in water-stressed areas (World Bank, 2006). In Nigeria, about 57 million people do not have access to safe water, the scarcity of water in Nigeria is taking a new dimension as residents of many urban and semi-urban areas do not have access to a readily available source of domestic water (Ojo, 2014). Nigeria is experiencing an increase in the rate of changes in her population coupled with urbanization and living standards.

Safe and clean water is important for socio-economic development and the ecosystem services. However, this resource is gradually diminishing in most parts of the country, including Idah. The current population growth rate of about 3.6% coupled with growth in development and urbanization may raise water demand and consumption pattern in the study area and may lead to deficit in the nearest future. Seasonal and spatial variation in water availability in Idah and the current demand that outstrip supply is an indication of water scarcity and diminishing improvement to water supply, which invariably may affect the consumption of these resources.

Several literatures abound in water study all over the world. For instance, Schleich and Hillenbrand (2007) carried out a study on the impact of economic, environmental and social determinants for the average per capita demand for water and sewage in about 600 water supply areas in Germany using econometric-analyzed prices, income, household size, the effects of population age, the share of wells and rainfall, and temperature during the summer months on water demand. The result suggests that the response of residential water demand in Germany is rather inelastic, and also found out that household size and the share of wells and summer rainfall have a negative impact on water demand. According to the study conducted by Ahmed and Smith (2007) in Bangladesh, water consumption per person per day for drinking, kitchen (cooking and utensil washing), bathing (bathing and washing clothes), sanitary. and other purposes were 2 L, 9 L, 20 L, and 8 L, respectively. Solley (2000) found in his study, "Global Water Quantity Supply And Demand Implication For Mega Cities," that structural changes affected supply and demand for domestic water due to federal laws controlling water pollution, technological changes in processes that use water as an input (including cooling towers and a movement away from once through cooling), and increase in recycling of water.

Mulwafu (2002) examined the status of water demand management in Malawi in Genta and in the Lake Chilva catchment in particular, the finding indicates that, while water demand management is highly advocated in the urban and semi-urban areas, very few aspects of water demand management are practiced in the rural areas. Ajadi (2003) carried out a study of water consumption pattern in Ilorin metropolis with the aid of questionnaire and interview method, the study found that increase in urbanization and rapid growth of population, coupled with ineffective management of water resources, have led to water crisis. Ijaiya (2000) investigated the impact of water shortage and depletion on the productive time of women in Ilorin, and the study made use of structured questionnaire, informal interview, and participant observation. The result of the study revealed that inadequate supply of water is due to inefficiency of the delivery agent and contamination of water due to poor sanitation (Singh *et al.*, 2023a).

From the foregoing, all the literatures reviewed have either carried out a study on consumption pattern or water supply pattern. There is therefore a salient point on assessing the factors of domestic water consumption pattern such as household size and composition, water accessibility, weather variability, and cost and price of water. It is on this premise that this present study intend to fill the gap. Therefore, this study seeks to assess the drivers of domestic water consumption pattern in Idah LGA of Kogi State.

3.2 Study Area

Idah lies between latitude 7°05'N and 7°83'N, and longitude 6°45'E and 6°75'E of the Greenwich meridian. Idah is a Local Government Area, a town in Kogi State, on the eastern bank of the Niger River in the North Central Region of Nigeria. It is one of the oldest local governments in Igala-land, created in 1979 alongside Dekina and Ankpa LGA. Idah has two main rock types, namely, basement complex rocks of the Precambrian age in the western half of the town and extending slightly eastwards beyond the lower Niger valley and the older sedimentary rocks in the eastern half. The various sedimentary rock groups extend along the banks of Rivers Niger and southeastwards through Enugu and Anambra states, to join the Udi Plateau. The drainage pattern in the area is dendritic. The area is well drained with major rivers such as Ocheche, Inachalo, Ofu, Emachi, and Ega, and streams occupying rather wide valleys. The rivers eventually drain into the River Niger. The tributaries are mostly seasonal, having their sources mainly within the area. This drainage configuration is thought to be the result of substantial degree of evolution from the primeval setting, *i.e.*, the Paleo drainage (Lahon *et al.*, 2023a).

Idah, like every other town in Kogi State, experiences a typical tropical climate with two distinct seasons—the wet/rainy season and the dry/summer season. The rainy season lasts from April to October and is accompanied by heavy humidity and strong rainfalls with annual rainfall in the range of 200–250 mm, heaviest rainfall occurs between June and July, reduces and gets heavy again in September (Iloeje, 2001).

The rain forest belt (selva type) covers Dekina, Ofu, Ankpa, Olamaboro, Idah, and Bassa local government areas with rich deciduous and occasional stunted trees including Palms, Iroko, Mahogany, Akee, apple, and other towering trees. Other Local Government Areas

(LGA) are in the Guinea Savannah or Parkland Savannah belt with tall grasses and some trees.

3.3 Methodology

The descriptive survey method of data collection was employed in the study. The survey method involves using a survey instrument, 'questionnaire,' to collect the required data. Nine communities were purposively and randomly selected forming the area of coverage for the purpose of this study. Idah was stratified into nine wards and one community was randomly selected from each of the ward. The communities selected are: Igalogba, Ede-Adejo, Inachalo, Ogenegu, Owoliapa, Oketecheje, Ega, Ojigagala, and Ofokpoju. Four hundred copies of questionnaires were administered in the study area. Forty-four copies of questionnaires were purposively administered to selected household in each of the community, and 48 questionnaires were administered in Igalogba because it is the major town in the local government and it comprises of people from different parts of Idah. The respondents were targeted to be the women of the house, because the women are majorly responsible for the

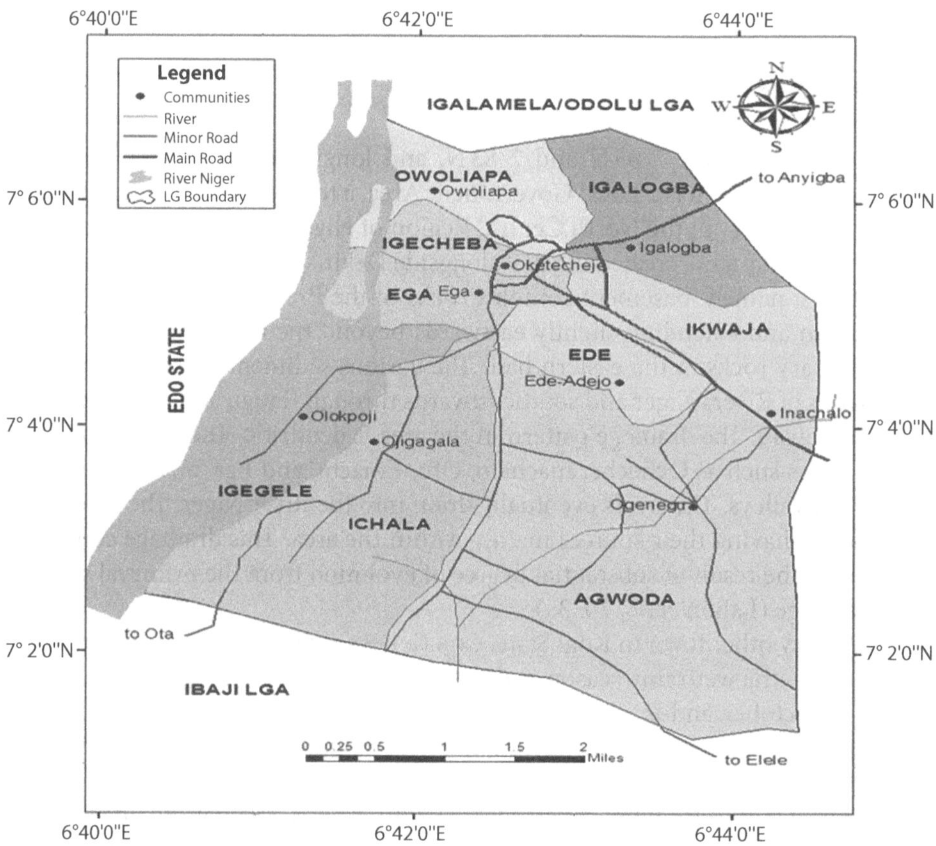

Figure 3.1 Study area source: Department of Geography K.S.U, Anyigba (2019).

collection of water in the household, and when the women are not around, the husbands were selected to answer to complete the questionnaire (Figure 3.1).

Descriptive and inferential statistics were used in the analysis of generated field data. The descriptive statistics employed in this study include frequency and percentage. The inferential test 'correlation analysis' was employed to examine the extent of water consumption and water supply in the study area (Figure 3.1).

3.4 Result and Discussion

3.4.1 Demographic Characteristics of Respondents

Table 3.1 reveals that 24% of the respondents were male while 76% were female. The investigation focused on heads of households especially women by virtue of their traditional role in water provision for home use; women are majorly responsible for collection of household's water. However, the proportion of males in the study was only accommodated when the female is not available.

On the distribution of respondents based on age, the study revealed there is uneven distribution of ages across the age groups with a majority 33.6% within the age range of 15–25 years, followed by 32.3% within the age range of 26–35 years. About 21.7% within the age range of 36–45 years, 7.8% were less than 15 years, and about 4.7% of the respondents were above 45 years.

The implication of the result is that, majority of the respondents are in their youthful age and they tend to use more water than the children and the aged. This component may be important especially where the respondent is within the age of working class. Such group of people is likely to use more water for various purposes which may not be relevant in homes with aged and teenager. Also, the distribution of respondents based on marital status: 72.6% are married, 12.4% are single, 10.3% are either widow or widower, while 4.7% are divorced. With respect to household sizes, the majority, 52.2%, of the households comprises of 6–10 persons, followed by 31.3% which comprises of 1–5 persons, 10.9% comprises of 11–15 persons, and 4.7% comprises of 16 and more persons. This finding is in agreement with that of Ayanshola, Sule and Salami (2010) where household size was observed to be one of the determinants of domestic water demand. The higher the household size, the higher the water consumption.

Table 3.1 further reveals the distribution of respondents based on income level, majority, 61%, earn above **N** 15,000 to **N** 20,000 per month, followed by **N** 5,000 to **N** 10,000 per month (33.6%), below **N** 5,000 per month (24%), above **N** 20,000 per month (13.4%), and 13.2% of the respondents earn above **N** 10,000 to **N** 15,000 per month. Majority earn between **N** 15,000 and **N** 20,000. By implication, more water will be demanded for because majority of the household earn an average income.

Results presented in Table 3.2 further reveals that 40.6% were traders, 16.5% were farmers, 9.6% were civil servants, 10.3% were ordinary house wives, and 15% had other occupations, they involved themselves in vocational jobs as means of their livelihood. This shows that majority of the respondents are business men and women. This is not unexpected as

Table 3.1 Socio-economic characteristics of respondents.

	Respondents background					**Total**
Sex	**Male**			**Female**		
Frequency	93			294		**387**
Percentage (%)	24.0			76.0		**100**
Age	**Below 15**	**15–25**	**26–35**	**36–45**	**Above 45**	
Frequency	30	130	125	84	18	**387**
Percentage (%)	7.8	33.6	32.3	21.7	4.7	**100**
Marital status	**Married**	**Single**	**Divorced**	**Widow/widower**		
Frequency	281	48	18	40		**387**
Percentage (%)	72.6	12.4	4.7	10.3		**100**
Household size	**1-5**	**6–10**	**11–15**	**Above 16**		
Frequency	121	202	42	22		**387**
Percentage (%)	31.3	52.2	10.9	5.7		**100**
Income level	**0–5000**	**>5000–10,000**	**>10,000–15,000**	**>15,000–20,000**	**>20,000**	
Frequency	93	130	51	61	52	**387**
Percentage (%)	24.0	33.6	13.2	61.0	13.4	**100**
Occupation	**Trading**	**Farming**	**Civil service**	**House wife**	**Others**	
Frequency	188	64	37	40	58	**387**
Percentage (%)	40.6	16.5	9.6	10.3	15.0	**100**
Educational qualification	**Tertiary**	**Secondary**	**Primary**	**No formal education**		
Frequency	34	100	140	113		**387**
Percentage (%)	8.8	25.8	36.2	29.2		**100**

Source: Field Work, 2020.

Table 3.2 Sources of water in the study area.

S/N	Sample location	Well		Borehole		Pipe-borne		River		Rainfall	
		Freq.	%	Freq.	%	Freq.	%	Freq.	%	Freq.	%
1	Igalogbo	0	0	43	100	0	0	0	0.0	0	0
2	Ede-Adejo	4	9.1	32	72.7	0	0	8	18.2	0	0
3	Ega	3	7.3	22	53.7	0	0	16	39.0	0	0
4	Oketecheje	4	9.1	23	52.3	0	0	17	38.6	0	0
5	Owoliapa	4	9.1	40	90.9	0	0	0	0.0	0	0
6	Ogenegu	0	0	0	0	0	0	43	100	0	0
7	Inachalo	0	0	23	56.1	0	0	18	43.9	0	0
8	Olokpoji	0	0	0	0.0	0	0	44	100	0	0
9	Ojigagala	0	0	5	11.6	0	0	38	88.4	0	0
Total		**15**		**188**		**0**		**184**		**0**	
Percentage (%)		**3.9**		**48.6**		**0.0**		**47.5**		**0**	

Source: Field Work, 2020.

Idah town is a major commercial center in Kogi East senatorial district. Due to the commercial activities in the town, there is an influx of movement from the rural areas to the town and also from the neighboring towns a common phenomenon observed in many developing countries of the world (Lahon *et al.*, 2023a, b). This tends to increase the rate of water demand in the area. This is in agreement with World Bank (2003) whereby occupation of the family is one of the most dominant factors affecting water consumption at the micro level.

Results presented in Table 3.2 also show that 8.8% of the respondents had tertiary education, 25.8% had secondary education, 36.2% had primary education, while 29.2% had no formal education. This implies that majority of the respondents were educated and education rate of people has influence on water usage. This study conforms to the study of Ifabiyi (2011) where literacy level has been found to have influence on domestic water use.

In reality, domestic water consumption is not only influenced by economic factors but also by socio-demographic factors such as population growth age (Kenney *et al.*, 2008), household size (House-Peters *et al.*, 2010), and income (Meraj *et al.*, 2021a).

3.4.2 Sources of Water for Domestic Uses within the Study Area

The sources of water to the people of Idah as revealed in Table 3.2 was mainly through boreholes and river/streams. This study revealed that in Igalogba, they mainly depend on borehole (100%); in Ede-Adejo, they depend majorly on borehole (72.7%), river (18.2), and well (9.1%); in Ega, they depend majorly on borehole (53.7%), river (39%), and well

(7.3%), in Oketecheje, they depend majorly on borehole (52.3%), river (38.6%), and well (9.1%); in Owoliapa, they depend majorly on borehole (90.9%) and well (9.1%); in Ogenegu and Olokpoji, they depend mainly on river (100%); in Inachalo, they depend on borehole (56.1%) and river (43.9%); and in Ojigagala, they depend majorly on river (88.4%) and borehole (11.6%).

In all sampled locations, a total of 188 (48.6%) source their water from borehole, 184 (47.5%) source their water from river or streams, while only 15 (3.9%) sources their water from wells.

None of the respondents sourced water from pipe-borne due to the non-functionality of the water board in Idah and also, none of the respondents sourced water from rainfall as the research was carried out during the dry season (December). However, some households still combined two or more water sources together. This finding is in line with the findings of various workers who have concluded that groundwater (borehole) is the major source of water among rural communities (Farooq *et al.*, 2022a; Gulati *et al.*, 2023).

Data presented in Table 3.3 reveals that three respondents representing 0.8% said the source is owned by government, 199 respondents representing 51.4% said it is owned by an individual, and 185 representing 47.8% said it is natural. It can be concluded that in Owoliapa community, the source is mainly owned by individuals as the households depend mainly on boreholes and wells because the river in the community is not good for use due to some traditional reasons, and household water sources in Ogenegu and Olokpoji community is owned by nature because they mainly depend on river as their water source. In general, majority of the sources are owned privately by an individual because majority of

Table 3.3 Distribution of respondents on ownership of water source.

S/N	Sample location	Government		Private individual		Community		Natural	
		Freq.	**%**	**Freq.**	**%**	**Freq.**	**%**	**Freq.**	**%**
1	Igalogbo	1	2.3	42	97.7	0	0	0	0
2	Ede-Adejo	0	0	36	81.8	0	0	8	18.2
3	Ega	2	4.9	23	56.1	0	0	16	39.0
4	Oketecheje	0	0	27	61.4	0	0	17	38.6
5	Owoliapa	0	0	44	100	0	0	0	0
6	Ogenegu	0	0	0	0	0	0	43	100
7	Inachalo	0	0	22	53.7	0	0	19	46.3
8	Olokpoji	0	0	0	0	0	0	44	100
9	Ojigagala	0	0	5	11.6	0	0	38	88.4
Total		**3**		**199**		**0**		**185**	
Percentage (%)		**0.8**		**51.4**		**0.0**		**47.8**	

Source: Field Work, 2020.

them depend on borehole. This conforms to the findings of World Bank (2006) where in Pakistan, the major contribution to water supply within the home comes from private hand, which boosts domestic water supply.

With respect to distance covered in all the sampled areas, Table 3.4 reveals that 118 respondents representing 30.5% trek a distance less than 51 m, 117 respondents representing 30.2% trek a distance between 51 and 100 m, 55 respondents representing 14.2% trek a distance between 101 and 150 m, 20 respondents representing 5.2% trek a distance between 151 and 200 m, and 77 respondents representing 19.9% trek a distance above 200 m. This shows that some people in some areas of the study (Olokpoji and Oketecheje) trekked long distances before they can get water because they majorly depend on river and streams, which are far from their homes.

Although, from the result, in the study area, majority (60.7%) trekked a distance below 100 m to their respective water sources, because boreholes and wells were dug in different places within the communities. This shows that majority trekked less than 50 m to water sources. By implication, this reduces the stress of trekking a long distance in search of water. This finding is in line with the recommended distance by the World Health Organization (WHO, 2000) which considered 200 m as a convenient distance for fetching water. Data presented in Table 3.5 reveals the mode of payment for household water among some the sampled communities. The result presented revealed that 180 of the respondents representing 46.5% pay for water, while 207 respondents representing 53.5% do not pay for water. Some communities in the study area such as Olokpoji, Ojigagala, and Ogenegu majorly depend on river, which is a natural source, therefore, they do not pay for water at all. It can

Table 3.4 Distribution of respondents on distance to water source(s).

S/N	Sample location	Less than 51 m		51–100 m		101–150 m		151–200 m		More than 200 m	
		Freq.	**%**	**Freq.**	**%**	**Freq.**	**%**	**Freq.**	**%**	**Freq.**	**%**
1	Igalogbo	20	46.5	23	53.5	0	0	0	0	0	0
2	Ede-Adejo	25	56.8	9	20.5	0	0	0	0	10	22.7
3	Ega	8	19.5	13	31.7	8	19.5	2	4.9	10	24.4
4	Oketecheje	13	29.5	8	18.2	0	0	4	9.1	19	43.2
5	Owoliapa	14	31.8	15	34.1	15	34.1	0	0	0	0
6	Ogenegu	20	46.5	10	23.3	13	30.2	0	0	0	0
7	Inachalo	14	34.1	17	41.5	7	17.1	3	7.3	0	0
8	Olokpoji	1	2.3	2	4.5	8	18.2	8	18.2	25	56.8
9	Ojigagala	3	7.0	20	46.5	4	9.3	3	7.0	13	30.2
Total		**118**		**117**		**55**		**20**		**77**	
Percentage (%)		**30.5**		**30.2**		**14.2**		**5.2**		**19.9**	

Source: Field Work, 2020.

Table 3.5 Distribution of respondents on payment for water.

S/N	Sample location	Yes		No	
		Freq.	%	Freq.	%
1	Igalogbo	39	9.7	4	9.3
2	Ede-Adejo	30	68.2	14	31.8
3	Ega	25	61.0	16	39.0
4	Oketecheje	23	52.3	21	47.7
5	Owoliapa	38	86.4	6	13.6
6	Ogenegu	0	0.0	43	100
7	Inachalo	20	48.8	21	51.2
8	Olokpoji	0	0	44	100
9	Ojigagala	5	11.6	38	88.4
Total		**180**		**207**	
Percentage (%)		**46.5**		**53.5**	

Source: Field Work, 2020.

be concluded that majority (53.5%) do not pay for water. This is because Idah is surrounded by water and inland streams, and this saves their money.

This study conforms to the study of Olajuyigba (2010) where a place with access to river pays no money to fetch water. On the amount paid for water, Table 3.6 reveals that 170 respondents representing 94.4% paid the sum of ten Naira (₦10) per 20 L of water fetched, and 10 respondents representing 5.6% paid fifteen Naira (₦15) per 20 L of water fetched. The results reveal that majority paid the sum of ₦10 to fetch water. This implies that majority spent about ₦55 to fetch water daily, and ₦1650 monthly for water. The few who pays ₦15 to fetch water is a result of the distance of the water source.

This is in line with Oghifo (2008) where he states, the farther away the source of water supplies, the higher the cost. Table 3.7 reveals the time taken to fetch water from the different respective sources. The results reveal that 167 respondents representing 43.2% spend less than 10 min to get water, 93 respondents representing 24% spend between 11 and 30 min to get water, 95 respondents representing 24.5% spend between 31 min and 1 h to get water, and 32 respondents representing 8.3% spend above 1 h to get water.

This shows that majority of the respondents spend less than 10 min to get water, by implication, it means water sources are close to their respective homes. It was observed during the investigation that most homes have their own boreholes or wells. This reduces the stress of trekking a long distance in search of water (Kotecha *et al.*, 2023).

Table 3.6 Distribution of respondents on amount (in Naira) paid per 20 liters.

S/N	Sample location	5 N		10 N		15 N		20 N	
		Freq.	**%**	**Freq.**	**%**	**Freq.**	**%**	**Freq.**	**%**
1	Igalogbo	0	0	37	94.9	2	5.1	0	0
2	Ede-Adejo	0	0	30	100	0	0	0	0
3	Ega	0	0	23	92.0	2	8.0	0	0
4	Oketecheje	0	0	23	100	0	0	0	0
5	Owoliapa	0	0	32	84.2	6	15.8	0	0
6	Ogenegu	0	0	0	0	0	0	0	0
7	Inachalo	0	0	20	100	0	0	0	0
8	Olokpoji	0	0	0	0	0	0	0	0
9	Ojigagala	0	0	5	100	0	0	0	0
Total		**0**		**170**		**10**		**0**	
Percentage (%)		**0**		**94.4**		**5.6**		**0**	

Source: Field Work, 2020.

Table 3.7 Distribution of respondents on time taken to fetch water from source(s).

S/N	Sample location	Less than 10 min		11–30 min		31 min–1 h		More than 1 h	
		Freq.	**%**	**Freq.**	**%**	**Freq.**	**%**	**Freq.**	**%**
1	Igalogbo	43	100	0	0	0	0	0	0
2	Ede-Adejo	27	61.4	7	15.9	10	22.7	0	0
3	Ega	11	26.8	10	24.4	20	48.8	0	0
4	Oketecheje	12	27.3	9	20.5	17	38.6	6	13.6
5	Owoliapa	30	68.2	14	31.8	0	0	0	0
6	Ogenegu	22	51.2	8	18.6	13	30.2	0	0
7	Inachalo	16	39.0	17	41.5	8	19.5	0	0
8	Olokpoji	2	4.5	10	22.7	22	50.0	10	22.7
9	Ojigagala	4	9.3	18	41.9	5	11.6	16	37.2
Total		**167**		**93**		**95**		**32**	
Percentage (%)		**43.2**		**24.0**		**24.5**		**8.3**	

Source: Field Work, 2020.

3.4.3 Magnitude of Household Water Demand in the Study Area

Data presented in Table 3.8 above reveals the quantity of water required for households per day. It was revealed that 11 respondents representing 2.8% need less than 50 L per day, 126 respondents representing 32.6% need between 51 and 100 L per day, 160 respondents representing 41.3% need between 101 and 150 L per day, 60 respondents representing 15.5% need between 151 and 200 L per day, and 30 respondents representing 7.8% need above 200 L per day. This shows that majority of the respondents need between 101 and 150 L of water per day for domestic uses. According to the WHO, between 50 and 100 L of water per person per day are needed to ensure that most basic needs are met and few health concerns arise. This implies that more water is required in the area, compared to the standard recommended by the WHO, and this will lead to water wastage, which is against sustainable development goal 6 (Farooq *et al.*, 2022b). This study disagrees with that of Shaban and Sharma (2007), which recommended 80 L per person per day and 100 L per person per day, respectively, as a basic quantity of water required for domestic use.

Table 3.9 reveals how often water is fetched in the study area. It was found that 235 respondents representing 60.7% fetch water daily, 117 respondents representing 30.2% fetch water every 2 d, 27 respondents representing 7.0% fetch water every 3 d, and 8 respondents representing 2.1% fetch water weekly. The result revealed that majority of the respondents fetch water on a daily basis. This implies that majority has to go in search of water daily and this affects their resourceful and relaxation time. This is in line with the observations Madagascar region, where women spend a lot of time collecting water on a daily basis.

Table 3.8 Distribution of respondents on water needed for household per day.

S/N	Sample location	Less than 50 L		51–100 L		101–150 L		151–200 L		More than 200 L	
		Freq.	**%**	**Freq.**	**%**	**Freq.**	**%**	**Freq.**	**%**	**Freq.**	**%**
1	Igalogbo	0	0	4	9.3	11	25.6	15	34.9	13	30.2
2	Ede-Adejo	1	2.3	4	9.1	31	70.5	3	6.8	5	11.4
3	Ega	0	0	3	7.3	28	68.3	8	19.5	2	4.9
4	Oketecheje	3	6.8	33	75.0	8	18.2	0	0	0	0
5	Owoliapa	0	0	15	34.9	24	54.5	5	11.4	0	0
6	Ogenegu	0	0	18	41.9	20	46.5	5	11.6	0	0
7	Inachalo	0	0	2	4.9	9	22.0	20	48.8	10	24.4
8	Olokpoji	3	6.8	20	45.5	18	40.9	3	6.8	0	0
9	Ojigagala	4	9.3	27	62.8	11	25.6	1	2.3	0	0
Total		**11**		**126**		**160**		**60**		**30**	
Percentage (%)		**2.8**		**32.6**		**41.3**		**15.5**		**7.8**	

Source: Field Work, 2020.

Table 3.9 Distribution of respondents on how often do you fetch water.

S/N	Sample location	Daily		Every 2 d		Every 3 d		Weekly	
		Freq.	**%**	**Freq.**	**%**	**Freq.**	**%**	**Freq.**	**%**
1	Igalogbo	20	46.5	19	44.2	3	7.0	1	2.3
2	Ede-Adejo	28	63.6	15	34.1	1	2.3	0	0
3	Ega	19	46.3	19	46.3	3	7.3	0	0
4	Oketecheje	30	68.2	9	20.5	5	11.4	0	0
5	Owoliapa	21	47.7	18	40.9	2	4.5	3	6.8
6	Ogenegu	33	76.7	5	11.6	5	11.6	0	0
7	Inachalo	23	56.1	9	22.0	5	12.2	4	9.8
8	Olokpoji	39	88.6	4	9.1	1	2.3	0	0
9	Ojigagala	22	51.2	19	44.2	2	4.7	0	0
Total		**235**		**117**		**27**		**8**	
Percentage (%)		**60.7**		**30.2**		**7.0**		**2.1**	

Source: Field Work, 2020.

On the collection of water, Table 3.10 reveals that 59.4% of water collectors were wives, 24.3% were children, 5.4% were men, and 10.9% represents collection by other people like water vendors. This indicates that women are majorly responsible for collection of water in the study area.

By implication, women spend most of their time and energy collecting water in the area since high quantity of water is required for domestic use. The water crisis is observed as a general problem for the rural population, women bear the greatest burden because of their social gendered roles, which involve looking for and collecting water for households (Kanga *et al.*, 2022a; Kotecha *et al.*, 2023).

Table 3.11 reveals the liters of water fetched in a day in the study area. The results reveal that 19 respondents representing 4.9% fetch less than 50 L per day, 146 respondents representing 37.7% fetch between 51 and 100 L per day, 165 respondents representing 42.6% fetch between 101 and 150 L per day, 48 respondents representing 12.4% fetch between 151 and 200 L per day, and 9 respondents representing 2.3% fetch above 200 L per day. This indicates that majority fetch between 101 and 150 L per day. From Tables 3.8 and 3.11, it can be deduced that the majority of the respondents do not get the demanded quantity of water that they need per day.

This is due to the inadequate distribution and the stress of having to go in search of water to buy every day, as well as the cost of getting water. This study is in conformity with that of Ojo (2014) where available sources of water is not enough to meet up with the demand of residents of many urban and semi-urban areas.

Table 3.10 Distribution of respondents on who is responsible for water collection.

S/N	Sample location	Wife		Children		Men		Others	
		Freq.	%	Freq.	%	Freq.	%	Freq.	%
1	Igalogbo	22	51.2	11	25.0	4	9.3	6	14.0
2	Ede-Adejo	25	56.8	10	22.7	2	4.5	7	15.9
3	Ega	23	56.1	11	26.8	2	4.9	5	12.2
4	Oketecheje	30	68.2	8	18.2	5	11.4	1	2.3
5	Owoliapa	27	61.4	9	20.5	4	9.1	4	9.1
6	Ogenegu	30	69.8	9	20.9	1	2.3	3	7.0
7	Inachalo	24	58.5	10	24.4	1	2.4	6	14.6
8	Olokpoji	25	56.8	11	25.0	2	4.5	6	13.6
9	Ojigagala	24	55.8	15	34.9	0	0	4	9.3
Total		**230**		**94**		**21**		**42**	
Percentage (%)		**59.4**		**24.3**		**5.4**		**10.9**	

Source: Field Work, 2020.

Table 3.11 Distribution of respondents on how many liters of water fetched a day.

S/N	Sample location	Less than 50 L		51–100 L		101–150 L		151–200 L		More than 200 L	
		Freq.	%	Freq.	%	Freq.	%	Freq.	%	Freq.	%
1	Igalogbo	0	0	4	9.3	21	48.8	15	34.9	3	7.0
2	Ede-Adejo	1	2.3	9	20.5	31	70.5	3	6.8	0	0
3	Ega	0	0	3	7.3	27	65.9	9	22.0	2	4.9
4	Oketecheje	3	6.8	33	75.0	8	18.2	0	0	0	0
5	Owoliapa	0	0	15	34.1	24	54.5	5	11.4	0	0
6	Ogenegu	0	0	23	53.5	15	34.9	5	11.6	0	0
7	Inachalo	0	0	6	14.6	21	51.2	10	24.4	4	9.8
8	Olokpoji	7	15.9	24	54.5	13	29.5	0	0	0	0
9	Ojigagala	8	18.6	29	65.9	5	11.6	1	2.3	0	0
Total		**19**		**146**		**165**		**48**		**9**	
Percentage (%)		**4.9**		**37.7**		**42.6**		**12.4**		**2.3**	

Source: Field Work, 2020

3.4.4 Consumption Pattern of Water in the Study Area

Table 3.12 reveals the average domestic water consumption pattern per day in the study area. For easy analysis, the various purposes for water consumption were classified into six, as shown in Table 3.12. For kitchen, it involves washing plates and utensils, washing vegetables and cooking of foods, and for personal hygiene, it involves brushing of teeth, ablution, washing of hands, feet, shaving, and others.

The daily household water use for drinking in all the nine sample areas as revealed in the table was 26.5 L per day, household water use for bathing in all the sample areas stood at 308 L per day, the daily household use for clothes washing was 448 L per day, the daily household use for kitchen stood at 81.5 L per day, the daily quantity used for personal hygiene was 32.5 L, and the daily water use for toilet in all the sampled areas was 52 L. This indicates that water use for washing clothes (448 L) and bathing (308 L) accounted for the largest quantity of household water consumption per day in the study area.

Also, Table 3.12 reveals the per capita water consumption in each sampled area. It was discovered that household in Igalogba consume 147 L of water per day. In Ede-Adejo, 102.5 L was used per household. In Ega, 131 L was used; in Oketecheje, 87.5 L was used; in Owoliapa, 105.5 L was used; in Ogenegu, 106.5 L was used; in Inachalo, 110 L was used; in Olokpoji, 73 L was used; and in Ojigagala, 85.5 L was used per household. From the result presented in Table 3.12, the communities used a high proportion of water for clothes washing and a small quantity was used for drinking.

The result of this study, however, revealed that residents of Igalogba, Ede-Adejo, Ega, Owliapa, Ogenegu, and Inachalo consumed more water than the recommended minimum requirement. Residents in Oketecheje and Ojigagala used 87.5 and 85.5 L per capita per

Table 3.12 Average water consumption pattern.

S/N	Sample location	Microcomponent uses (L)						
		Drinking	Bathing	Clothes washing	Kitchen	Personal hygiene	Toilet	Total
1	Igalogbo	3	43	71	11	4	15	**147.0**
2	Ede-Adejo	2.5	30	50	8	3	9	**102.0**
3	Ega	4	41	60	12	10	4	**131.0**
4	Oketecheje	3	28	35	10	3.5	8	**87.5**
5	Owoliapa	3	40	42	8	3.5	9	**105.5**
6	Ogenegu	2.5	42	51	7	2	2	**106.5**
7	Inachalo	3	33	61	8	3	2	**110.0**
8	Olokpoji	2.5	23	35	10	1.5	1	**73.0**
9	Ojigagala	3	28	43	7.5	2	2	**85.5**
Total		**26.5**	**308**	**448**	**81.5**	**32.5**	**52**	

Source: Field Work, 2020.

day, respectively, with only about 7.5 L and 5.5 L, respectively, were considered as waste. In Olokpoji, the residents used below the required quantity of water. These findings disagreed with UNDP (2008) which says that the minimum absolute daily water need per person per day is 50 L (13.2 gallons) which include: 5 L for drinking, 20 L for sanitation and hygiene, 15 L for bathing, and 10 L for preparing food.

This study conforms to the international consumption figures released by the 4th World Water Forum (2006), which indicates that a person living in an urban or semi-urban area uses an average of 250 L per day.

3.4.5 Factors Influencing Consumption Pattern of Household Water in the Study Area

Table 3.13 reveals the factors influencing domestic water consumption in the study area. It reveals that household size has the highest proportion at 33.6%. The relevance of household size in domestic water use cannot be overlooked. It implies that the larger the household, the higher the domestic water use. Also, the quantity of water supplied by the supplier stood at 22.7%. This implies that the more the water supplied by the fetchers, the higher the household water use. Also, household preference for a particular water source also influences domestic water use in the study area with 10.9% of the respondents. This observation

Table 3.13 Distribution of respondents on main determinants of domestic water consumption.

Variables	Frequency	Percentage (%)
Household size	130	33.6
Quantity supplied by fetchers	88	22.7
Preference for a source	42	10.9
Religious use of water	22	5.7
Gender composition	29	7.5
Family income	10	2.6
Cost of water	22	5.7
Cultural influences	16	4.1
Others	28	7.2
Total	**387**	**100**

Source: Field Work, 2020.

is the central focus of Vasques (2011). This factor is relevant, where family prefers a particular source for a given home use.

Another component is gender composition, which stood at 7.5%, the more the females in a given household, the higher the domestic water use. Females have been found to use more water than their male counterparts.

This observation is similar to that of Xinming *et al.* (2000). Another component is religious use of water, which stood at 5.7%. This component becomes important because of the presence of Muslims in Idah who use water for ablution purposes. Cost of water and level of income are also a determinant affecting domestic water consumption in the study area. Other factors include: age range of the supplier, closeness to water sources, and age component. Households dominated by young adults are more likely to have more supply of water than those homes dominated by aged or little children. Also, the closer the water source, the more water was supplied for household uses. A similar observation was made by Environment Agency (2008). The age compositions also affect water consumption of a particular household. This component may be important especially where the respondent is within the age of working class. Such group of people is likely to use more water. The study investigated the relationship between water supply and water consumption using the Pearson Moment Correlation Analysis technique. The results are shown in Table 3.14.

$$\text{Water Consumed} = \frac{\text{Total water consumed}}{\text{Number of respondents}} = \frac{40{,}672}{387} = 105.1 \text{ L per person per day}$$

$$\text{Water Supplied} = \frac{\text{Total water supplied}}{\text{Number of household}} = \frac{42{,}725}{387} = 110.4 \text{ L per household}$$

$$r_{xy} = \frac{n\sum xy - \sum y \sum x}{\sqrt{n\sum x^2 - (\sum x)^2 (n\sum y^2 - (\sum y)^2}}$$

Where, r_{xy} = Pearson r correlation coefficient between x and y

n = number of observations (9)

x = value of x (water consumed)

y = value of y (water supplied)]

Table 3.14 Correlation analysis of water consumed and water supplied.

S/N	Sample location	Questionnaire administered	Questionnaire recovered	Total water consumed (X)	Total water supplied (Y)	X^2	Y^2	XY
1	Igalogbo	48	43	4519.3	4747.2	20424072.49	22535907.84	21454020.96
2	Ede-Adejo	44	44	4624.4	4857.6	21385075.36	23596277.76	22463485.44
3	Ega	44	41	4309.1	4526.4	18568342.81	20488296.96	19504710.24
4	Oketecheje	44	44	4624.4	4857.6	21385075.36	23596277.76	22463485.44
5	Owoliapa	44	44	4624.4	4857.6	21385075.36	23596277.76	22463485.44
6	Ogenegu	44	43	4519.3	4747.2	20424072.49	22535907.84	21454020.96
7	Inachalo	44	41	4309.1	4526.4	18568342.81	20488296.96	19504710.24
8	Olokpoji	44	44	4624.4	4857.6	21385075.36	23596277.76	22463485.44
9	Ojigagala	44	43	4519.3	4747.2	20424072.49	22535907.84	21454020.96
Total		**400**	**387**	**40,673.7**	**42,724.8**	**183949204.53**	**202969428.48**	**193225425.12**

Source: Field Work, 2020.

$$r_{xy} = \frac{9(193225425.12) - (40673.7)(42724.8)}{\sqrt{(9(183949204.53)^2 - (40673.7)^2 (9(202969428.48)^2 - (42724.8)^2}}$$

$$r_{xy} = \frac{1739028825 - 1737775697.8}{\sqrt{1655542840.8 - (1654349871.7)(1826724856.3) - 1825408535}}$$

$$r_{xy} = \frac{1253127.2}{\sqrt{1192969.1)(1316321.3}}$$

$$r_{xy} = \frac{1253127.2}{\sqrt{(1192969.1)(1316321.3)}}$$

$$r_{xy} = \frac{1253127.2}{\sqrt{1570330650000}}$$

$$r_{xy} = \frac{1253127.2}{1253128.454}$$

$$r_{xy} = 0.999999$$

From the result of the correlation coefficient (0.999999), there is high positive linear relationship between the household's water supply and water consumed in the study area.

Hypothesis:

H_0 = There is no significant relationship between household water supply and consumption in the study area.

Level of significance = 0.05

Significance level = 2 degree of freedom

$$\text{Statistics} = t_{cal} = \frac{r\sqrt{n-2}}{\sqrt{1-r}}$$

Where - r = correlation coefficient (0.999999)

n = number of observation (9)

$$t_{cal} = \frac{0.999999\sqrt{9-2}}{\sqrt{1-0.999999}}$$

$$t_{cal} = \frac{0.999999(2.64575)}{\sqrt{0.000001}}$$

$$t_{cal} = \frac{2.6457}{0.001}$$

$$t_{cal} = 2645.7$$

$$t_{tab} = 1.67$$

The decision rule is that reject null hypothesis if the calculated value is greater than the critical value at 0.05 significance level. Since t_{cal} (2645.7) is greater than t_{tab} (1.67), we reject null hypothesis (H_0) and accept alternative hypothesis (H_1). Therefore, it can be concluded that there is a significant relationship between household water supply and consumption in the study area.

3.5 Conclusion and Recommendations

Findings from the study indicate that the demand for domestic water varies in terms of household size, which shows that the number of people found in a household determines the quantity of water needed for domestic purpose. This study further shows that variation in domestic water use pattern in the study area could be attributed to the availability of water and economic status of residents. The high rate of domestic water wastage in the high income and middle-income groups like Igalogba, Ega, Inachalo, and Ede-Adejpo in the study area was a function of their economic status as well as the easy access to domestic water. Although residents of the rural areas belonged to the low income group, like Olokpoji and Ojogagala, they consumed less water because they are not able to afford water-consuming appliances or engage in water-guzzling activities. The study also revealed that the government played very little role in the supply of water in the study area because majority of the water sources are owned by private individuals.

The study concluded that there is need for the intervention of Government in the provision of adequate water to every individual in the study area. If we are to meet the sustainable development goal 6 set by the international bodies, there is a need for enforcement of water legislation and policies related to water resource planning, development, and management (Kanga *et al.*, 2022). Also, it is evidenced from the study that domestic water use on a daily basis in Idah is not static but rather under the influence of certain variables. In order for policy makers and water planners to effectively manage water resources, they must consider the factors influencing water demand, supply, and consumption in the area. (Farooq *et al.*, 2022a, b; Sajan *et al.*, 2022; Debnath *et al.*, 2023a, b, c; Sud *et al.*, 2023; Lavane *et al.*, 2023; Upadhyay *et al.*, 2023).

References

Farooq, M., Meraj, G., Rishabh, Kanga, S., Nathawat, R., Singh, S. K., & Ranga, V. (2022a). Slum Categorization for Efficient Development Plan—A Case Study of Udhampur City, Jammu and Kashmir Using Remote Sensing and GIS. In *Geospatial Technology for Landscape and Environmental Management: Sustainable Assessment and Planning* (pp. 283-299). Cham: Springer International Publishing.

Kanga, S., Meraj, G., Farooq, M., Singh, S. K., & Nathawat, M. S. (Eds.). (2022a). Disaster Management in the Complex Himalayan Terrains. *Geography of the Physical Environment*. Springer, Cham.

Debnath, J., Sahariah, D., Lahon, D., Nath, N., Chand, K., Meraj, G., Kumar, P., Singh, S. K., Kanga, S., & Singh, S. K. (2022a). Geospatial modeling to assess the past and future land use-land cover changes in the Brahmaputra Valley, NE India, for sustainable land resource management. *Environ. Sci. Pollut. Res.*, 1-24.

Meraj, G., Kanga, S., Kranjčić, N., Đurin, B., & Singh, S. K. (2021a). Role of natural capital economics for sustainable management of earth resources. *Earth*, 2(3), 622-634.

House-Peters, L., Pratt, B., & Chang, H. (2010). Effects of urban spatial structure, socio-demographics, and climate on residential water consumption in Hillsboro, Oregon. *J. Am. Water Resour Assoc.*, *46*(3), 461-472.

Ifabiyi, I.P. (2011) Willingness to Pay for Water at Household Level in Ilorin Kwara State: *Glob. J. Hum. Soc. Sci.*, 11(2):14 -19.

Debnath, J., Sahariah, D., Lahon, D., Nath, N., Chand, K., Meraj, G., Kumar, P., Singh, S. K., Kanga, S., & Farooq, M. (2023b). Assessing the impacts of current and future changes of the planforms of River Brahmaputra on its land use-land cover. *Geosci. Front.* https://doi.org/10.1016/j.gsf.2023.101557.

Kenney, S. D., Goemans, C., Klein, R., Lowery, J., & Reidy, K. (2008). Residential water demand management: Lessons from Aurora, Colorado. *J. Am. Water Resour. Assoc.*, *44*(1), 192-207. http://dx.doi.org/10.1111/j.1752-1688.2007.00147.x

Debnath, J., Sahariah, D., Saikia, A., Meraj, G., Nath, N., Lahon, D., Annayat, W., Kumar, P., Chand, K., Singh, S. K., & Kanga, S. (2023c). Shifting Sands: Assessing Bankline Shift Using an Automated Approach in the Jia Bharali River, India. *Land*, 12(3), 703.

Kotecha, M. J., Kanga, S., Pankhaniya, S. K., Agrawal, S., Meraj, G., & Singh, S. K. (2023). Urban Heat Island (UHI) Resilience Plan in Varying Climatic Conditions Using Geospatial Approach: A Case Study of Rajkot City. In *Climate Change, Agriculture and Society: Approaches Towards Sustainability* (pp. 307-338). Cham: Springer International Publishing.

Mohammed, A,H and Sanullah, P. (2017). An Empirical Analysis of domestic Water Sources, Consumption and Association Factors in Kandahar City. Afghanistan. *Environ. Nat. Resour. J.*, 17(2):49 to 61

Lahon, D., Sahariah, D., Debnath, J., Nath, N., Meraj, G., Farooq, M., Kanga, S., Singh, S. K., & Chand, K. (2023a). Growth of water hyacinth biomass and its impact on the floristic composition of aquatic plants in a wetland ecosystem of the Brahmaputra floodplain of Assam, India. *PeerJ*, 1-32. http://dx.doi.org/10.7717/peerj.14811

Lahon, D., Sahariah, D., Debnath, J., Nath, N., Meraj, G., Kumar, P., Hashimoto, S. and Farooq, M. (2023b). Assessment of Ecosystem Service Value in Response to LULC Changes Using Geospatial Techniques: A Case Study in the Merbil Wetland of the Brahmaputra Valley, Assam, India. *ISPRS Int. J. Geo-Inf.*, 12(4), p.165.

Lavane, K., Kumar, P., Meraj, G., Han, T. G., Ngan, L. H. B., Lien, B. T. B., Ty, T. V., Thanh, N. T., Downes, N. K., Nam, N. D. G., Minh, H. V. T., Singh, S. K., & Kanga, S. (2023). Assessing the Effects of Drought on Rice Yields in the Mekong Delta. *Climate*, 11(1), 13.

Farooq, M., Rashid, H., Meraj, G., Kanga, S., & Singh, S. K. (2022c). Assessing the Microclimatic Environmental Indicators of Climate Change of a Temperate Valley in the Western Himalayan region. In A. R. Siddiqui and Avijit Sahay (Eds.), *Climate Change, Disaster and Adaptations - Contextualizing human responses to ecological change* (pp. 47–61). https://link.springer.com/chapter/10.1007/978-3-030-91010-5_5

Farooq, M., Mushtaq, F., Meraj, G., Singh, S. K., Kanga, S., Gupta, A., Kumar, P., Singh, D., & Avtar, R. (2022b). Strategic Slum Upgrading and Redevelopment Action Plan for Jammu City. *Resources*, 11(12), 120.

Omvir, S. and Sushila, T. (2013). A Survey of Household Domestic Water Consumption Patterns in Rural Semi-arid Village, India. *GeoJournal, Spatially Integrated Social Sciences and Humanities*, 78 (5): 777-790.

Sajan, B., Mishra, V. N., Kanga, S., Meraj, G., Singh, S. K., & Kumar, P. (2022). Cellular automata-based artificial neural network model for spatial-temporal dynamics and land use/land cover change prediction. *Agronomy*, 12, 2772.

Singh, S., Singh, S. K., Prajapat, D. K., Pandey, V., Kanga, S., Kumar, P., & Meraj, G. (2023a). Assessing the Impact of the 2004 Indian Ocean Tsunami on South Andaman's Coastal Shoreline: A Geospatial Analysis of Erosion and Accretion Patterns. *J. Mar. Sci. Eng.*, 11(6), 1134.

Sud, A., Kanga, R., Singh, S. K., Meraj, G., Kanga, S., Kumar, P., Ramanathan, A. L., & Bhardwaj, V. (2023). Simulating Groundwater Potential Zones in Mountainous Indian Himalayas—A Case Study of Himachal Pradesh. *J. Geophys. Res.: Earth Surf.*, 10(3), 65.

Shaban, A. And Sharma, R.N. Sharma (2007). Water Consumption Patterns in Domestic Households in Major Cities of India. *Econ. Polit. Wkly*. June 9,

Upadhyay, R. K., Tripathi, G., Đurin, B., **Šamanović**, S., Cetl, V., Kishore, N., Sharma, M., Singh, S. K., Kanga, S., Wasim, M., Rai, P. K., & Bhardwaj, V. (2023). Groundwater Potential Zone Mapping in the Ghaggar River Basin, North-West India, Using Integrated Remote Sensing and GIS Techniques. *Water*, 15(5), 961.

Vasquez WF (2011). Household preferences and governance of water services: A Hedonic analysis from rural Guatemala. *International food Policy Research Institute (IFRRI) Discussion Paper 01152.*

WHO. (2003). *Domestic water quantity, service level and health.* Geneva: World Health Organization. World Bank, 2006

World Bank. (2004). *World Development Indicators.* Washington, DC: The World Bank.

World Health Organization (2000). *Guidelines For Drinking Water Quality*, Volume 1. Recommendations 2nd Edition.

World Water Forum (2006); *The right to Water: A right for all citizens*(www.academic-eau.org/img/pdf/02-ADEO6-EN-4fev.pdf).

Xinming M, Whittington D, Briscoe J (2000). Modeling Village Water Demand Behaviour: A Discrete Choice Approach. *Water Res.* 26(4):521-529.

Young, G. (2006). *World Water Development Report* Willrid Lauries University Waterloo Ontario.

Gulati, B., Sharma, R., Kanga, S., Singh, S. K., Sajan, B., Meraj, G., Kumar, P., & Ramanathan, A. L. (2023). Unraveling the Relationship Between Stubble Burning and Air Quality Degradation in Punjab: A Temporal and Spatial Analysis (2019-2022). *J. Clim. Change*, 9(2), 43-53.

4

Disaster Risk Reduction and Risk Management: A Conceptual Framework

Fouzia Rahman

University Department of Geography, V.K.S. University, Arrah, Bhojpur, Bihar, India

Abstract

Disaster is a common phenomenon in the world. It is of different types but it creates havoc and damages life and resources in common. Nature and man themselves are responsible for these disasters. In other words, human development and disasters are two faces of a coin. With the advancement of human civilization, the nature and frequency of disasters somehow change. Humans have to live on this earth and disaster occurrences, unfortunately, have negative impacts on this earth. Therefore, it is necessary to disseminate the knowledge and concept of disasters, so as to manage its impact and risks. It is now mandatory to know the nature of different disasters, their frequency of occurrences, how to be resilient amid these disasters, and so on, to minimize or reduce the disaster risk. The author has tried to disseminate the conceptual knowledge about this with the present chapter.

***Keywords*:** Disaster, human civilization, resilience, risk reduction, risk management

4.1 Introduction

A disaster is a sudden catastrophic event or phenomenon that is intimately associated with calamity, environmental losses, human casualties, economic losses, social order, and much more (Debnath *et al.*, 2022). In other words, disasters are events which disrupt the functioning of a societyor community. Sometimes, hazards are taken in the same context of disasters but when hazard affects the vulnerable people, then disaster occurs. Generally, disasters occur due to natural causes but somehow, they have human origin. The development of human civilization and disasters are the two faces of a coin. So, when humans cross their limits in resource utilization, and when man and nature's relationship gets imbalanced, then disaster may occur.

Disasters are multifaceted events that significantly disrupt the social, economic, and political fabric of society. They are not merely incidents of physical destruction and death

Email: Fouziarahman87@gmail.com

Sanjay Kumar, Suraj Kumar Singh, Shruti Kanga, Gowhar Meraj, Majid Farooq and M. S. Nathawat (eds.)
Disaster Management and Environmental Sustainability, (61–74) © 2024 Scrivener Publishing LLC

but embody the intricate interplay between human vulnerability and environmental threats. Disasters challenge the conventional order, compelling individuals and communities to deviate from established norms and adopt new behaviors to cope with the aftermath. This deviation underscores the essence of disasters as both a real and perceived threat, emphasizing the critical need for long-term recovery strategies due to the significant irreversible loss and damage they cause. The literature on disasters highlights their ability to disrupt routines fundamentally, reflecting a broader hazards perspective that geographers and researchers adopt. This perspective focuses on the interaction between society's susceptibility to natural hazards—such as earthquakes, floods, and tornadoes—and the socio-economic systems' capacity to withstand and recover from these events. Thus, defining disasters extends beyond their immediate impact, incorporating the societal response to unexpected life conditions and the overarching necessity to re-evaluate human resilience in the face of environmental extremes (Farooq *et al.*, 2022a-c; Debnath *et al.*, 2022, 2023a, b; Meraj *et al.*, 2021a, 2022; Gulati *et al.*, 2023; Kanga *et al.*, 2022a, b; Lavane *et al.*, 2023).

Usually, disasters occur due to natural causes, but on the basis of their origin, they can be described in different terms. There are many agencies and research centers across the world that are associated with the task of disaster reduction and management, have classified disasters into different types (Figure 4.1) such as natural, man-made, and hybrid (Kotecha *et al.*, 2023).

4.2 Types of Disasters

1) **Natural disaster:** natural disasters are devastating in nature and are outcomes of natural causes such as cyclones, volcanic eruptions, hurricanes/tornadoes (Figure 4.2), etc. Natural disasters are often termed as "Act Of God".
2) **Man-made Disaster:** man-made disasters are the cataclysmic result of anthropogenic activities. The International Federation of Red Cross and described that, 'it is an impetuous or long term calamity where natural catastrophic causes are absent.' Where impetuous disasters are involved with structural, building, and mine collapses, and disasters are associated with land, air, and sea. On the other hand, long-term, man-made disasters include national and international conflicts (Figure 4.3).
3) **Hybrid Disaster:** this is the occurrence of both human error and natural causes. For example, heavy landslides resulting in the clearing of villages and ruining the transport system. We can understand this with the recent example which happened in Uttarakhand in the month of February 2021.

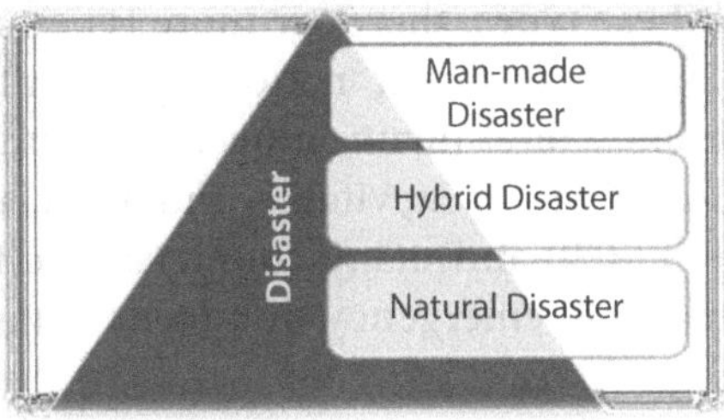

Figure 4.1 Types of disaster.

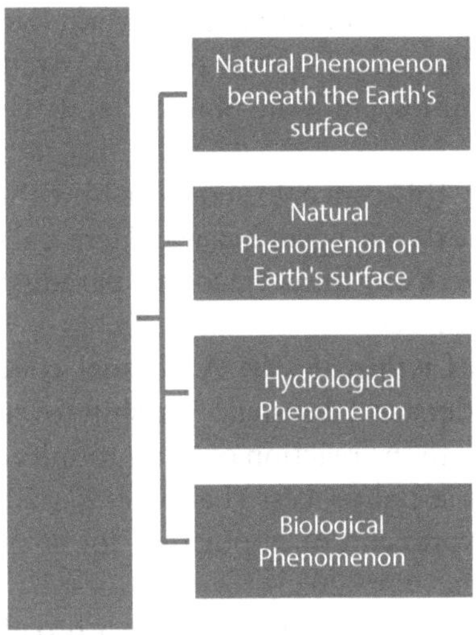

Figure 4.2 Natural disaster. Source: (Kanga *et al.*, 2023a): Disaster Prevention and Management.

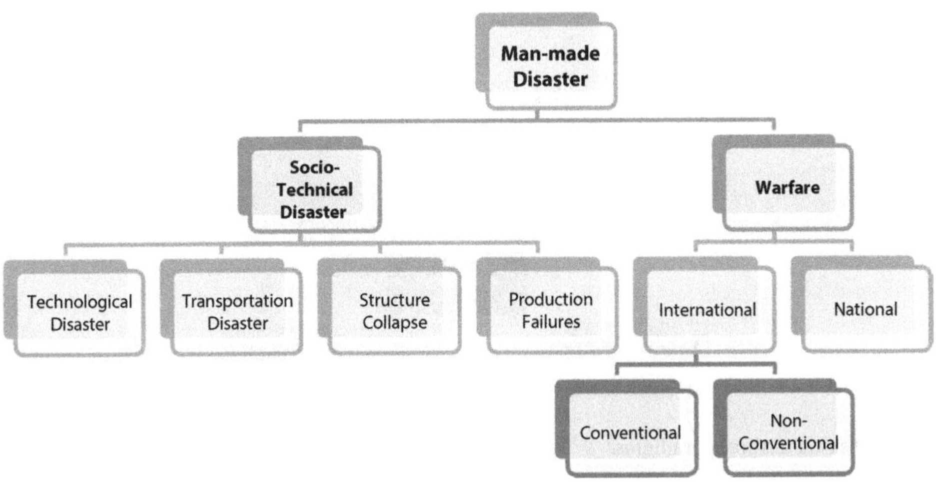

Figure 4.3 Man-made disaster. Source: M. Shaluf (2007): Disaster Prevention and Management.

Apart from all these types, there are more types of disasters that have been described by the different national and international agencies and research centers. Here, mentioning the recent outbreak of global pandemic, *i.e., COVID-19* is inevitable. India declared it as a 'Notified Disaster' on March 2020 and many countries like the USA and others also declared a national emergency due to its catastrophic impacts on human lives.

4.3 Disasters in Different Paradigms

When we try to keep disasters in a conceptual framework, there are three major paradigms that help to understand the disasters. These three main paradigms are: duplication of war, expression of social vulnerabilities, and state of uncertainty (Figure 4.4).

a) ***Duplication of war***: This concept is concerned with external agent, just like a bomb which is an external agent in a war. Here, calamity acts as an external agentfor occurrence of any disaster, so this paradigm keeps disasters in the same category of war.
b) ***Expression of social vulnerability***: A gradual conceptual shift should be seen in order to understand disasters. The shift towards social perspective is that disasters deal with the reaction of a community on serious disturbances in social order and social process (Lahon *et al.*, 2023a,b). In other words, disasters are the expression of population for social vulnerability (Kanga *et al.*, 2022b; Kotecha *et al.*, 2023).
c) ***State of uncertainty***: This paradigm shows a broad concept, *i.e.*, disasters also explain the impossible things which never have been before. So, new events and imagination may happen in reality, which is uncertain at present time. These uncertain events, which are imaginary for now, can be measured when disaster happens due to its calamitous impact on the community and society.

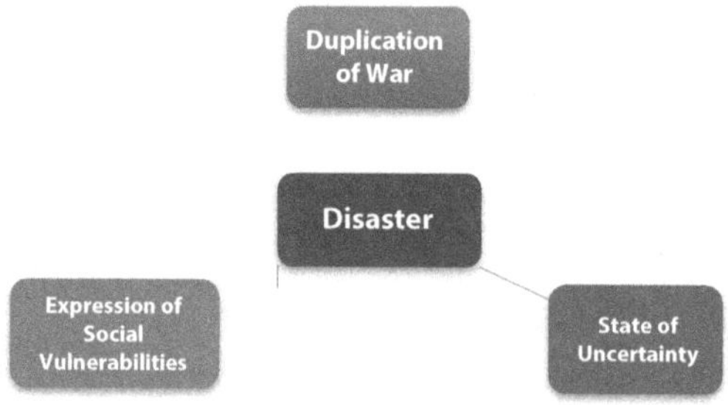

Figure 4.4 Disaster in three paradigms.

4.4 Criteria for Disasters

Every event that disrupts the social order is categorized as disasters; there are some criteria for disasters. In order to identify a phenomenon as disaster, the *United Nations Environment Programme and the Center for Research* have proposed some criteria which are as follows: index according to their exposure and vulnerability risk to a particular physical hazard. It is very helpful for planners and policy makers to identify countries which are more vulnerable to disasters. Here, hazard can be of different types, such as cyclone, earthquake, flood, and drought (UNDP). Disaster Risk index (DRI) helps policy makers to perform the task of disaster risk reduction (DRR) (Gulati *et al.*, 2023; Kanga *et al.*, 2022a).

1. Number of casualty is 10 or more people,
2. 100 people reported affected,
3. A call for international assistance,
4. Declaration of an emergency situation.

4.5 Disaster Risk

The disastrous physical events along with vulnerable conditions which act as an agent of the catastrophe can be described as disaster risk (Farooq *et al.*, 2022c). In other words, these events and phenomenon facilitates disasters. "Disaster risk is defined as the loss of life, destruction and damage of assets in specific period of time". It is associated with three main terms: hazards, exposure, and vulnerability. Generally, disasters seem like external physical phenomenon but risk is the outcome of complex developmental processes that favor or trigger the condition of hazard, exposure, and vulnerability.

a) ***Hazard:*** it is a phenomenon or activity which favors the catastrophic events that may cause casualties, destructions, and damage of assets and infrastructure, for example, flood, drought, cyclone, earthquake, and tsunami. Population or society settled in any geographical territory is exposed to such hazards.
b) ***Exposure:*** it is involved with the number of people living in an area along with their assets, infrastructure, housing, and other capabilities (Farooq *et al.*, 2022b).
c) ***Vulnerability:*** it is a multidimensional term and associated with social, physical, economic, and environmental factors. There are more factors that can be added in this list suchas cultural and institutional factors. These factors trigger the surroundings vulnerable to hazard and then that may be a cause of a disaster.

These terms collectively generate risk and there is an intimate relationship between development and risk. Therefore, to understand this relationship, the United Nation Development Program (UNDP) developed the disaster risk index (DRI). This list is prepared through measurement and comparison of vulnerability and risk with respect to different countries. Countries are enlisted to risk reduction has been an inevitable part of policy making process. Policy makers describe it (DRR) not only as post disaster healing or reconstruction but also with development planning and policy which can minimize the vulnerability. Aside from this, there is a need to promote investment in vulnerability reduction. Enhancing people awareness, their knowledge about the vulnerability, and promoting their capacity to manage risk, they all collectively form a disaster risk reduction system (Lavane *et al.*, 2023).

4.6 Disaster Risk Reduction (DRR)

The *United Nations Disaster Risk Reduction* program defines disaster risk reduction with the help of three keywords: preventing, reducing and managing (Figure 4.5). Preventing new disaster risk, reducing existing disaster risk, and managing residual risk are three measures through which DRR can be performed. These three actions are further associated with restoring, resilience, and sustainable development. DRR is the policy objective of disaster risk management. Disaster risk is intimately associated with development processes and these developmental projects/work trigger the risk. This is one aspect of development and disaster risk, another aspect is that man can contribute to reduction in disaster risk with the help of his development strategies and planning as well (Farooq *et al.*, 2022a).

The disaster risk can be reduced by minimizing the level of vulnerability. The level of vulnerability can be reduced through different preventive measures such as mitigation and preparedness. Through displacement and relocation of community and population, exposure can be kept far away from hazard and its destructive effects, shown in Figure 4.6. The figure shows how disaster risks can be reduced by reducing the level of vulnerability and shifting exposure far from hazard. So, the triangle of disaster risk shown here is smaller than that shown in Figure 4.5.

With the help of DRI, disaster risk reduction (DRR) involves strategies, policies, and planning that are used to reduce vulnerabilities and disaster risk in the community. Sometimes, it is used in the same context as disaster risk management (DRM). In other words, disasters and development are intimately connected to each other. Therefore, disaster risk walks side-by-side with development. We cannot say 'no' to development, rather, we can reduce the risk by adopting disaster-resilient developmental projects. Here, assessment, adaptation, and mitigation become necessary practices in development projects (Figures 4.7, 4.8).

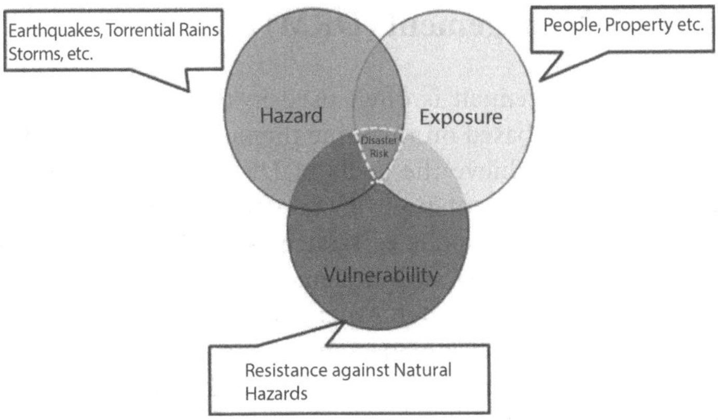

Figure 4.5 Emergence of natural disaster. Source: Asian Disaster Reduction Center.

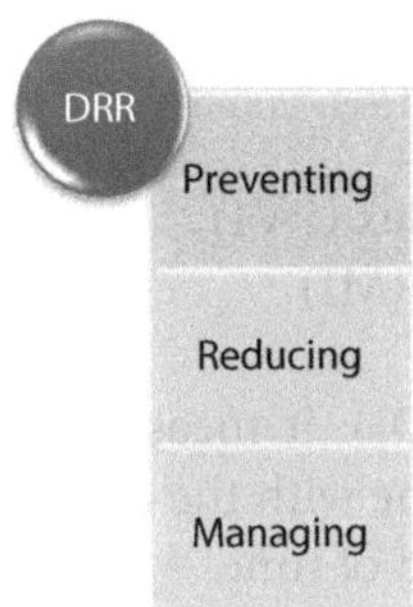

Figure 4.6 Measures for reduction of disaster risk.

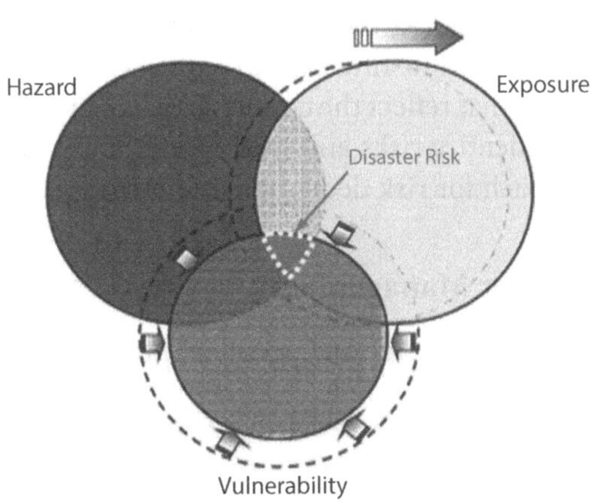

Figure 4.7 Mechanism of natural disaster reduction. Source: Asian Disaster Risk Reduction Center.

4.7 Disaster Risk Management (DRM)

The term disaster risk management is often used in the same context as disaster risk reduction. Both are generally based on the same theme but the basic difference between them is that DRM helps to achieve the goals of DRR. In other words, DRM is more focused on the practical implementation of preventive measures/initiatives. Basically, DRR considers the issues regarding policies, strategies, institutional and operational matters while DRM deals with practices and implementation of these strategies at the local and national level. For this purpose, Inter-American Development Bank (IDB) proposed a system of indicators for disaster risk management. The aim is to provide national policy makers with information that identify risk and suggest the suitable strategy to risk management. These indicators identify the socio-economic factors affecting the risk and risk management (IDB, 2010).

4.7.1 Indicators of Disaster Risk Management

Indicators of disaster risk management can be classified as:

a) Disaster deficit index (DDI)
b) Local disaster index (LDI)
c) Prevalent vulnerability index (PVI)
d) Risk management index (RMI)

a) Disaster deficit index (DDI): it measures countries' macroeconomic and financial capabilities to cope with the disaster risk or any destructive physical events in a given period of time.
b) Local disaster index (LDI): this index measures the impact of social and environmental risks at the local level, which is collectively responsible for a catastrophe and damage at the national level.
c) Prevalent vulnerability index (PVI): it is involved with a series of indicators that consider common vulnerable circumstances. Socio-economic weaknesses and lack of social resilience are very much related to this index.
d) Risk management index (RMI): actions taken for minimizing risks by different organizations and institutions are considered for this index. It has a group of indicators that reflect the country's risk management performance. So, these indices identify risks and their causes and also measure/monitor actions and approach for risk deduction and measurement (IDB, 2010).

4.7.2 Measures of Disaster Management

There are two main measures of disaster management: prevention and mitigation.

a) Prevention: in order to explain it, the question that comes to mind is why is there a need to consider disaster prevention? So, for interrupting or preventing the occurrence of any disastrous phenomenon, preventive measures become essential.

Source: Internet

For example, in constructing a dam to control floods, prevention as a measure of risk management should be considered in different areas of planning such as national development planning, counter-disaster planning, disaster legislation, and national disaster management policy. It can be implemented through different organizations, agencies, and authorities. Many government departments, NGO's, military forces, and international disaster assistance agencies are considered as resources that implement the preventive measures. There are public awareness programs and workshops that can also enrich the process of prevention.

b) Mitigation: it is a measure of risk management that helps in reducing the impact of a natural or man-made disaster in any society. It is very much similar to prevention. Preventive measures can also be used in the mitigation process such as strengthening the building to be more resistant to earthquakes so that the impact of earthquakes can be reduced. Adopting a crop pattern that can resist the impact of flood/drought is another example. With appropriate mitigation programs, a comprehensive national disaster policy, and adequate public awareness and education programs, the task of mitigation can be performed very well. The process of disaster management can be understood through the disaster management cycle. The traditional approach of risk management is based on "Disaster Cycle". It is divided into two phases, *i.e.*, before and after disaster. Each phase has different steps, such as mitigation, preparedness, response, and recovery.

1. **Factor prior to disaster:** as mentioned earlier, the management cycle has two phases, *i.e.*, before and post disaster. So there are some important factors

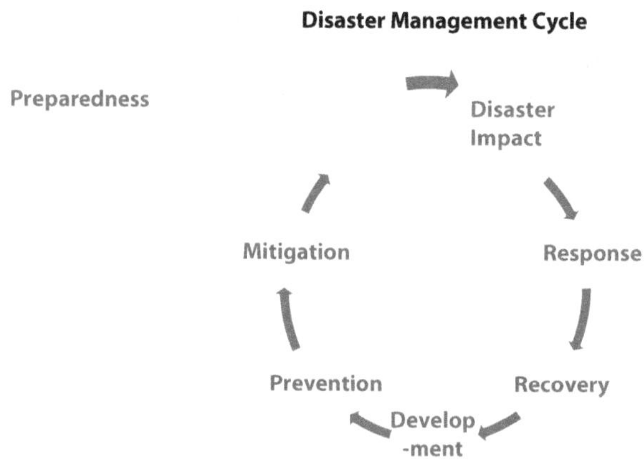

Figure 4.8 Disaster management cycle. Source: Disaster Management Handbook (Asian Development Bank, 2008).

in each phase. Phase one (before cycle) has many factors, ***preparedness*** is most important of them. It plays a vital role in disaster management and associated with almost all aspects of disaster management. It is quite different from prevention/mitigation. Preparedness is comparatively more strongly oriented towards action by an individual organization. For example, in a flood-prone area, people move from the area to be affected or they shift their household appliances to the upper portion of the house or somewhere else. But for effective preparedness, a strong forecast system should bein place at the local and national level. Also, activities like training activities, exercises and tests, public awareness activities, publicity, and education in schools have become essential for preparedness (Figure 4.8).

2. **Post disaster factor:** there are many factors that affect the post disaster impact. ***Recovery*** is one of them. It is a process by which society try to go back to normal. Restoration and reconstruction are the main parts of recovery. It takes 5–10 years and more to recover from disaster loss. Recovery requires national and international assistance because it is a very complex and extensive process. It generates various problems and challenges for the government and other agencies. But at the same time, we should keep in mind that the recovery process generates opportunity or improvements in context with national/international development. Sometimes, it is called the "disaster as a benefit syndrome." Recovery sometimes takes place as an important part of national development planning, and it may be in the form of reconstruction and restoration projects. The strategy for recovery is generally taken at the top level of government, which may include areas of recovery need, an effective time frame of recovery action, resource assessment, and monitoring by the government.

Besides these approaches, planners felt a need for a holistic approach to disaster management. In order to do this, the United Nations accepted three frameworks for action, which are very well known for this purpose:

1) **Yokohama strategy for a safe world:** it was accepted by the United Nations at the 1st *World Conference on Disaster Reduction* in May 1994 in Yokohama, Japan. This conference provided ten principles, such as:
   - Risk assessment is a required step.
   - Disaster prevention and preparedness are of primary importance.
     In the process of national planning and policy, prevention and preparedness should be included.
   - Strengthening the capacities to prevent, reduce, and mitigate disasters should bethe top priority.
   - Early warning system is the key factor to effective disaster prevention andpreparedness.
   - Efficient and proper training and education can reduce vulnerability.
   - Technology transfer should be done at the international level.
   - Sustainable development, environment, and poverty alleviation is compulsory inprevention and mitigation.
   - Political determination should be present at the national and international level.
2) **Hyogo framework for action 2005–2015:** it was accepted by the United Nations at the 2nd *World Conference on Disaster Reduction* in Kobe, Japan in January 2005. This conference emphasized these points:
   - Ensuring DRR as a national institutional system for risk management assessment and early warning.
   - To reduce main factors affecting the process of DRR.
   - Disaster preparedness and effective responses.
3) **Sendai framework for disaster Risk reduction 2015–2030:** It was approved by the United Nations at the 3rd *World Conference on Disaster Reduction* in Sendai, Japan in 2015. This conference described four priorities for risk reduction and management.
   - Understanding disaster risk.
   - Strengthening disaster risk governance.
   - Investing for resilience.
   - Preparedness, *"Build back better"* in recovery, rehabilitation, and reconstruction.

The above discussion describes measures of disaster management, their different indices, factors, etc., but the implementation of disaster management programs is more complex task atall levels. Basically, local community and people cannot understand and cope with the vulnerability of catastrophic situations, thus, what the solution will be, they don't know until they are trained. So, there are many equipment that can be used for this purpose, such as training, public awareness, and research.

a) ***Training:*** here, the question is why do we need training for disaster management? The answer is it is for planning, day-to-day management, counter disaster operations for crisis management, logistic functions, recovery management, first aid, assessment, emergency relief, prevention, mitigation, and so on. Every country has different training needs based on its capacity, technical training system, and intensity of disaster. Training programs include government officials, NGO workers, and other agency personnel. There may be disaster management training, skill training, coordination training, and specialized training programs. Public education programs, awareness workshops related to preparedness, rescue, and prevention are the main parts of the training process.

b) ***Public awareness:*** public awareness is complementary to all planning and policy. If public awareness programs are successfully done then it is understood that half of the task has been completed It helps the governmental agencies and trigger the process of management and development. The aim of these programs is to promote and inform, alert and make the community capable of taking suitable precautionary steps so that they can be less affected by disaster. Awareness makes people be able to contribute their efforts and support in management processes and they cooperate with authorities to fulfill the task. It includes notices, posters, cartoons, films, videos, talks, and so on.

c) ***Research:*** it is an inevitable part of any policy and planning. Research shows the new areas of development which can be applied or opted for management. It helps to identify previous mistakes and not to repeat them. Research contributes to improvements in ongoing disaster management capability. It also shows how to reduce vulnerability and stimulate forward-looking concepts for the future.

4.8 Conclusion

Disaster is a sudden catastrophic event or phenomenon that is intimately connected to calamity, environmental losses, human casualties, economic losses, social order, and much more. It generates hindrances in developmental processes and creates new challenges for the community. It can be a result of natural phenomenon, man-made causes, and hybrid (both). Apart from these types, any other event that is fatal to mankind in a large context is a disaster. For example, since last year, the entire world has been facing the pandemic, *i.e.*, COVID-19. The agent of disastrous phenomenon is stated as disaster risk. A disaster risk reduction process has become imperative in that case. It is a process of reducing, preventing, and managing new disaster risks. Disaster risk management is the next step in this context. Disaster management deals with the implementation of policies through different measures such as prevention and mitigation. It includes various indices that help to identify the problem area and indexing them. More vulnerable areas can then be identified and future planning and policies of prevention can be made efficiently.

The occurrence of any disaster cannot be stopped completely as it will surely happen. But we can reduce the impact of disaster on the community with the help of strong disaster management mechanisms. For this purpose, disaster management should be taken as an integral part of national and international policies and strategies as the United Nations says. There should be a strong early warning system to alert people and prepare them for upcoming situations. Strengthen the capacities at local, national and international levels.

People should receive training and also public awareness programs should be incorporated by government agencies and here, the role of NGOs has become very important. More and more research and developmental projectsshould be promoted in the field of disaster management and risk reduction so that new ideas and dimensions can be added. For this, a proper system of aid assistance is required so the government should create and promote disaster funding agencies. There should be political determination at the national and international levels and corruption and negligence of task should be avoided for the sake of safety and for the sake of humanity.

References

Debnath, J., Sahariah, D., Lahon, D., Nath, N., Chand, K., Meraj, G., Kumar, P., Singh, S. K., Kanga, S., & Singh, S. K. (2022). Geospatial modeling to assess the past and future land use-land cover changes in the Brahmaputra Valley, NE India, for sustainable land resource management. *Environ. Sci. Pollut. Res.*, 1-24.

Debnath, J., Sahariah, D., Lahon, D., Nath, N., Chand, K., Meraj, G., Kumar, P., Singh, S. K., Kanga, S., & Farooq, M. (2023a). Assessing the impacts of current and future changes of the planforms of River Brahmaputra on its land use-land cover. *Geosci. Front.*, https://doi.org/10.1016/j.gsf.2023.101557.

Debnath, J., Sahariah, D., Saikia, A., Meraj, G., Nath, N., Lahon, D., Annayat, W., Kumar, P., Chand, K., Singh, S. K., & Kanga, S. (2023b). Shifting Sands: Assessing Bankline Shift Using an Automated Approach in the Jia Bharali River, India. *Land*, 12(3), 703.

Farooq, M., Meraj, G., Rishabh, Kanga, S., Nathawat, R., Singh, S. K., & Ranga, V. (2022a). Slum Categorization for Efficient Development Plan—A Case Study of Udhampur City, Jammu and Kashmir Using Remote Sensing and GIS. In *Geospatial Technology for Landscape and Environmental Management: Sustainable Assessment and Planning* (pp. 283-299). Cham: Springer International Publishing.

Farooq, M., Mushtaq, F., Meraj, G., Singh, S. K., Kanga, S., Gupta, A., Kumar, P., Singh, D., & Avtar, R. (2022b). Strategic Slum Upgrading and Redevelopment Action Plan for Jammu City. *Resources*, 11(12), 120.

Farooq, M., Rashid, H., Meraj, G., Kanga, S., & Singh, S. K. (2022c). Assessing the Microclimatic Environmental Indicators of Climate Change of a Temperate Valley in the Western Himalayan region. In A. R. Siddiqui and Avijit Sahay (Eds.), *Climate Change, Disaster and Adaptations - Contextualizing human responses to ecological change* (pp. 47–61). https://link.springer.com/chapter/10.1007/978-3-030-91010-5_5

Gulati, B., Sharma, R., Kanga, S., Singh, S. K., Sajan, B., Meraj, G., Kumar, P., & Ramanathan, A. L. (2023). Unraveling the Relationship Between Stubble Burning and Air Quality Degradation in Punjab: A Temporal and Spatial Analysis (2019-2022). *J. Clim. Change*, 9(2), 43-53.

Kanga, S., Meraj, G., Farooq, M., Singh, S. K., & Nathawat, M. S. (Eds.). (2022a). Disaster Management in the Complex Himalayan Terrains. *Geography of the Physical Environment*. Springer, Cham.

Kanga, S., Singh, S. K., Meraj, G., Kumar, A., Parveen, R., Kranjˇ, N., & Đurin, B. (2022b). Assessment of the Impact of Urbanization on Geoenvironmental Settings Using Geospatial Techniques: A Study of Panchkula District, Haryana. *Geographies*, 2(1), 1-10. ISSN No. 2673-7086.

Kotecha, M. J., Kanga, S., Pankhaniya, S. K., Agrawal, S., Meraj, G., & Singh, S. K. (2023). Urban Heat Island (UHI) Resilience Plan in Varying Climatic Conditions Using Geospatial Approach: A Case Study of Rajkot City. In *Climate Change, Agriculture and Society: Approaches Toward Sustainability*, (pp. 307-338). Cham: Springer International Publishing.

Lahon, D., Sahariah, D., Debnath, J., Nath, N., Meraj, G., Farooq, M., Kanga, S., Singh, S. K., & Chand, K. (2023a). Growth of water hyacinth biomass and its impact on the floristic composition of aquatic plants in a wetland ecosystem of the Brahmaputra floodplain of Assam, India. *PeerJ*, 1-32. http://dx.doi.org/10.7717/peerj.14811

Lahon, D., Sahariah, D., Debnath, J., Nath, N., Meraj, G., Kumar, P., Hashimoto, S. and Farooq, M. (2023b). Assessment of Ecosystem Service Value in Response to LULC Changes Using Geospatial Techniques: A Case Study in the Merbil Wetland of the Brahmaputra Valley, Assam, India. *ISPRS Int. J. Geo-Inf.*, 12(4), p. 165.

Lavane, K., Kumar, P., Meraj, G., Han, T. G., Ngan, L. H. B., Lien, B. T. B., Ty, T. V., Thanh, N. T., Downes, N. K., Nam, N. D. G., Minh, H. V. T., Singh, S. K., & Kanga, S. (2023a). Assessing the Effects of Drought on Rice Yields in the Mekong Delta. *Climate*, 11(1), 13.

Meraj, G., Farooq, M., Singh, S. K., Islam, M. N., & Kanga, S. (2022). Modeling the sediment retention and ecosystem provisioning services in the Kashmir valley, India, Western Himalayas. *Model. Earth Syst. Environ.*, 8(3), 3859-3884.

Meraj, G., Kanga, S., Kranjčić, N., Đurin, B., & Singh, S. K. (2021a). Role of natural capital economics for sustainable management of earth resources. *Earth*, 2(3), 622-634.

Gulati, B., Sharma, R., Kanga, S., Singh, S. K., Sajan, B., Meraj, G., Kumar, P., & Ramanathan, A. L. (2023). Unraveling the Relationship Between Stubble Burning and Air Quality Degradation in Punjab: A Temporal and Spatial Analysis (2019-2022). *J. Clim. Change*, 9(2), 43-53.

5

Impact of Environmental Degradation and Disaster Happenings on Human Health

Kishor Kumar Anand[1]* and Sanjay Kumar[2]

[1]*DMWO, District Administration, Arwal, Bihar, India*
[2]*P.G. Department of Geography, Maharaja College, Ara, V.K.S.U, Ara, Bihar, India*

Abstract
In our solar system, the Earth is the only planet where life exists. Unfortunately, the quality of the environment is deteriorating day by day. It is human activities which are basically responsible for this deterioration. The term "environmental degradation" refers to the deterioration in the physical component brought into biological processes mainly by human activities to such an extent that it cannot be set right by self-regulatory mechanism or homeostatic mechanism of the environment. Nature is generous and tolerant, but when the limits are overstepped, it retaliates many times more destructive than the initial assault upon it. Environmental degradation consists of extreme events or hazards and various types of pollutions like air pollution, water pollution, soil pollution, noise pollution, and nuclear pollution. The main causes of theses pollutions are human activities. These are affecting various arena of human life especially human health, for example, many diseases are spread by drinking water, polluted air, noisy pressure horns, and nuclear radiation, etc. Environmental degradation and its impact on human health is now a major problem. It has various dimensions. These problems can be solved by understanding our environment, environmental degradation, and environmental degradation process. Therefore, environmental education is becoming more important. For this, "education about environment" and "education through environment" will be essential. The measurement of pollution is another aspect, but it is not easy. So, it is also important to develop simple measuring techniques to measure pollution at the root level with environmental education. It will be very beneficial to choose the appropriate "remedies and healthcare" for a particular area.

Keywords: Environmental degradation, human health, environmental education, pollution, healthcare

5.1 Introduction

We are living in the 21st year of the 21st century with a lot of technological advancement, but even after that, we are still facing the problems of environmental degradation. Environmental degradation is one of the ten threats officially cautioned by the high level threat panel of the United Nations. It affects various domains of life, especially human health. The effect

**Corresponding author*: kishorkumaranand@gmail.com

Sanjay Kumar, Suraj Kumar Singh, Shruti Kanga, Gowhar Meraj, Majid Farooq and M. S. Nathawat (eds.)
Disaster Management and Environmental Sustainability, (75–84) © 2024 Scrivener Publishing LLC

is not only direct but indirect, too. To understand these effects is the first step towards its remedies. Environmental education can play a very important role here.

Significant research has been conducted on the impact of environmental degradation on human health, highlighting the intricate relationship between the destruction of natural habitats and the emergence of diseases. Scholars have identified how the disruption of natural environments can provide opportunities for disease-causing agents to emerge and spread, underlining the critical need for comprehensive environmental health policies. The degradation of ecosystems has been shown to significantly affect human health, necessitating a deeper understanding and the implementation of new solutions. Moreover, the interplay between ecological values and human health has been explored, emphasizing the importance of integrating environmental considerations into health practices. Despite a consensus on the detrimental effects of environmental degradation on health, there is a noted gap in strategies aimed at enhancing grassroots awareness of these issues, pointing to an area that requires further attention and action (Debnath *et al.*, 2022a; Kotecha *et al.*, 2023; Farooq *et al.*, 2022a, b; Lahon *et al.*, 2023a, b; Kanga *et al.*, 2022a. b).

This chapter is concentrated on direct and indirect impacts of environmental degradation on human health and the role of environmental education, especially "education through environment" as its remedies.

Aims and Objectives

- To understand the nature of environmental degradation, human health, and impact of environmental degradation on human health.
- To find out the suitable remedies/suggestions for the given problem.

Research Hypothesis

- There are direct and indirect impacts of environmental degradation on human health.
- Environmental education is the best remedy for both environmental degradation and human health.

5.2 Methods, Results and Discussion

This research chapter was prepared with the help of observation, secondary data from various reports and different sources, print and electronic media, and library work.

5.2.1 Environmental Degradation

Environment can be defined as (1) the circumstances or conditions that surround an organism or group of organisms, or (2) the complex of social or cultural conditions that affect an individual or community (Debnath *et al.*, 2023c; Farooq *et al.*, 2022c). Unfortunately, the quality of the environment is deteriorating day by day. It is human activities which are basically responsible for this deterioration and we are facing the problems of environmental degradation.

When natural habitats are destroyed or natural resources are depleted, the environment is degraded. Environmental degradation is the deterioration of the environment through depletion of resources such as air, water, and soil, the destruction of ecosystems, and the extinction of wildlife. It is defined as any change or disturbance to the environment perceived to be deleterious or undesirable. This undesirable change cannot be set right by nature itself (Gulati *et al.*, 2023; Nath *et al.*, 2023). So, the term "environmental degradation" refers to the deterioration in physical component brought into biological processes mainly by human activities to such an extent that it cannot be set right by self-regulatory mechanism or homeostatic mechanism of the environment. Therefore, the quality of the environment decreases. So, The United Nations International Strategy for Disaster Reduction defines environmental degradation as "the reduction of the capacity of the environment to meet social and ecological objectives, and needs." Nature is generous and tolerant, but when the limits are overstepped, it retaliates many times more destructive than the initial assault upon it (Lahon *et al.*, 2023b; Rafiq *et al.*, 2022).

In the recent past, the world has faced the problem of the pandemic of the corona virus/Covid 19. It is worth mentioning here the death toll due to the pandemic has been substantial globally. Still, we are under the threat of it and still doubtful about its origin.

Environmental degradation consists of extreme events or hazards and various types of pollutions (Lavane *et al.*, 2023; Sajan *et al.*, 2022; Upadhyay *et al.*, 2023) (Figure 5.1).

Pollution is an undesirable change in physical, chemical, and biological characteristics of environment. Pollutions are of various types like air pollution, water pollution, land pollution, noise pollution, nuclear pollution, and thermal pollution. Some important types of pollution with their causes are as follows.

In another classification, pollution is of two types: physical pollution and social pollution. Air, water, noise, and land pollution are the parts of physical pollution. Social pollution consists of economical pollution (poverty), religious pollution (religious violence), Political pollution (war), ethnic pollution (riot), and sociological pollution (crime, educational & social backwardness) (Figure 5.2).

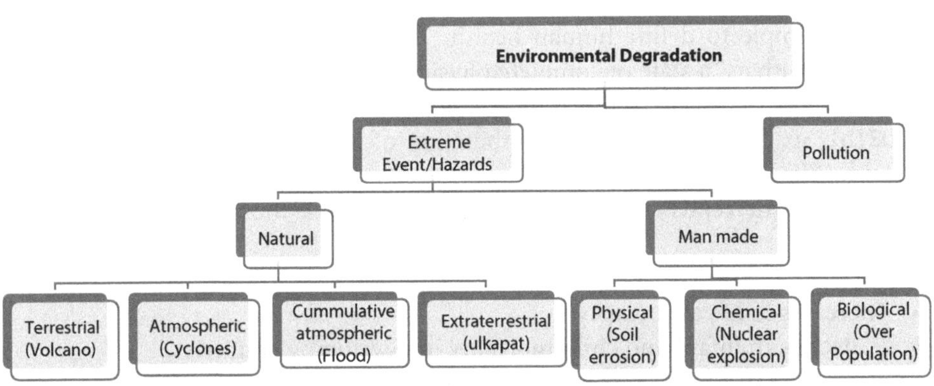

Figure 5.1 Shows the various categories of environmental degradation.

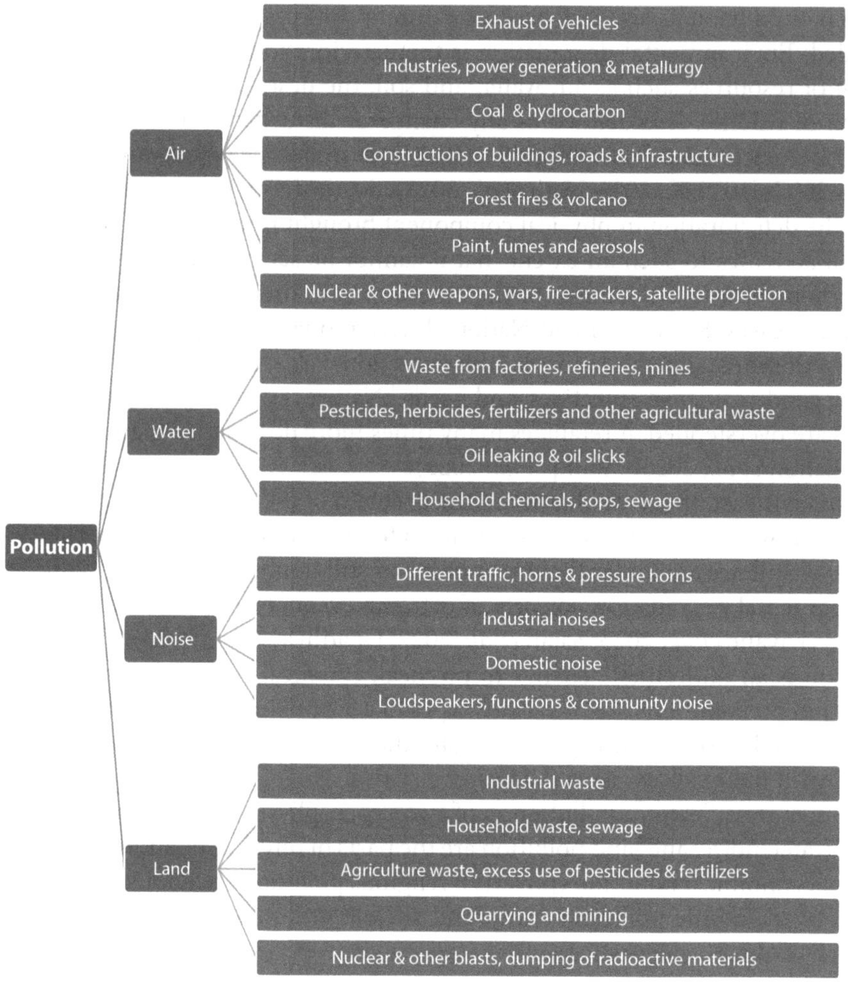

Figure 5.2 Various categories of pollution and their sub-types.

5.2.2 Human Health

It is not very simple to define human health. The World Health Organization (WHO) characterizes health as "a state of complete physical, mental, and social well-being and not merely the absence of disease or infirmity" (Pandey *et al.*, 2010; Meraj *et al.*, 2021a, b; Meraj *et al.*, 2021b). So, human health is a combination of physical health, mental health, and social health (Figure 5.3).

The major current trends of human health globally include the following features (WHO, 1996):

- Increases in life expectancy,
- A decline in infant and child mortality in developing countries,
- Reduction of certain vaccine-preventable diseases, and
- Increased incidences of chronic non-infectious diseases and HIV/AIDS.

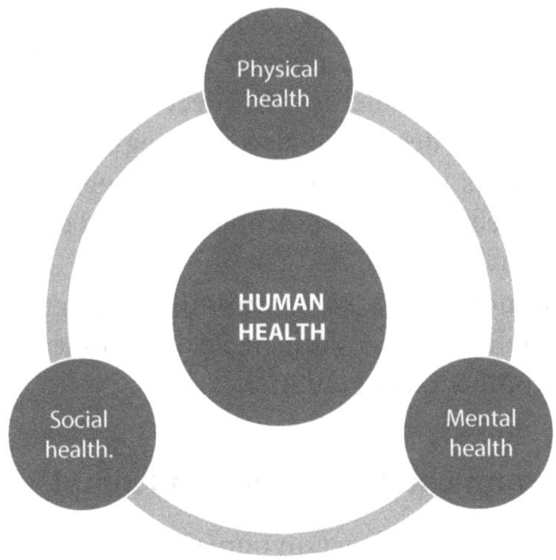

Figure 5.3 Shows set of physical health, mental health and social health.

5.2.3 Impact of Environmental Degradation on Human Health

The WHO estimates that 24% of global disease burden and 23% of all deaths can be attributed to environmental factors. The impact of environmental degradation on human health is categorized as follows:

The following pictorial diagram shows the impact of physical pollution on human health (Figure 5.4).

With the help of the above diagram, we have two results:

1. The effect of environmental degradation is both direct and indirect. For example, air pollution causes suffocation and breathing problems directly and acid rain-related problems indirectly. Physical pollution affects human health

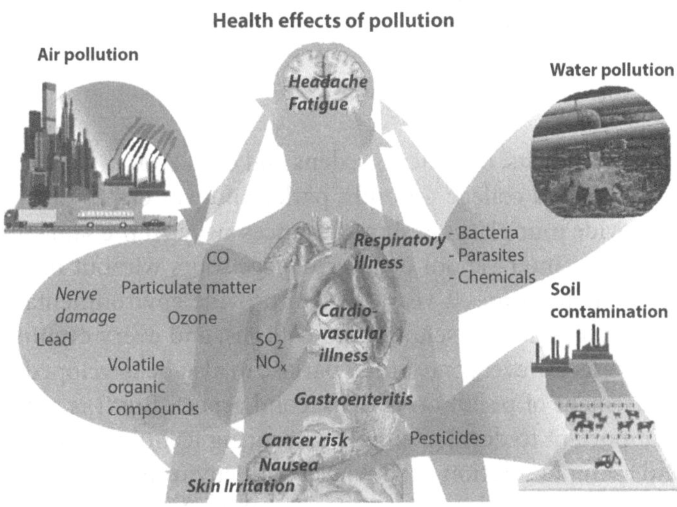

Figure 5.4 Impact of pollution on human health.

both directly and indirectly, while social pollution basically affects human health indirectly. For example, poverty produces lack of nutrition which results in various health problems. The effect of an extreme event/hazards is instantaneous and even death, but also has long-term effects. Instantaneous and short-term effects are mainly direct and long-term effects are mainly indirect. Indirect effects may be of various steps and complications increased with increasing steps.

2. The presence of more types of environmental degradation causes more health-related problems. Thus, human health problems are positively correlated with environmental degradation.

5.2.4 Remedies: Environmental Education and Simple Measuring Techniques

Environmental degradation and its impact on human health is now a major problem. It has various dimensions. The effect of environmental degradation is both direct and indirect (Kanga *et al.*, 2022b; Meraj *et al.*, 2022; Singh *et al.*, 2022a, b; Sud *et al.*, 2023). Indirect effects may be of various steps and complications increased with increasing steps. At the same time, the nature of the health problems are also changing in nature. In our view, awareness is the best way and the first step towards remedies for different health problems. The significant changes in health conditions are also posing a demand for knowledge. But how to make people aware at the root level is a big question facing society.

Environmental education can play a very important role here. These problems can be solved by understanding our environment, environmental degradation, and environmental degradation process (Singh *et al.*, 2023a). Environmental education consists of two dimensions: "education about environment" and "education through environment." We can say that these two are the theoretical and practical approaches of environmental education, respectively (Singh *et al.*, 2022). Practical approach is more beneficial because it is real-life experiences and student-centered learning. The Government of Bihar provides a special package for school students for an educational tour (named '*Shakshanik paribhraman*'). This is related to 'education through environment.' However, the tours are a long distance away, only once a year, and insufficient. At the same time, it is for school students only and not for all people.

At this situation, there are two suggestions:

1. To develop green belts, botanical gardens and even zoos at the local level. We can use schools and college grounds, parks, and wasteland for this purpose. This will provide tour places at short distances away. At the same time, we can also think about inter-school or inter-college tours without much cost and will be possible ten times a year. We can provide environmental awareness programs on weekends in schools and colleges, and every evening in public places like parks for everyone. Study parks can also be developed here.
2. The measurement of pollution is the second aspect. But it is not easy. Thus, it is also important to develop simple measuring techniques to measure pollution and a simple solution at root level with environmental awareness. It is also not impossible. We have two major examples from Bihar.

Eklabhya Prasad from "Megh payin abhiyan" is working on spreading awareness about water problems. He suggests a very simple and beneficial technique which is basically given by Mo. Murtaja, a villager from Supaul in 2007. This is the technique to find the presence of iron in water. Excess iron is a slow poison which does not kill directly but causes various illnesses like gastric problems, indigestion, etc. Guava leaves is crushed by hand and dropped into the water sample. If the color of the water becomes green, there is no iron, if it turns violet, there is iron, and if it is black, there is excess of iron (Sud *et al.*, 2023). The international society "*Serving in Mission*" is using this technique to make people of Bangladesh aware. Dr. Saiyd Mumtazuddin, principal investigator, major research project UGC, with his research scholar, Abdul Kalam Azad and other students have developed the methods to eliminate arsenic and excess copper from water with the help of ashes and gumba plant (Lukas spera), respectively. These techniques will be very beneficial to choose as an appropriate "remedies and health care" for a particular area.

5.3 Conclusions

There are direct and indirect impacts of environmental degradation on human health. The problems related to human health are positively correlated with environmental degradation and awareness with special reference to 'education through environment' is very beneficial. Conclusions also justify the validity and relevance of the formulated hypotheses for the chapter.

References

Debnath, J., Sahariah, D., Lahon, D., Nath, N., Chand, K., Meraj, G., Kumar, P., Singh, S. K., Kanga, S., & Singh, S. K. (2022a). Geospatial modeling to assess the past and future land use-land cover changes in the Brahmaputra Valley, NE India, for sustainable land resource management. *Environ. Sci. Pollut. Res.*, 1-24.

Debnath, J., Sahariah, D., Lahon, D., Nath, N., Chand, K., Meraj, G., Kumar, P., Singh, S. K., Kanga, S., & Farooq, M. (2023b). Assessing the impacts of current and future changes of the planforms of River Brahmaputra on its land use-land cover. *Geosci. Front.*, https://doi.org/10.1016/j.gsf.2023.101557.

Debnath, J., Sahariah, D., Saikia, A., Meraj, G., Nath, N., Lahon, D., Annayat, W., Kumar, P., Chand, K., Singh, S. K., & Kanga, S. (2023c). Shifting Sands: Assessing Bankline Shift Using an Automated Approach in the Jia Bharali River, India. *Land*, 12(3), 703.

Farooq, M., Meraj, G., Rishabh, Kanga, S., Nathawat, R., Singh, S. K., & Ranga, V. (2022a). Slum Categorization for Efficient Development Plan—A Case Study of Udhampur City, Jammu and Kashmir Using Remote Sensing and GIS. In *Geospatial Technology for Landscape and Environmental Management: Sustainable Assessment and Planning* (pp. 283-299). Cham: Springer International Publishing.

Farooq, M., Mushtaq, F., Meraj, G., Singh, S. K., Kanga, S., Gupta, A., Kumar, P., Singh, D., & Avtar, R. (2022b). Strategic Slum Upgrading and Redevelopment Action Plan for Jammu City. *Resources*, 11(12), 120.

Farooq, M., Rashid, H., Meraj, G., Kanga, S., & Singh, S. K. (2022c). Assessing the Microclimatic Environmental Indicators of Climate Change of a Temperate Valley in the Western Himalayan region. In A. R. Siddiqui and Avijit Sahay (Eds.), *Climate Change, Disaster and Adaptations - Contextualizing human responses to ecological change* (pp. 47–61). https://link.springer.com/chapter/10.1007/978-3-030-91010-5_5

Gulati, B., Sharma, R., Kanga, S., Singh, S. K., Sajan, B., Meraj, G., Kumar, P., & Ramanathan, A. L. (2023). Unraveling the Relationship Between Stubble Burning and Air Quality Degradation in Punjab: A Temporal and Spatial Analysis (2019-2022). *J. Clim. Change*, 9(2), 43-53.

Kanga, S., Meraj, G., Farooq, M., Singh, S. K., & Nathawat, M. S. (Eds.). (2022a). Disaster Management in the Complex Himalayan Terrains. *Geography of the Physical Environment*. Springer, Cham.

Kanga, S., Singh, S. K., Meraj, G., Kumar, A., Parveen, R., Kranjˇ, N., & Đurin, B. (2022b). Assessment of the Impact of Urbanization on Geoenvironmental Settings Using Geospatial Techniques: A Study of Panchkula District, Haryana. *Geographies*, 2(1), 1-10. ISSN No. 2673-7086.

Kotecha, M. J., Kanga, S., Pankhaniya, S. K., Agrawal, S., Meraj, G., & Singh, S. K. (2023). Urban Heat Island (UHI) Resilience Plan in Varying Climatic Conditions Using Geospatial Approach: A Case Study of Rajkot City. In *Climate Change, Agriculture and Society: Approaches Toward Sustainability* (pp. 307-338). Cham: Springer International Publishing.

Lahon, D., Sahariah, D., Debnath, J., Nath, N., Meraj, G., Farooq, M., Kanga, S., Singh, S. K., & Chand, K. (2023a). Growth of water hyacinth biomass and its impact on the floristic composition of aquatic plants in a wetland ecosystem of the Brahmaputra floodplain of Assam, India. *PeerJ*, 1-32. http://dx.doi.org/10.7717/peerj.14811

Lahon, D., Sahariah, D., Debnath, J., Nath, N., Meraj, G., Kumar, P., Hashimoto, S. and Farooq, M. (2023b). Assessment of Ecosystem Service Value in Response to LULC Changes Using Geospatial Techniques: A Case Study in the Merbil Wetland of the Brahmaputra Valley, Assam, India. *ISPRS Int. J. Geo-Inf.*, 12(4), p. 165.

Lavane, K., Kumar, P., Meraj, G., Han, T. G., Ngan, L. H. B., Lien, B. T. B., Ty, T. V., Thanh, N. T., Downes, N. K., Nam, N. D. G., Minh, H. V. T., Singh, S. K., & Kanga, S. (2023). Assessing the Effects of Drought on Rice Yields in the Mekong Delta. *Climate*, 11(1), 13.

Meraj, G., Farooq, M., Singh, S. K., Islam, M. N., & Kanga, S. (2022). Modeling the sediment retention and ecosystem provisioning services in the Kashmir valley, India, Western Himalayas. *Model. Earth Syst. Environ.*, 8(3), 3859-3884.

Meraj, G., Kanga, S., Kranjčić, N., Đurin, B., & Singh, S. K. (2021a). Role of natural capital economics for sustainable management of earth resources. *Earth*, 2(3), 622-634.

Meraj, G., Singh, S. K., Kanga, S., Islam, M. N., & Singh, S. K. (2021b). Modeling on comparison of ecosystem services concepts, tools, methods and their ecological-economic implications: A review. *Model. Earth Syst. Environ.*, 1-20.

Nath, N., Sahariah, D., Meraj, G., Debnath, J., Kumar, P., Lahon, D., Chand, K., Farooq, M., Chandan, P., Singh, S. K., & Kanga, S. (2023). Land Use and Land Cover Change Monitoring and Prediction of a UNESCO World Heritage Site: Kaziranga Eco-Sensitive Zone Using Cellular Automata-Markov Model. *Land*, 12(1), 151.

Rafiq, M., Meraj, G., Kesarkar, A. P., Farooq, M., Singh, S. K., Kanga, S., & Bhardwaj, V. (2022). Hazard Mitigation and Climate Change in the Himalayas–Policy and Decision Making. In S. Kanga, G. Meraj, M. Farooq, S. K. Singh, & M. S. Nathawat (Eds.), Disaster Management in the Complex Himalayan Terrains. *Geography of the Physical Environment*. Springer, Cham. https://doi.org/10.1007/978-3-030-89308-8_12.

Sajan, B., Mishra, V. N., Kanga, S., Meraj, G., Singh, S. K., & Kumar, P. (2022). Cellular automata-based artificial neural network model for spatial-temporal dynamics and land use/land cover change prediction. *Agronomy*, 12, 2772.

Singh, A., Singh, S. K., Meraj, G., Kanga, S., Farooq, M., Kranjcic, N., Đurin, B., & Sudhanshu. (2022b). Designing Geographic Information System Based Property Tax Assessment in India. *Smart Cities*, 5, 364–381.

Singh, H., Meraj, G., Singh, S., Shrivastava, V., Sharma, V., Farooq, M., Kanga, S., & Kumar, P. (2022a). Status of Air Pollution during COVID-19-Induced Lockdown in Delhi, India. *Atmosphere*, 13(12), 2090.

Singh, S. K., Gupta, S. K., Meraj, G., Kumar, P., & Kanga, S. (2023b). Uncovering the hydrometeorological drivers responsible for forest fires utilizing geospatial techniques. *Theor. Appl. Climatol.* https://doi.org/10.1007/s00704-023-04497-y

Singh, S. K., Meraj, G., Mondal, N., Bera, A., Verma, M. K., Tomar, J. S., & Kanga, S. (2022). Assessment of seasonal vegetation dynamics over parts of thar desert using geospatial techniques. *ANGRAU*, 105.

Singh, S., Singh, S. K., Prajapat, D. K., Pandey, V., Kanga, S., Kumar, P., & Meraj, G. (2023a). Assessing the Impact of the 2004 Indian Ocean Tsunami on South Andaman's Coastal Shoreline: A Geospatial Analysis of Erosion and Accretion Patterns. *J. Mar. Sci. Eng.*, 11(6), 1134.

Sud, A., Kanga, R., Singh, S. K., Meraj, G., Kanga, S., Kumar, P., Ramanathan, A. L., & Bhardwaj, V. (2023). Simulating Groundwater Potential Zones in Mountainous Indian Himalayas—A Case Study of Himachal Pradesh. *J. Geophys. Res.: Earth Surf.*, 10(3), 65.

Upadhyay, R. K., Tripathi, G., Đurin, B., Šamanović, S., Cetl, V., Kishore, N., Sharma, M., Singh, S. K., Kanga, S., Wasim, M., Rai, P. K., & Bhardwaj, V. (2023). Groundwater Potential Zone Mapping in the Ghaggar River Basin, North-West India, Using Integrated Remote Sensing and GIS Techniques. *Water*, 15(5), 961.

6

Impact of Development on Environmental Degradation: An Indian Diaspora

Rekha Vishwakarma

Department of Geography, M. L. K. (P.G.) College, Balrampur, Uttar Pradesh, India

Abstract
Environment and development are considered as two sides of a coin. The environment degradation, in fact, started with the propagation of the human race, *e.g.*, considering the mythological suggestion that the story of Adam and Eve, who ate an apple and threw away the peels, led to the beginning of the environment degradation.

The development of a large-scale canal system led to deforestation; water-logging around the canal system led to barren lands. It is clear that on the one side, agriculture was intensified with the input of fertilizers and irrigation, and on the other side, the environment was degrading. Major issues like global warming, greenhouse effect, and air and water pollution are increasing day by day. The eradication of poverty and increase in economic development is mandatory. Without economic development, nations cannot come out of this vicious circle of poverty. Development without environment is meaningless; we need to save our resources for our future generation so that they can live in a healthy way/environment. The natural environment directly influences children's well-being by playing significant roles in food and nutrition, water and sanitation, disease and immunity, physical and mental development, and hope and security of successful long-term development.

***Keywords*:** Environment, degradation, pollution, economic growth, urbanization

6.1 Introduction

Development can cause environmental degradation, in turn, environmental degradation can undermine development. This can be expressed more simply by saying that poverty is both a cause and an effect of environmental degradation (Debnath *et al.*, 2022a; Farooq *et al.*, 2022a). The circumstances, objects, or conditions by which one is surrounded, the complex of physical, chemical, and biotic factors (such as climate soil and living things) that act upon an organism or an ecological community ultimately determine its form and survival (Debnath *et al.*, 2023c; Farooq *et al.*, 2022b).

Email: drrekhavishwakarma1969@gmail.com

Sanjay Kumar, Suraj Kumar Singh, Shruti Kanga, Gowhar Meraj, Majid Farooq and M. S. Nathawat (eds.)
Disaster Management and Environmental Sustainability, (85–96) © 2024 Scrivener Publishing LLC

Environment is considered as the physical world consisting of the land masses, oceans, and atmosphere. In other words, the surrounding in which an organism lives and work is considered as their environment (Debnath *et al.*, 2023b; Farooq *et al.*, 2022c; Gulati *et al.*, 2023).

The 2021 World Environment Day was observed at the most difficult of times with India battling on two fronts—the second wave of Covid-19 has caused devastation, and evidence shows that the country has not augured well on environmental norms (Kanga *et al.*, 2022b; Kotecha *et al.*, 2023). Water and air continues to be as polluted and forest degradation continues, as India tried to revive its economy (Lavane *et al.*, 2023; Meraj *et al.*, 2022; Meraj *et al.*, 2021a).

Development is the process of qualitative change in humans, social, political, and economic system is known as development.

Environmental degradation is a process through which the natural environment is compromised in the same way biological diversity and the general health of the environment are reduced. This process can be entirely natural in origin or it can be accelerated or caused by human activities (Kanga *et al.*, 2022a; Lahon *et al.*, 2023a).

The rapid population growth and economic development in country are degrading the environment through the uncontrolled growth of urbanization and industrialization, expansion and intensification of agriculture, and the destruction of natural habitats (Lahon *et al.*, 2023b).

Population size and growth and related consumption patterns are critical elements in the many environmental degradation and social problems we currently face. The population issues should be urgently addressed by education, empowering women, including in the workforce, in rights, and ownership and inheritance, health care of children and the elderly, and making modern contraception accessible to all.

6.2 Environmental Degradation

Environmental degradation is a result of socio-economic, technological, and institutional activities. Degradation occurs when the earth's natural resources are depleted. These affected resources include water, year, soil, wildlife, plants, animals, and microorganisms.

The cause for environmental degradation is a result of the dynamic interplay of socio-economic and technological activities; a full list of environmental changes that may be driven by manufacturers include economic growth, population growth urbanization, intensification of agriculture, rising energy use, and transportation (Sajan *et al.*, 2022).

It includes increased poverty, overcrowding famine, weather, loss of species, acute and chronic medical illness, war and human right abuse, and increasingly unstable global situations that portends Malthusian chaos disasters.

Causes

1. Social factors
2. Economic factors

Environmental changes are based on many factors including:

- Urbanization
- The population explosion
- Intensification of agriculture
- Increase in energy use
- Increase in transportation

Social Factors

- **Population Pressure**

Population growth related to natural resources depletion is an environmental degradation. Human population growth and environmental sustainability are related since more humans require more natural resources. This in turn, can cause a demand of natural resources that is unsustainable. Some suggested that rapidly growing population not only increases pressure on marginal lands, over-exploitation of soil, over-grazing, over-cutting of wood, soil-erosion, silting flooding but also increases excess use of pesticides, fertilizers, causing land degradation, and water pollution (Figure 6.1).

There is a difficult balance between people and their natural environment. Rapid economic development has raised increasing pressure on India's natural environment. The scale of pollution has a big impact on the health of the population and has contributed to global climate change.

The force exerted by a growing population upon its environment resulting in dispersal and reduction of the population. The sum of the factors (increase in number or excessive food consumption) within a population that reduce the ability of an environment to support the population and that, therefore, tends to result in migration and expansion of range or in extinction or decline of the population.

The increase of population has been trending towards an alarming situation. India has 18% of the world's population on 2.4% of its land area, and places a great deal of pressure on

Figure 6.1 Heavily populated areas in Indian sub-urbs.

its all-natural resources. Water shortages, soil exhaustion, deforestation, and air and water pollution afflict many areas.

However, during this, the population increased from 235.9 million in 1891 to 361 million in 1951, and further to about 800 million in 1988. One out of every six people on the planet live in India and between 2001 and 2011 census, the country grew by 17.7%, adding 181.5 million people. The country has doubled in size in just 40 years. In the present scenario, the total population of our country is 1,395,618,865 as of 1 August 2021.

The excess population is itself a symptom of overpopulation. This problem is becoming more acute every day, due to rapidly increasing population by about 22 million persons a year. So, India is overpopulated. However, a small minority of people see that India is not really overpopulated because it is a vast country with plenty of natural resources. These resources have not been fully used yet.

Rapidly growing population and economic development are the most important issues leading to environmental degradation in India. Uncontrolled and unsustainable growth of urbanization, industrialization expansion, massive increase of agriculture, description of forests, and extraction of natural resources are the important factors for environmental degradation. Environmental concerns which affect the ecosystem are multidisciplinary in nature and vary in the scale of the problem.

- **Poverty**

While people living in poverty are seldom the principal creators of environment damage, they often bear the brunt of environmental damage and are often caught in a downward spiral whereby the poor are forced to deplete resources to survive, and this degradation of the environment further impoverishes people. There is a strong association between environmental health problems and urban poverty and the confusion between environmental health risk and environmental degradation. It may explain why urban poverty is thought to contribute to environmental degradation. But the two should not be confused. Most environmental health risks pose no threats to environmental degradation (Figure 6.2).

Environmental degradation increases poverty:

1. Environmental degradation reduces the stock of natural capital and the poor are disproportionately affected
2. Pollution increases morbidity (ill health and mortality, *e.g.*, exposure to pesticides, contaminated water, indoor air pollution)
3. Environmental degradation increases vulnerability

Figure 6.2 Environment status in a slum area.

4. Deforestation and upper pumping increase the cost of basic goods such as firewood and drinking water.

- **Urbanization**

Due to uncontrolled urbanization, environmental degradation has been occurring very rapidly and causing many problems like land insecurity, worsening water quality, excessive air pollution, noise, and other problems of waste disposal.

As cities grow in number, spatial extent, and density, their environmental and ecological footprints increase. Urban expansion that takes place in forest wetlands and agricultural systems lead to habitat clearing, degradation, and fragmentation of the landscapes. Similarly, the growing urbanization affects the environment by shifting the levels of contaminating emissions as a result of the change in population and shifts in the population behavior pattern after relocating from rural areas to urban areas (Figure 6.3).

Economic Factors

1. Growing transport activities
2. Expansion of chemical-based industry
3. The manufacturing technology adopted by most of the industries, which generally is based on intensive resource and energy use
4. Non-existent or poorly functioning market for environmental services
5. Market distortion created by price controls and subsidies.

Development vs. Environment

Development and environment are two sides of the same coin. On one hand, people are concerned about the environment in which they live. Major issues like global warming, greenhouse effect, air, and water pollution are increasing day by day. On the other hand, to eradicate poverty and increase growth in the economy, development is mandatory. Without economic development, a nation cannot come out of the vicious circle of poverty. As our

Figure 6.3 Heavily urbanized areas due to population in-migration.

population grows, there should be a balance between economic advancement and consumption of natural resources for our future generations so that they can live healthy.

Rapid industrialization and urbanization are inevitable to bring in desired levels of economic development. However, these income generating activities are sure to have a negative environmental quality is being such as pollution. Environment quality is being compromised for the goals of mass employment generation and poverty reduction. The reality is that the continued growth generation activities only increases the deterioration of the environmental quality. This is also believed to be essential to substantially increase the per capita income.

India comprises 2.4% of the land of the world and comprises 18% of the world's population. There is a difficulty in balancing between people and the natural environment. The scale of pollution has great impact over health and contributed to global climate change, too.

Deforestation

Deforestation is the clearance of forest by logging and burning (Figure 6.4). Deforestation occurs for several reasons: trees are used for cooking among villagers and cutting down of trees on a large scale by timber merchants, derived charcoal are used or sold for fuel or as a commodity, while cleared land is used as pasture for livestock, plantation of commodities, and settlements. The removal of trees without sufficient reforestation has resulted in damage to habitat and biodiversity loss. It has an adverse impact on bio-sequestration of atmospheric carbon dioxide. Deforested regions typically incur significant adverse soil erosion and frequently degrade into wasteland.

Deforestation causes a range of serious problems such as flooding, loss of biodiversity, soil erosion, and climate change. Desertification means that the land is becoming increasingly dry, and therefore, like a desert, this means it will be hard to grow food. Deforestation is

Figure 6.4 A scene of the deforestation.

cutting down trees, which is major, with the ever-growing demand for forest-based products. Things causing deforestation:

- Commercial logging
- Urban expansion
- Converting forest to agriculture land
- Mining
- Overgrazing
- Forest fires

Impact of Deforestation

Deforestation affects the water cycle, reduces soil quality, and results in soil erosion and flooding. Land's capacity to hold the groundwater is shrinking with the depleting forest cover. The absence of trees leads to increased salinity in the soil cover, and affects the agricultural activity. Forests are an important natural resource of India. They have moderate influence against floods and thus protect the soil erosion. Forests also play an important role in enhancing the quality of environment by influencing the ecological balance and life support system.

Deforestation vs. Desertification

Deforestation is the term that describes the cutting down or clearing of trees from a wooded area. There are a number of reasons woodlands or forests are cut down. In some cases, the trees themselves are the desired resource. Trees have been a source of fuel for many generations and continue to be used in this way today. They are also turned into timber for use in building and carpentry, as well as used in the production of paper products. India, deforestation rate: GFW (Global Forest Watch) in 2010, India has 31.3 Mha of natural forest extending over 11.1% of its land area in 2020; it lost 132 Katha of natural forest equivalent to 67.3 mt of CO_2 of emissions. In other cases, trees are cleared because they are in the way of progress. Forests may be cleared to make room for farmland or for grazing land for cattle, or they may be removed to make room for new houses, neighborhoods or expanding cities. Some deforestation is not intentional and can result from natural causes, such as wildfires.

Desertification refers to the persistent degradation of dry land ecosystems by climatic variations and human activities. It occurs on all continents (except Antarctica) and affects the livelihoods of millions of people, including a large proportion of the poor in dry lands.

Biodiversity

India is one of the world's most biodiverse countries with 7% of the world's animal and plant species and 21% of its geographical area under forest cover. To pursue higher economic growth and meet the aspiration of people, India is scarifying a part of its biodiversity. And measuring the health of forest is the best way to understand the country's biodiversity status. India has over 45,000 species of plants, and 91,000 species of animals. Out of the world's top 35 biodiversity hotspots, four are found in India. Less than 10% of these hotspots are protected vegetation and this hotspot has fallen by up to 50% in the last four decades or so.

Pollution

Pollution is the introduction of harmful material into the environment. These harmful materials are called pollutants. Pollutants damage the quality of air, water, and land. Pollution is a global problem. They can also be created by human activity, such as trash or runoff produced by factories.

Air Pollution

The World Health Organization (WHO) found that 13% of the world's top polluted cities are present in India. Delhi is listed at the top. Here, air pollution has reduced life expectancy by 3.2 y for the 660 million Indians who live in cities and is the fifth biggest cause of death in India as a whole .But it is not only in just the urban areas, in the rural areas as well, the pollutions caused by fire flue by burning paraffin stoves may kill more than 1,000,000 people per year.

Air pollution also affects the lifestyle of many workers who suffer from heart and lung disease. The poorest population who live on the streets are at the greatest risk, with 6.67 million deaths due to area pollution in 2019 (Figure 6.5).

Water Pollution

Most of the Indian rivers are highly polluted because the citizens throw flowers, idol of God and goddess into the rivers, and they bath and wash their dirt in the rivers. This practice turns the fresh water of the rivers into poison and thus, rivers become polluted. Increased level of sewage is the main cause; less than one of three sewage from the urban areas is treated, the rest flows directly into the water resources as industrial waste, and water running off field agricultural run-off are other causes of the water pollution (Figure 6.6). The number of polluted rivers rose from 121 to 275 between 2010 and 2015, respectively. Around 80% of India's water is severely polluted because the people dump raw sewage and garbage into the country's rivers and lakes. This has led to undrinkable and polluted water, and having to rely on illegal and expensive sources. Each year, more than 1.5 million Indian children die because of water pollution-related diseases. Water pollution is a serious problem in India. Almost 70% of its surface water and a growing percentage of its groundwater reserves are contaminated by biological, toxic organic and inorganic pollutants.

Figure 6.5 Factories emitting gaseous pollutants into the environment.

Figure 6.6 Sewage being directly drained into a natural water body.

Land Pollution

It is one of the most destructive forms of agriculture that result in large scale deforestation. It involves burning of large scales of land and subsequent plantation of crops in the same soil despite the practice being abandoned by the several countries.

Mining sometimes leaves the soil contaminated with dangerous chemicals. Pesticides and fertilizers from the agricultural field are blown by the wind. They can harm plants, animals, and sometimes people. Some fruits and vegetables absorb the pesticides that help them to grow. When people consume these fruits and vegetables, the pesticides enter their bodies. One pesticide can cause cancer and other diseases.

Landfills

Landfills are where waste is kept once the waste is crushed into very small pieces. Waste is buried but without oxygen, a dangerous gas called methane is then created. This process is called anaerobic digestion. In some countries, the methane from landfills are used to generate energy. Landfill sites, also known as tip, dump, and dumping ground, are sites for the disposal of waste material. Landfill is the oldest and most common form of waste disposal, although the systematic burial of waste with daily, intermediate, and final covers only began in the 1940's. Urban India generates 62 million tons of waste annually, and it has been predicted that this will reach 165 million tons in 2030. Forty-three million tons of municipal solid waste is collected annually, out of which, 31 million is dumped in landfill sites and just 11.9 million is treated. The Deonar dumping ground is a waste dumping ground or landfill in the city of Mumbai (Figures 6.7). In 2019, the state of Madhya Pradesh had 378 landfills. It was the highest number in any state that year.

Consequences

By destroying the environment, we risk our own quality of life, gamble with the climate and weather, and the existence of other species. The continuing loss of the natural system could make human activities increasingly vulnerable to ecological surprises in the future.

Figure 6.7 Waste deposition in a unscientific landfill site.

Preventions

There are thousands of ways to protect or save the environment from destructions but there is the need to practice these ways:

- Change the way you get around
- Grow your own food or buying it locally
- Replace standard product with the energy efficient version
- Spread the word
- Stop using plastic water bottles
- Be mindful of eating habits
- Stop second-hand shopping embarrassment
- Buy recycled products.

Environmental Problem in India

Air pollution, power management out of solid waste, water scarcity, falling of ground water table, water preservation and quality of forest, biodiversity loss, and land/soil degradation are the major environmental issues India is facing today.

There are 30 most polluted cities counted from all over the world, and among them, 21 are found in India. Fifty-one percent of pollution is caused by industries, 27% of pollution are by vehicles, 70% of pollution is because of crop burning on agricultural land, and the remaining 5% of pollution is from increase in fireworks. Air pollution contributes to the premature death of the 2,000,000 Indians every year.

In recent years, the urbanization in India has increased rapidly and it was determined, in many ways, that an increase in the country's population is linked to inadequate infrastructure and have rendered many Indian cities with unhealthy living conditions. With existing challenges of poverty, depletion natural resources, poor water qualities, and lack of sanitation, many citizens bear the burden of the toxic and unhealthy living conditions.

Sustainable Development

Sustainable development is needed at the present time not only for the survival of mankind but also for its future protection. Unlike the other great revolution in Indian history by humans. **The green revolution** and the **industrial revolution** are sustainable developments or we can see **sustainable revolutions** will have to take place rapidly, consciously, and on many different levels in many different spheres simultaneously. On a technical level, for example, it will involve sustainable technologies based upon the use of non-renewable fossil fuels for technologies that take advantage of renewable energies like the sun, wind, and biomass; adaptation of conservation, recycling practice on a larger scale, and transition of cleaner and more energy efficient technologies in the countries in to the developing world.

6.3 Conclusion

Evidence used for this chapter is from India. Over the years, rapidly growing populations and economic development are most important issues leading to environmental degradation in India. The uncontrolled and unsustainable growth of urbanization and industrialization expansion, as well as the massive increase of agricultural destructions of forests and extraction of natural resources, are the most important factors for environmental degradation. Environmental concerns which affect the ecosystem are multidisciplinary in nature.

Environmental pollution is multidisciplinary in nature as some problems are global and regional while local ones affect the most of the well being of nature. The chapter is an attempt to study the local environmental problem in India and establish their linkage with environmental degradation and human well-being. It also lists India's major environmental problems and suggests changes in the policies for sustainable development in India.

References

1. Debnath, J., Sahariah, D., Lahon, D., Nath, N., Chand, K., Meraj, G., Kumar, P., Singh, S. K., Kanga, S., & Singh, S. K. (2022a). Geospatial modeling to assess the past and future land use-land cover changes in the Brahmaputra Valley, NE India, for sustainable land resource management. *Environ. Sci. Pollut. Res.*, 1-24.
2. Debnath, J., Sahariah, D., Lahon, D., Nath, N., Chand, K., Meraj, G., Kumar, P., Singh, S. K., Kanga, S., & Farooq, M. (2023b). Assessing the impacts of current and future changes of the planforms of River Brahmaputra on its land use-land cover. *Geosci. Front.*, https://doi.org/10.1016/j.gsf.2023.101557.
3. Debnath, J., Sahariah, D., Saikia, A., Meraj, G., Nath, N., Lahon, D., Annayat, W., Kumar, P., Chand, K., Singh, S. K., & Kanga, S. (2023c). Shifting Sands: Assessing Bankline Shift Using an Automated Approach in the Jia Bharali River, India. *Land*, 12(3), 703.
4. Farooq, M., Meraj, G., Rishabh, Kanga, S., Nathawat, R., Singh, S. K., & Ranga, V. (2022a). Slum Categorization for Efficient Development Plan—A Case Study of Udhampur City, Jammu and Kashmir Using Remote Sensing and GIS. In *Geospatial Technology for Landscape and Environmental Management: Sustainable Assessment and Planning* (pp. 283-299). Cham: Springer International Publishing.

5. Farooq, M., Mushtaq, F., Meraj, G., Singh, S. K., Kanga, S., Gupta, A., Kumar, P., Singh, D., & Avtar, R. (2022b). Strategic Slum Upgrading and Redevelopment Action Plan for Jammu City. *Resources*, 11(12), 120.
6. Farooq, M., Rashid, H., Meraj, G., Kanga, S., & Singh, S. K. (2022c). Assessing the Microclimatic Environmental Indicators of Climate Change of a Temperate Valley in the Western Himalayan region. In A. R. Siddiqui and Avijit Sahay (Eds.), *Climate Change, Disaster and Adaptations - Contextualizing human responses to ecological change* (pp. 47–61). https://link.springer.com/chapter/10.1007/978-3-030-91010-5_5
7. Gulati, B., Sharma, R., Kanga, S., Singh, S. K., Sajan, B., Meraj, G., Kumar, P., & Ramanathan, A. L. (2023). Unraveling the Relationship Between Stubble Burning and Air Quality Degradation in Punjab: A Temporal and Spatial Analysis (2019-2022). *J. Clim. Change*, 9(2), 43-53.
8. Kanga, S., Meraj, G., Farooq, M., Singh, S. K., & Nathawat, M. S. (Eds.). (2022a). Disaster Management in the Complex Himalayan Terrains. *Geography of the Physical Environment*. Springer, Cham.
9. Kanga, S., Singh, S. K., Meraj, G., Kumar, A., Parveen, R., Kranjˇ, N., & Đurin, B. (2022b). Assessment of the Impact of Urbanization on Geoenvironmental Settings Using Geospatial Techniques: A Study of Panchkula District, Haryana. *Geographies*, 2(1), 1-10. ISSN No. 2673-7086.
10. Kotecha, M. J., Kanga, S., Pankhaniya, S. K., Agrawal, S., Meraj, G., & Singh, S. K. (2023). Urban Heat Island (UHI) Resilience Plan in Varying Climatic Conditions Using Geospatial Approach: A Case Study of Rajkot City. In *Climate Change, Agriculture and Society: Approaches Toward Sustainability* (pp. 307-338). Cham: Springer International Publishing.
11. Lahon, D., Sahariah, D., Debnath, J., Nath, N., Meraj, G., Farooq, M., Kanga, S., Singh, S. K., & Chand, K. (2023a). Growth of water hyacinth biomass and its impact on the floristic composition of aquatic plants in a wetland ecosystem of the Brahmaputra floodplain of Assam, India. *PeerJ*, 1-32. http://dx.doi.org/10.7717/peerj.14811
12. Lahon, D., Sahariah, D., Debnath, J., Nath, N., Meraj, G., Kumar, P., Hashimoto, S. and Farooq, M. (2023b). Assessment of Ecosystem Service Value in Response to LULC Changes Using Geospatial Techniques: A Case Study in the Merbil Wetland of the Brahmaputra Valley, Assam, India. *ISPRS Int. J. Geo-Inf.*, 12(4), p.165.
13. Lavane, K., Kumar, P., Meraj, G., Han, T. G., Ngan, L. H. B., Lien, B. T. B., Ty, T. V., Thanh, N. T., Downes, N. K., Nam, N. D. G., Minh, H. V. T., Singh, S. K., & Kanga, S. (2023a). Assessing the Effects of Drought on Rice Yields in the Mekong Delta. *Climate*, 11(1), 13.
14. Meraj, G., Farooq, M., Singh, S. K., Islam, M. N., & Kanga, S. (2022). Modeling the sediment retention and ecosystem provisioning services in the Kashmir valley, India, Western Himalayas. *Model. Earth Syst. Environ.*, 8(3), 3859-3884.
15. Meraj, G., Kanga, S., Kranjčić, N., Đurin, B., & Singh, S. K. (2021a). Role of natural capital economics for sustainable management of earth resources. *Earth*, 2(3), 622-634.
16. Sajan, B., Mishra, V. N., Kanga, S., Meraj, G., Singh, S. K., & Kumar, P. (2022). Cellular automata-based artificial neural network model for spatial-temporal dynamics and land use/land cover change prediction. *Agronomy*, 12, 2772.

7

Food Crisis During Covid-19 Pandemic Among Migrants: A Study With Reference to Rohtak City (Haryana)

Vinod Kumar[1]* and Nitin[2]

[1]*Geography, Sakshi Malik Govt College for Girls Mokhra Rohtak-Haryana, Haryana, India*
[2]*Department of Geography, Delhi School of Economics, University of Delhi, Delhi, India*

Abstract

The Covid-19 pandemic has brought many and diverse challenges to human society, among these challenges, issues about safety, security, and wellness of migrants were at the forefront. As many countries, including India, adopted lockdowns as a potent instrument to curb the spread of the infectious virus, the impact of lockdowns on migrants was also most visible in all parts of the country. In India, millions of migrants migrated to different areas of the country, including to Maharashtra, Gujarat, and Haryana to earn their and their families' livelihood. Due to uncertainties caused by unforeseen economic lockdown, migrants and their families had to suffer a lot. In this chapter, an attempt is made to understand the nature of the suffering of migrants and their families during the period of uncertainties and sufferings. In this chapter, Rohtak an emerging city of NCR- Haryana, is taken as the study area. As Rohtak has emerged as a very important site for migrants, industrial and agricultural laborers from different parts of the country have always preferred Rohtak as it has the potential to provide year-round employment. The present study is based on data collected from interviews with head of the laborers' families who were present in the study area just before lockdown. And their interviews were conducted once they were back after lockdowns were over. In this chapter, an attempt is made to understand the qualitative aspects of their life rather than dealing with issues quantitatively.

***Keywords*:** Pandemic, lockdowns, food crisis, migration, mobility

7.1 Introduction

Migration, along with fertility and mortality, is the main component that brings population changes in an area. The nature of migration is different from fertility and mortality, as both fertility and mortality are biological phenomena. Migration is caused by the combined effect of social, cultural, and economic conditions (Debnath *et al.*, 2022a). It is an age-old phenomenon, which has remained prevalent throughout history for man's survival and existence (Debnath *et al.*, 2023b; Farooq *et al.*, 2022a). Migration can be defined as

**Corresponding author*: vinodkumar7262@gmail.com

Sanjay Kumar, Suraj Kumar Singh, Shruti Kanga, Gowhar Meraj, Majid Farooq and M. S. Nathawat (eds.)
Disaster Management and Environmental Sustainability, (97–108) © 2024 Scrivener Publishing LLC

the geographic movement of people from one region to another in a permanent or semi-permanent manner. Migration is a universal phenomenon of population redistribution. World over, poor people migrate to ensure their survival as their native places (in most cases rural areas) are devoid of resources and facilities required for quality of life. In the Indian case, most of the migrants move away due to miseries and sufferings at their source/origin place. Rural areas in India have been trapped in the problems of the vicious cycle of low agricultural productivity, underemployment, low income, and poverty (Debnath *et al.*, 2023c). These problems leave no choice for the poor other than to migrate in search of better opportunities for themselves and their families (Farooq *et al.*, 2022b).

The world is witnessing unprecedented events in the wake of the Covid-19 pandemic as its spread is unparalleled by any other disease in human history. In light of this, different nations responded to the crisis in different manners. The coronavirus that is believed to have originated from Wuhan in China spread across nations as it is highly infectious; most scientists believe that migration and mobility of people contributed to its spontaneous spread (Farooq *et al.*, 2022c; Gulati *et al.*, 2023). Its fatality and adverse health impacts made nations use activities such as lockdowns, home isolation, and social distancing methods to curb the spread of the deadly virus. Many countries, including India, adopted "lockdowns" as the most preferred mechanism to minimize the spread of the virus by reducing community gatherings. Severe restrictions on normal mobility and economic activities brought many hardships to most individuals, especially to migrants and their families (Kanga *et al.*, 2022a).

In fact, the life of every individual was hard during Covid-19 related lockdowns; the migrants were the most sufferers as their capacity to cope with any form of unforeseen event was minimal. As it is often said, in any calamity poor people suffer most; it was also true in the context of the Covid-19 crisis. Most of the migrants had to suffer the dual burden of being in poverty and migrants have very few social, political, and economic rights in the destination. In the absence of any political, social, and economic strength, the plight of the migrant can be well understood given the responsibilities on them, as on one hand, they have to earn for themselves, and on the other, they have to send remittances to their families back home (Kanga *et al.*, 2022a; Upadhyay *et al.*, 2023).

As per Kumar (2021), along with many other countries, India also initiated lockdowns to curtail the spread of the Corona virus, but still, India ranked among the most affected country in the world. The sudden and prolonged lockdowns in India led to panic, chaos, insecurity, and fear among people, particularly migrants. The sudden lockdown pushed already marginalized migrants into deeper distress as they lost their job and had to face a food crisis in the absence of any earnings (Kotecha *et al.*, 2023; Lahon *et al.*, 2023a; Meraj *et al.*, 2022). Many migrants were thrown out of their rental houses as they were not able to pay rent and other charges. In India, most of the migration is due to the push factors or negative factors of the source region including unemployment, poverty and other forms of human suffering. Many studies have found that regional variations in the level of development and migration streams are highly interlinked and most of the time, male migrants pour into developed regions in search of better employment opportunities (Lavane *et al.*, 2023; Meraj *et al.*, 2021a).

Migrants, after losing their jobs and living days without food, were not even able to return to their native places during lockdowns as all forms of transport facilities were also curtailed, to restrict the spread of the virus. Job losses, non-payments of old due salaries by companies, no savings, and unable to pay rents of rental houses, were the few common issues that had to be confronted by every migrant the world over (Meraj *et al.*, 2021b). In general, it is found that the rate of in-migration is higher in higher-income group states; whereas low-income group states witness a higher rate of out-migration (Bhagat *et al.*; 2020; Nath *et al.*, 2023). Migrants move out of economically underdeveloped regions towards areas of better economic opportunities in the form of gainful employment. Over the last few decades, Delhi, Haryana, Punjab, and Maharashtra have remained the focus of in-migration. As per the NSSO, 2010 study, only one-third of total migrants are employed as irregular or are working in informal sectors. Migrants working as informal workers had to suffer even more as compared to other workers from economic lockdowns and related events.

Scholars see lockdowns as barriers in the natural flow of migrants, restriction in the movement of people, and stoppage of mass modes of transport, and has brought much

Table 7.1 Haryana immigration from different states 2011.

States	**Number of migrants reached**
Uttar Pradesh	378,173
Rajasthan	179,562
Bihar	149,701
Punjab	146,075
Delhi	120,001
Chandigarh	36,937
West Bengal	35,377
Uttaranchal	28,046
Chhattisgarh	26,609
Himachal Pradesh	20,179
Madhya Pradesh	16,268
Maharashtra	10,229
Jharkhand	9227
Orissa	7090
Jammu and Kashmir	6291
Assam	5231
Others	20,700

Source: Calculated from Census of India 2011, D-2 Migration Table (www.censusindia.gov.in).

suffering to poor migrants and their families. Migrants being poor and marginalized, were the worst affected community during Covid-19-related lockdowns. Migrants died due to reasons like starvation, suicides, exhaustion, road and rail accidents, police brutality, and denial of timely medical care (Guha *et al.*, 2020). Geographically, long lockdowns caused panic among migrants, triggering reverse migration from urban centers to their native places (Singh *et al.*, 2022b). This reverse migration, too, was not free from suffering as migrants had to return to their place on foot in the absence of any transport mode (Singh *et al.*, 2022a), carrying loads of household materials, small children without food or money. Some died while returning and some had so much physical, mental, and economic suffering that this period became unforgettable for them. Apart from economic problems, the migrants also had to encounter various psychosocial issues like high degrees of anxieties and fears due to various concerns related to the Covid-19 pandemic (Chaudhary, 2020; Pandey *et al.*, 2010). Occasionally, they also faced harassment and adverse reactions from the local community (Das 2016).

Moreover, the lockdown and resulting economic conditions in India also changed the food status of most of the migrants. Therefore, in the light of the above, this chapter is an attempt to look into the food status of migrants coming to Haryana from different parts of the country. It is a matter of great curiosity at how a food surplus state of India had handled the food issues of migrants during the lockdown.

7.2 Study Area

The present study is based upon the Rohtak district; it lies in the state of Haryana that is the agriculturally and industrially most prosperous state of India. Having a special position in terms of economic development in Haryana as per the census of India, out of total migrants, 2.17% of migrants come to Haryana (Table 7.1). Migrants from different states of India come to Haryana in search of better economic opportunities and higher wages. Haryana is a state where the green revolution began; with the start and diffusion of green revolution technology from core areas to non-core areas, employment opportunities also increased tremendously. The need for labor in different agricultural processes was met by migrant workers and local workers were not able to meet the needs of a large number of manual labor requirements.

7.3 Patterns and Reasons for Migration

Man has remained mobile ever since its origin on earth and man tends to migrate to different places for different causes. In all, geographical studies based on migration, analysis of migration patterns, and their causes are very important. Haryana being an agriculturally and industrially prosperous area, large number of migrants get attracted to it. As far as migration pattern is concerned, the study found that a large number of migrants from NCT (National Capital Territory) of Delhi, Uttar Pradesh, Bihar, Punjab, Rajasthan and Uttarakhand migrate towards Rohtak. Its location near Delhi and better connectivity both via train and road route, along with agricultural prosperity and industrial development, make Rohtak a hot destination for migrants. Moreover, over the year, many new industries have emerged in and around Rohtak, which has added to the attraction for migrants.

Another dimension that is very crucial in the analysis of migration is to know the causes for which migration is taking place. As per different migration theories in population geography such as Ravenstein's Law of Migration (1885), Stouffer's theory of Mobility (1940), and Everett Lee's theory (1966), migration is a short distance, majority of migrants tend to travel a short distance, and it is always age, sex-selective; it underlines that majority of migrants will be male in working age group venturing out for economic causes. Data obtained from the census of India, 2011, also shows a similar picture as shown by the above-discussed theories. As per data, most of the migrants are males migrating to Rohtak for economic causes that include better job opportunities.

7.4 Objectives

The main objective of this chapter is to analyze the impact of lockdowns on the food status of migrants and their families. At the same time, this chapter also attempts to present some short- and long-term policy suggestions to overcome any such future event.

7.5 Data Source and Methodology

The present chapter is based on primary and secondary data sources. The primary data was collected from field interviews of migrants who were in Haryana during the starting of lockdown; and later, they have returned on foot or through any other mode to their native places. During the field survey, migrants were interviewed in a structured manner to know their sufferings and experiences of Covid-19-related lockdowns and economic crisis. Infield interviews removed any regional biases. Fifty migrants from different states were selected and interviewed openly and transparently. The secondary data was also utilized to know the number of migrants, and some other characteristics of migrants were collected from the Census of India 2011, D-2 migration (Table 7.2). In terms of methodology, the present study attempts to analyze the issues of migrants, especially issues on food in a qualitative manner. As the suffering of migrants was much more than just numbers, therefore, to reach the crux of the issue, a qualitative approach was adopted.

7.6 Results and Discussions

During the field interviews with respondents, questions were asked related to their experiences of Covid-19-related lockdown in the state of Haryana when they were doing labor just before the lockdown was announced. They all had some experiences and the present study is based on sharing those experiences.

Sudden stoppage of all forms of economic activities was the cause of all the other worries and sufferings migrants had to face. As in today's world the availability of money drives every decision. And all the respondents replied that economic causes were the cascading cause of all the other worries. In most cases, migrants had job losses and with a few, they were paid at a reduced salary. Although there was no data from the government on the unemployment caused during the said time, some prominent economists predict that nearly 2–3 crore people lost their jobs. In the case of migrants, the situation was even more critical as most of them were working in the unorganized sector having little social security net. Moreover, the absence of any compensation or expectation on the revival of the economy, they left their workspaces in the absence of any hope for a bright future.

Similarly, all the migrants that were on daily wages were under severe economic and other forms of stress due to Covid-19-related economic lockdown. Their informal nature of jobs suddenly came to a halt, leaving no scope for their and their families' survival. Women narrated how they had to feed their children by begging for food and milk as they had to

Table 7.2 Inter-state migrants to Rohtak district from the major states of India 2011.

Source of migration	Number of migrants
NCT of Delhi	19,086
Uttar Pradesh	18,937
Bihar	6365
Punjab	4679
Rajasthan	4219
Uttarakhand	1485
Madhya Pradesh	1186
West Bengal	1176
Maharashtra	665
Assam	632
Chandigarh	540
Himachal Pradesh	445
Chhattisgarh	296
Jammu and Kashmir	267
Gujarat	223
Odisha	219
Jharkhand	214
Tamil Nadu	147
Andhra Pradesh	104

Source: Calculated from Census of India 2011, Migration Table (www.censusindia.gov.in).

walk around for 10 d without any help. These kinds of suffering by migrants put a serious question mark on the claim of India of being the 5th largest economy, when we are not able to provide just basic needs to children and women. Some brave-hearted migrants decided not to return and instead, kept working in whatever work was available. They were aware that the situation in terms of food and employment was not so rosy back in their home. They kept doing whatever work was offered. Respondents narrated their experience that they had to work as home cleaners, as factories had closed, on well below the normal salaries.

A similar sorrowful study was presented by migrants who were on some form of formal jobs that their salaries were also reduced as their factories were not functioning. Daily wage workers and contractual workers shared their views that they were not paid full salaries. Uncertainties in their future and a very high level of insecurity made them suffer. Parents were not able to pay school fees, resulting in schools closing their doors for online classes for such students. The worst sufferers were the female students as they are marginalized even in normal days and now during the lockdown, they had no scope of getting back to school. The economic crisis in families led to an increase in dropouts in the general contest, and girl children especially. Long periods of lockdowns wiped out their savings and they were funding their basic needs using savings in the absence of jobs and no earnings scenario.

One more long-lasting consequence was the economic slowdown, which resulted in a total halt in production. This, combined with higher fuel prices, led to a phenomenal, very high rate of inflation. This time, inflation was different from all other previous inflation levels, as people were suffering from job losses and they were finding it hard to stay alive, and much higher prices of essential commodities such as edible oil, pulses, and petroleum have pushed many to below subsistence level (Figures 7.1, and 7.2).

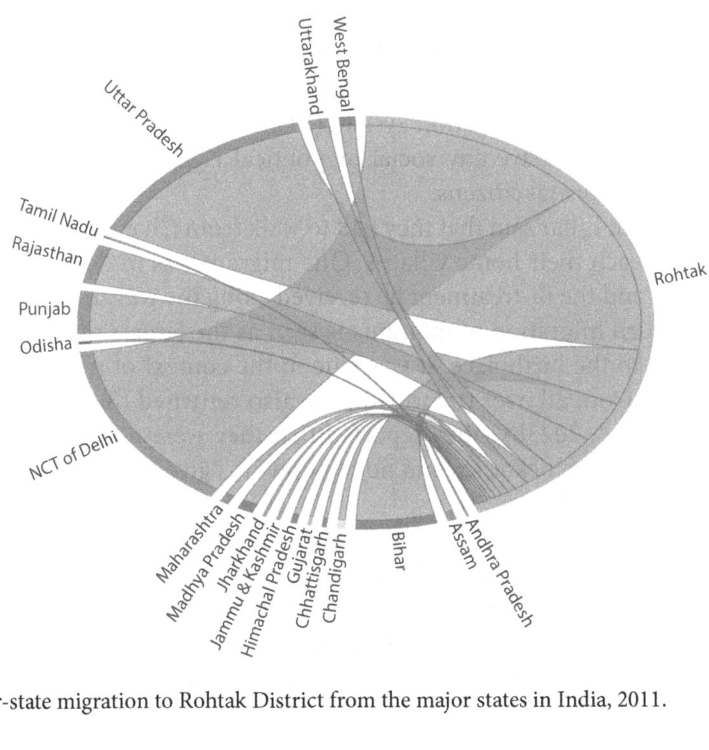

Figure 7.1 Inter-state migration to Rohtak District from the major states in India, 2011.

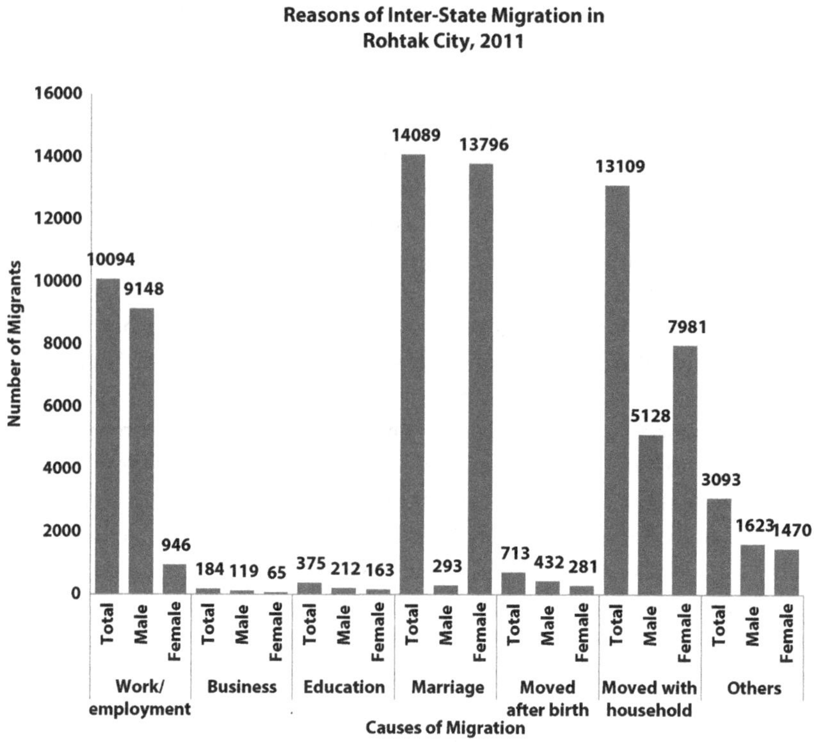

Figure 7.2 Aerial distribution of inter-state migration in Rohtak City, 2011.

Abject poverty has altered the food status of millions as some people are finding it hard to support their families; especially in the case of migrants. One other aspect needs to be understood, that by residing away from their home state, most of the migrants do not have any entitlement under the food security net as most of them do not have any access to ration facilities from Public Distribution System (PDS). Therefore, most of the respondents replied that as they do not have any social or political rights, their sufferings are much higher as compared to normal citizens.

A few migrants from Bihar said that they had to walk from Chapra in Bihar for hundreds of miles on foot to reach their home villages. One migrant was in deep sorrow while narrating his long walk and the ill-treatment he received along the way in between Rohtak and back home. The return migration of migrants is seen as the largest mass return migration from the workplace to the birthplace. It was true in the context of migrants within India, but Indian migrants from all over the world have also returned (Sajan *et al.*, 2022; Rafiq *et al.*, 2022; Singh *et at.*, 2023b). They reported that they were not given any prior information about the period of lockdown and no window was given to them so that they could move back to their native places.

One other sorrowful picture was about the behavior of their state government and police administration where they were treated badly and even trashed as they were seen as the potential carrier of infection. Even after returning to their native places, migrants were thrown in camps for 20 d without many basic facilities. Some of the respondents were those who got infected by corona, their voices became weak while narrating their experience at the workplace and at the hospital where they had spent 3–4 weeks. Fear of getting infected, doctors and other staff members of the hospital were behaving in such a manner as if they are not human (Singh *et al.*, 2022).

Some complaint that governments halted transportation overnight to contain the spread of the virus; migrants from large urban centers started travelling on foot from Haryana, Uttar Pradesh, and Maharashtra towards their native states, therefore, in the absence of any prior information, they had to suffer a lot. Some migrants expressed their gratitude to NGOs and the Haryana government as they arranged food, water, and medicines as and when it was required. Being a welfare state, states offered helping hands to migrants by sheltering them, water, food, and medicines to needy. During this hour of need, many individuals and NGOs also came forward with a helping hand. The Haryana government, going one step further, also helped stranded migrants to travel back by registering migrants digitally.

7.7 Policy Suggestions and Way Forward

Migration will always play an important role in the Indian economic scenario. As most of the migrants migrate due to economic causes, they always have low coping strength with unforeseen events, like Covid-19 and lockdowns. Therefore, in case of any such future scenarios, the following issues should be kept in mind while dealing with the issue at the government or societal levels.

Better information dissemination methods and mechanisms should be followed so that rumors may not spread about the safety of migrants. Moreover, as migration is bound to happen, it is for the betterment of our economy that the free flow of migration takes place (Singh *et al.*, 2023a; Sud *et al.*, 2023). Therefore, to safeguard the interests of migrants, a proper return policy should be in place. In this regard, the current need is to have better coordination between centers and states. Keeping in mind the interests of migrants, a better framework for safe migration should be created to avoid unnecessary panic and problems. Moreover, the government should implement policies for better food opportunities at subsidized rates through PDS. Similarly, the one nation–one ration card policy is rolled out for all included migrants so that they can have legal entitlement for ration under PDS.

7.8 Conclusions

Conventionally, migration in India is not considered a positive event, most of the time, migration is caused by prevalent socio-economic compulsions. Aiming for higher wages and quality of life, poor migrants migrate to economically developed regions of the country. At the time, due to some natural or economic emergencies, their dreams and aspirations fail to bear fruit. Covid-19 and related lockdowns provided such conditions that were very hard for all the people, especially for migrants and their families as they stood last in terms of resources and capacity to cope with such unforeseen events. Poor and deprived sections of society had to bear the brunt of the suffering. Field surveys and interviews of migrants suggested that they suffered on more than one front, on one hand, economic losses due to job losses and reduction in salaries. On the other, due to economic losses, many other related issues cropped up in their life, such as food insecurity, reduction in expenses in the household, and higher dropouts from schools; these all seriously affected the quality of life in the family. Covid-19-related lockdowns were not just economic losses as they had effects ranging from social and psychological, along with others. All these also had a profound impact on the food status of migrants that included a reduction in expenditure on food items, reduction in meals per day, uncertainties of availabilities of food articles especially milk and vegetables, dependence on NGOs or other social groups for daily needs, and borrowing from other relatives had changed their food status to highly insecure during said time.

It emerged from the study that lockdown was required for containing the spread of the virus. Problems emerged due to faulty procedure of the lockdowns and keeping information, including its probable period, a secret from migrants was not justified. Ration to all, arrangement for child education, essential daily items and medicines should have remained at the priority for policy measures. It emerged from the study that the Haryana government did relatively better as compared to most of the other states. Among many steps taken by the state government for the welfare of migrants during the that period, the most appreciated was provision for free ration, psychological counseling of migrants, and arrangements done for migrants to return to their native places. Similarly, providing information on lockdowns, steps to reduces job losses, and sufferings for those who tried to go to their native villages on foot needs to be rectified in the future, so that the sufferings of poor migrants can be minimized.

References

Bhagat, R.B., Reshmi R.S., Sahoo, Harihar, Roy, Archana K. and Govil, Dipti: (2020): The COVID-19, Migration, and Livelihood in India: Challenges and Policy Issues, *Migrat. Lett.*, Debnath, J., Sahariah, D., Lahon, D., Nath, N., Chand, K., Meraj, G., Kumar, P., Singh, S. K., Kanga, S., & Singh, S. K. (2022a). Geospatial modeling to assess the past and future land use-land cover changes in the Brahmaputra Valley, NE India, for sustainable land resource management. *Environ. Sci. Pollut. Res.*, 1-24.

Debnath, J., Sahariah, D., Lahon, D., Nath, N., Chand, K., Meraj, G., Kumar, P., Singh, S. K., Kanga, S., & Farooq, M. (2023b). Assessing the impacts of current and future changes of the planforms of River Brahmaputra on its land use-land cover. *Geosci. Front.*, https://doi.org/10.1016/j.gsf.2023.101557.

Debnath, J., Sahariah, D., Saikia, A., Meraj, G., Nath, N., Lahon, D., Annayat, W., Kumar, P., Chand, K., Singh, S. K., & Kanga, S. (2023c). Shifting Sands: Assessing Bankline Shift Using an Automated Approach in the Jia Bharali River, India. *Land*, 12(3), 703.

Farooq, M., Meraj, G., Rishabh, Kanga, S., Nathawat, R., Singh, S. K., & Ranga, V. (2022a). Slum Categorization for Efficient Development Plan—A Case Study of Udhampur City, Jammu and Kashmir Using Remote Sensing and GIS. In *Geospatial Technology for Landscape and Environmental Management: Sustainable Assessment and Planning* (pp. 283-299). Cham: Springer International Publishing.

Farooq, M., Mushtaq, F., Meraj, G., Singh, S. K., Kanga, S., Gupta, A., Kumar, P., Singh, D., & Avtar, R. (2022b). Strategic Slum Upgrading and Redevelopment Action Plan for Jammu City. *Resources*, 11(12), 120.

Farooq, M., Rashid, H., Meraj, G., Kanga, S., & Singh, S. K. (2022c). Assessing the Microclimatic Environmental Indicators of Climate Change of a Temperate Valley in the Western Himalayan region. In A. R. Siddiqui and Avijit Sahay (Eds.), *Climate Change, Disaster and Adaptations - Contextualizing human responses to ecological change* (pp. 47–61). https://link.springer.com/chapter/10.1007/978-3-030-91010-5_5

Gulati, B., Sharma, R., Kanga, S., Singh, S. K., Sajan, B., Meraj, G., Kumar, P., & Ramanathan, A. L. (2023). Unraveling the Relationship Between Stubble Burning and Air Quality Degradation in Punjab: A Temporal and Spatial Analysis (2019-2022). *J. Clim. Change*, 9(2), 43-53.

Kailash C.Das and Subhasis Saha (2013): Inter-state migration and regional disparities in India, *International Union for the Scientific Study of Population (IUSSP)*, PP 1-27

Kanga, S., Meraj, G., Farooq, M., Singh, S. K., & Nathawat, M. S. (Eds.). (2022a). Disaster Management in the Complex Himalayan Terrains. *Geography of the Physical Environment*. Springer, Cham.

Kanga, S., Singh, S. K., Meraj, G., Kumar, A., Parveen, R., Kranjˇ, N., & Đurin, B. (2022b). Assessment of the Impact of Urbanization on Geoenvironmental Settings Using Geospatial Techniques: A Study of Panchkula District, Haryana. *Geographies*, 2(1), 1-10. ISSN No. 2673-7086.

Kotecha, M. J., Kanga, S., Pankhaniya, S. K., Agrawal, S., Meraj, G., & Singh, S. K. (2023). Urban Heat Island (UHI) Resilience Plan in Varying Climatic Conditions Using Geospatial Approach: A Case Study of Rajkot City. In *Climate Change, Agriculture and Society: Approaches Toward Sustainability* (pp. 307-338). Cham: Springer International Publishing.

Kumar Shailendra, Choudhury Sanghamitra (2021): "Migrant workers and human rights: A critical study on India's COVID-19 lockdown policy. *Soc. Sci. Humanit. Open*, 3 (2021) 100130.

Lahon, D., Sahariah, D., Debnath, J., Nath, N., Meraj, G., Farooq, M., Kanga, S., Singh, S. K., & Chand, K. (2023a). Growth of water hyacinth biomass and its impact on the floristic composition of aquatic plants in a wetland ecosystem of the Brahmaputra floodplain of Assam, India. *PeerJ*, 1-32. http://dx.doi.org/10.7717/peerj.14811

Lahon, D., Sahariah, D., Debnath, J., Nath, N., Meraj, G., Kumar, P., Hashimoto, S. and Farooq, M. (2023b). Assessment of Ecosystem Service Value in Response to LULC Changes Using Geospatial Techniques: A Case Study in the Merbil Wetland of the Brahmaputra Valley, Assam, India. *ISPRS Int. J. Geo-Inf.*, 12(4), p.165.

Lavane, K., Kumar, P., Meraj, G., Han, T. G., Ngan, L. H. B., Lien, B. T. B., Ty, T. V., Thanh, N. T., Downes, N. K., Nam, N. D. G., Minh, H. V. T., Singh, S. K., & Kanga, S. (2023a). Assessing the Effects of Drought on Rice Yields in the Mekong Delta. *Climate*, 11(1), 13.

Malhotra, Neena and Devi, Pushpa (2016): "Analysis of Factors Affecting Internal Migration in India", *Amity Journal of Economics*, 1(2), PP. 34–51

Meraj, G., Farooq, M., Singh, S. K., Islam, M. N., & Kanga, S. (2022). Modeling the sediment retention and ecosystem provisioning services in the Kashmir valley, India, Western Himalayas. *Model. Earth Syst. Environ.*, 8(3), 3859-3884.

Meraj, G., Kanga, S., Kranjčić, N., Đurin, B., & Singh, S. K. (2021a). Role of natural capital economics for sustainable management of earth resources. *Earth*, 2(3), 622-634.

Meraj, G., Singh, S. K., Kanga, S., Islam, M. N., & Singh, S. K. (2021b). Modeling on comparison of ecosystem services concepts, tools, methods and their ecological-economic implications: A review. *Model. Earth Syst. Environ.*, 1-20.

Nath, N., Sahariah, D., Meraj, G., Debnath, J., Kumar, P., Lahon, D., Chand, K., Farooq, M., Chandan, P., Singh, S. K., & Kanga, S. (2023). Land Use and Land Cover Change Monitoring and Prediction of a UNESCO World Heritage Site: Kaziranga Eco-Sensitive Zone Using Cellular Automata-Markov Model. *Land*, 12(1), 151.

National Sample Survey Office (2010) *Migration in India 2007-08*, Ministry of Statistics and Programme Implementation, Govt. of India, New Delhi

Rafiq, M., Meraj, G., Kesarkar, A. P., Farooq, M., Singh, S. K., Kanga, S., & Bhardwaj, V. (2022). Hazard Mitigation and Climate Change in the Himalayas–Policy and Decision Making. In S. Kanga, G. Meraj, M. Farooq, S. K. Singh, & M. S. Nathawat (Eds.), Disaster Management in the

Complex Himalayan Terrains. *Geography of the Physical Environment.* Springer, Cham. https://doi.org/10.1007/978-3-030-89308-8_12.

Rajan, S. Irudaya, P. Siva Kumar and Aditya Srinivasan (2020): "The COVID-19 Pandemic and Internal Labor Migration in India: A Crisis of Mobility", *Indian Society of Labor Economics* 2020, Vol. 63, PP. 1021–1039

Sahni, Sakshi and Aulakh, Rawal Singh (2020): "Impact of COVID-19 on Rural Migrants in India", book edited by R.K Sinha entitled *Impact of COVID 9 on Rural India*, Kalpaz Publications, New Delhi

Sajan, B., Mishra, V. N., Kanga, S., Meraj, G., Singh, S. K., & Kumar, P. (2022). Cellular automata-based artificial neural network model for spatial-temporal dynamics and land use/land cover change prediction. *Agronomy*, 12, 2772.

Singh, A., Singh, S. K., Meraj, G., Kanga, S., Farooq, M., Kranjcic, N., Đurin, B., & Sudhanshu. (2022b). Designing Geographic Information System Based Property Tax Assessment in India. *Smart Cities*, 5, 364–381.

Singh, H., Meraj, G., Singh, S., Shrivastava, V., Sharma, V., Farooq, M., Kanga, S., & Kumar, P. (2022a). Status of Air Pollution during COVID-19-Induced Lockdown in Delhi, India. *Atmosphere*, 13(12), 2090.

Singh, S. K., Gupta, S. K., Meraj, G., Kumar, P., & Kanga, S. (2023b). Uncovering the hydrometeorological drivers responsible for forest fires utilizing geospatial techniques. *Theor. Appl. Climatol.* https://doi.org/10.1007/s00704-023-04497-y

Singh, S. K., Meraj, G., Mondal, N., Bera, A., Verma, M. K., Tomar, J. S., & Kanga, S. (2022). *Assessment of seasonal vegetation dynamics over parts of thar desert using geospatial techniques.* ANGRAU, 105.

Singh, S., Singh, S. K., Prajapat, D. K., Pandey, V., Kanga, S., Kumar, P., & Meraj, G. (2023a). Assessing the Impact of the 2004 Indian Ocean Tsunami on South Andaman's Coastal Shoreline: A Geospatial Analysis of Erosion and Accretion Patterns. *J. Mar. Sci. Eng.*, 11(6), 1134.

Sud, A., Kanga, R., Singh, S. K., Meraj, G., Kanga, S., Kumar, P., Ramanathan, A. L., & Bhardwaj, V. (2023). Simulating Groundwater Potential Zones in Mountainous Indian Himalayas—A Case Study of Himachal Pradesh. *J. Geophys. Res.: Earth Surf.*, 10(3), 65.

Upadhyay, R. K., Tripathi, G., Đurin, B., Šamanović, S., Cetl, V., Kishore, N., Sharma, M., Singh, S. K., Kanga, S., Wasim, M., Rai, P. K., & Bhardwaj, V. (2023). Groundwater Potential Zone Mapping in the Ghaggar River Basin, North-West India, Using Integrated Remote Sensing and GIS Techniques. *Water*, 15(5), 961. Volume: 17, No: 5, pp. 705 - 718.

8

Crime Against Women in Patna and Its Environs: Degradation in Social Environment

Vimlesh Kumar

P.G. Department of Geography, V.K.S.U, Ara, Bihar, India

Abstract

Crime is a problem reported in almost every society. However, it is not evenly distributed. This fact gives an opportunity to geography and geographers to explore the variable characters of this social problem and locate the clusters and 'hotspots' of crime. Unfortunately, geography of crime—a relatively recent sub-discipline of social geography and geography of social well-being, in general, has not been able to attract the mainstream professional geographers much.

In tune with Patna City's identity and image, as "cultural capital of India," the city and its surroundings has undergone various transformations involving remarkable economic, political, and social changes, especially during the last quarter century. Consequently, decrease in poverty, more educated people, increased literacy rates, and improved gender equality are reported. The outcomes however are not optimistic as they also include increased economic inequality, environmental destruction, differences between urban and rural classes, and continuing social problems including crime, as revealed by the District Crime Record Bureau (DCRB). Together, they increase tensions and create gender inequalities as well. The tension can then appear in crimes especially against women as an attempt by men to maintain the traditional gender power structures in transition today.

Scholars believe that cities are becoming "safest and most scared ever" simultaneously. In a city like Patna, crime and violence have a known history. The literary, cinematic, and media representations attest to them, too. This is how it becomes an interesting research issue to explore, the nature of urban crime and fear which scholars term 'complex' as it entails not only several layers but also lacks simple cause and effect kind of relationships. However, there is no denying the fact that there exists a clear cut gender difference in (urban) fear perception, particularly in traditional societies like Patna. This study aims at analyzing the nature of crimes against women and developing an understanding of women's perception of feared places and locations, their victimization experience, and vulnerability in Patna City, from the point of women-safety. Accordingly, the qualitative approach and methods (narratives, case studies, ethnography, and phenomenology) were employed in this study to collect the primary data and analyze them. In addition, the secondary data sources were used to compliment them.

***Keywords*:** Crime, cultural, unfortunately, complex, compliment

Email: 87vimleshkumar@gmail.com

Sanjay Kumar, Suraj Kumar Singh, Shruti Kanga, Gowhar Meraj, Majid Farooq and M. S. Nathawat (eds.) Disaster Management and Environmental Sustainability, (109–120) © 2024 Scrivener Publishing LLC

8.1 Introduction

Crime is a surface expression of discontents which lie deeply embedded in the social system. Like any surface manifestation, crime can provide clue(s) to the particular forces which causes it, these in turn, may be traced to the deeper contradictions which churn in the guts of the social and economic system (Peet, 1975). Crime leaves deep scars in society; therefore, we need to study the nature, causes, and patterns of crime affecting people and their property to find out the reasons behind them and prescribe measures to curb such incidences. The geography of crime is the interplay between crime, space, and society through the analysis of offences, offenders, and the end effects. Crime involves rule breaking while violence involves intentional harm-doing using physical means. Violent crime involves both crime and violence.

Fear of crime is a cause of women's non-participation in many spheres of life. Further, it has far-reaching consequences for women, their children, and society as a whole. Victims suffer a range of physical, mental, and health problems, and their ability to earn a living and to participate in public life is diminished. Their children are more at risk of health problems, poor school performance and behavioral disturbances. Crime thus impoverishes women, their families, communities, and nations (Ibid.). Crimes against women are broadly classified under two categories: the first category includes crimes identified under the Indian Penal Code (IPC), such as, rape, kidnapping and abduction, dowry homicides, torture-physical and mental, molestation, eve teasing or sexual harassment, importation of girls, and murder (other than dowry homicides); and, the second category covers crimes identified under the special laws (SL), such as immoral trafficking, demanding dowry, and indecent representation of women. Taking into account both IPC and Special and Local Laws (SLL), the incidences of crime have consistently risen during 2015–2019. A total of 2,44,270 incidents of crime against women were reported in the country during the year 2020 as compared to 2,28,650 in the year 2019, recording an annual increase of 6.4% in the same period. As per the National Crime Records Bureau (NCRB) data, Bihar, accounting for nearly 16.7% of the country's women population, recorded 10.5% of total cases of crimes against women in the country. Eastern Bihar and Patna, too, show increasing trend in this regard.

Rape is the fastest growing crime and reports indicate that rape of women in India has increased considerably especially in recent years (Crime in India, 2012–19). The issue of rape has received increased attention as the number of victims is increasing and it is considered as a major social and human rights issue. The elimination of crimes against women remains one of the serious challenges especially in India as the prevalence of rape has exceeded to a point where even children of five years of age are being tortured and raped (CNN, April 22, 2019). Rape cases recorded in the various police stations of Patna City and its environs has consistently risen between 2015 and 2019. However, experts claim that the actual number of instances of rape is far from being recorded since the unreported number of cases is extremely high (Human Rights Watch: December 29, 2019). Rape remains one of the highly underreported crimes of the world due to various reasons such as fear of retribution from their abusers, lack of remedies for the victim's situation, fear of skepticism, and societal stigmatization. Patna City is one of the traditional societies of India where socio-cultural factors play an important role in the pattern of reporting the serious sexual crimes against women like rape.

"Dowry deaths" comprise a unique category of deaths in India. The custom of payment of dowry by the bride's family to the prospective bridegroom's family is ancient and widely prevalent. When demands for cash, jewelry or goods remain unfulfilled in arranged marriages, or when the dowry is deemed unsatisfactory (Banerjee, 2013). The resulting tensions may lead to the husband or his extended family harassing the woman, sometimes to the extent of killing her or creating such intolerable conditions that she decides to take her own life. Such deaths are termed "dowry deaths" in the Indian Penal Code (defined in section 304B). A conundrum for classification purposes is that both homicides and suicides can constitute 'dowry deaths. "Head wise crime report" of Patna City shows that the number of recorded dowry deaths has been reducing very slowly, but it is unclear whether this is because the numbers have remained stable or reporting practices have remained unchanged. Nevertheless, the numbers are sufficiently high to generate concern.

Sexual harassment and molestation is also serious social problems which affect the lives of many women. It is likely to occur more commonly in cultures that foster beliefs of perceived male superiority and social and cultural inferiority of women. Although culture is an important factor to understand crimes like sexual harassment and molestation in its entirety, we need to look at, as well as beyond cultural structures, their strengths and weaknesses. Patna City, being one of the culturally rich cities in India, it becomes important to explain the harsh reality of problems related to sexual harassment of women as the city is an example of under reporting of sexual crimes against women. The study analyzed the nature of crimes against women and developed an understanding of women's perception of feared places and locations, their victimization experience, and vulnerability in Patna City from the point of women-safety. Accordingly, the qualitative approach and methods (narratives, case studies, ethnography, and phenomenology) were employed in this study to collect the primary data and analyze them. In addition, secondary data sources, too, were used to compliment them. The results provided many interesting and important facts about crimes against women in the city.

8.2 Literature Review

Crime is an outcome of the negative effect of a society. As held by some, the concept of crime is a relative one and it varies in accordance with the socio-economic development of the society and is reflected in the moral and social values of a community. Although many geographers in the past have contributed to the crime literature, their work is still very small when compared to the vast literature on crime and delinquency by sociologists, criminologists, and psychologists. As a geographical paradigm, environmentalism might have provided the most logical link to geography of crime, that it did not do so was in part a function of scale but more particularly, of its view of the natural environment as the habitat of man. There have been studies of the relationship between the natural environment and crime.

Geography and Crime

Climate and seasons affect the crime rate. The crimes against persons are always more numerous in the summer and the crimes against property are more numerous in winter.

The geography of a region plays a significant role in crime against women. Crime and justice vary significantly according to the size and location of a particular area and furthermore, the descriptive and analytical tools of the geographer can help legal scholars to understand crime and its causes. After a thorough analysis of crime on these varying geographic levels, Lavane proceeds to a short discussion of the geography of justice, concentrating on regional variations in sentencing and penalty patterns. Here, too, his analysis proves useful, reinforcing commonly held perceptions of the inadequacies of the criminal justice system. Problems arise only when social reproductions are frustrated, retarded, or prevented. The focus should be on the entire economic and social formation for it is only in this context that problems can be identified and understood.

Valentine (1992) explained the issue of mental maps that women create in their minds of places which they feel would be dangerous. This study gave emphasis on the concept of use of space by women, both in the context of fear and the constrained space, which is available to her, due to patriarchy. It demonstrated that crime reflects the organization of the built-up environment, the spatial structure of social relations and also the distribution of power and wealth. Through the previous studies, it is evident that crime is both spatial as well as temporal in nature. However, crime also differs from region to region. So, it must be noted that "no correct scientific theory of criminal behavior is possible."

Few researches have also been carried out in the Indian context to show the spatiality of crime. In his district level study, he mapped important crime heads and tried to analyze the socio-economic correlates of crime. Sharma and Chowdhary (2012) analyzed the spatial pattern of crime in Shimla city. The purpose of spatial analysis is to identify pattern in geographic data and attempt to explain these patterns. This study reveals the importance of geographical phenomena in the analysis of crime pattern. In their study, they extracted the inference that the development variables do not play an effective and significant role in inhibiting the criminal instances.

A number of works had been done on crime against women but not many of them attempted to analyze the spatiality of the problem. Different reasons and theories have been hypothesized about this problem like the patriarchal system of society giving greater powers to men over women is a major cause of various types of crime committed against (Hiller, 1948). Among the types of crimes against women, rape and dowry death are of much significance in the Indian context. According to Moynihan (1969), rape tends to become acts of passion among intimates and acquaintances. According to the study conducted by a U.S. scholar, majority of offenders and victims of rape come from the same age group and offenders are mostly from lower occupational and lower social status. On the other hand, authors have found that in India, the victims and the offenders are not from the same age group. Victims are much younger than the offenders and both come from mixed economic groups but poorer women run greater risk of being raped. It has also been observed that out the place of occurrence of rape and found that 24% of crimes occurred in the victim's house, 17% in the offender's house, 12% in the offender's relative's house, and the remaining 47% occurred in various distant places. On the issue of bride burning, in India it has been found that it is a typical upper-class problem but statistical analysis did not reveal a positive trend in terms of income. According to him, not only the joint structure of the family but its composition is also important in bride burning cases.

Geography of Fear of Crime

The geographic study of fear of crime reveals the spatial manifestation of institutions of power which operate within a specific setting: that is, who has power, how it is expressed, its material ramifications, and everyday manifestations. Fear of crime is intricately associated with the sense of place ingrained in the minds of women and is in turn anchored in the way public space is constructed and perceived by them. Therefore, the public sphere is much beyond physical forms and structures. It is perceptual and experienced through the senses; it is symbolic and experienced through a spatial symbolic representation; it is representational and lived through symbols and images; and it is subjective and shaped by experiences of the mental realm. The meanings associated with public space emanate from socialization processes.

Koskela *et al.* (2000) examined the relationship between the built environment and women's fear of crime, based on qualitative studies in two European cities. While particular environments are often identified when women talk about the threat of attack, this reflects much broader processes operating to create fear.

Fear influences our experience of places, as much as places influence our experiences of fear. Pain (2001) took this fear-of-crime discourse a step further by reviewing and exploring the differences in people's experience of fear through the lens of social identities. She came to the conclusion that differences in age, gender, and race can lead to differential experiences of fear in the same social and geographic setting. Further, she discusses the ambiguity concerning the term fear of crime. She defines fear of crime as "the wide range of emotional and practical responses to crime and disorder individuals and communities may make." This definition takes into account more aspects than simply knowledge of crime locations, yet it is still heavily dependent on crime as an agent of fear. Some more recent studies show an interest in examining the social, economic, and built environments for evidence that fear is not solely influenced by crime. Many of these studies attribute a place-based quality to personal fears. It is in these studies that the interest in examining the influence of mediating factors affecting perceptions and feeling of fear really emerges as a dominant area of interest within the geographic study of fear. Ranade (2007) observes that women in Mumbai while moving through public spaces, avoid unrespectable (bars, lottery shops) or unsafe (isolated and dark lanes) places. Further, women also tend to avoid low-income squatters, fearing "unfriendly men" who are likely to impart a feeling of discomfort. It is because of this fear that urban spaces are experienced differently by women and men.

Ranade (2007) points out that women can access public space legitimately only when they can manufacture a sense of purpose and are seldom found wandering in public spaces without any purpose. Further, women tend to occupy public spaces as a transit between one private space and another. Phadke (2007, p. 1511) enlisted certain possible "risks" in relation to women in the public space. She talks about four potential risks as:

> First, the potential physical assault,
> second, the risk to "reputation" of accessing public space against the stereotypical construction that defines women's place as the private spaces of the home,

third, the risk of being blamed for being in the public space for physical assault or sexual harassment, if any, and,
fourth, the risk of loss of opportunity to "engage city space and the loss of the experience of public spaces."

This culminates in the adoption of certain precautionary steps, such as avoiding certain parts of the city or not going out after dark. Daigle *et al.* (2008) found that a small proportion of college women experience a large proportion of violent and sexual victimizations. Women are more likely to experience repeat sexual victimization than repeat violence incidents. Repeat victimization tends to happen in the same month of the initial victimization, and the most likely next type of victimization is by far the same type of victimization. Comparing incident-level characteristics of repeat incidents to single incidents, there are few differences, with the exception that, in a larger proportion of single incidents, women took self-protective action. Implications for prevention and educational programs are discussed. Owens (2015) attempted to understand (a) whether or not, compared with men, women are more likely to report being frightened by stalking behaviors, when controlling for correlates associated with fear, and (b) whether or not men and women have significantly different (gendered) reactions to factors associated with increased levels of fear. The author found women were exponentially more likely than men to report fear, despite controlling for indications that women had experienced the more serious stalking incidents. Furthermore, these variable factors affected both men and women in similar ways. Arguably, the fear requirement present in most states' definitions of stalking is inherently gender-biased and should be removed, as no other type of crime is defined by an emotional response.

Building on these conclusions, this research demonstrates not only the trend, pattern, and spatiality of crimes against women but also explores the linkages between gendered space and crime. Fear can be more accurately attributed to individuals' reaction to visible signs of disorder and not merely crime or knowledge of crime statistics.

Urban Crime

The concentration of a huge population due to urbanization often leads to squalid and congested settlements. Congestion due to urbanization encourages deviant behavior, spreads diseases, creates conditions of mental illness, crime alcoholism, etc. Such settlements have a large unemployed and underemployed population, which in turn leads to poverty, low social status, crime and other vices. Urban crime is a result of deteriorating social fabric in families, material deprivation, frustrated aspirations, boredom, and anger (Meraj *et al.*, 2021a). Modern cities present a meeting point of people from different walks of life having no affinity with one another.

When we talk about the challenges of urbanization, we cannot ignore the changing lifestyle of people in urban areas and the role this "change" has to play in their behavior. An individual's position in the society puts several limitations on him and he resorts to crime in a bid to adapt to these limitations. Other than the option of resorting to crime, the person also needs to have opportunities to learn how he can commit crime. This works on the basis of Edwin Sutherland's differential association theory, which states that a person learns the values, attitudes, and techniques required for criminal behavior on interacting with others. While things are not very different in rural areas, the intensity of certain factors differs there. This difference can be attributed to the difference in urban and rural living. In rural areas, cultural factors have a much more dominant role to play as compared to other factors. The connection between crime and city size is hardly a new fact. Criminologists have discussed the urban tendency towards crime for decades. The observed connection between crime and urbanization is evidenced by the theory of "urbanism as a way of life." In fact, social observers have for centuries assailed cities for their supposed concentration of immoral activities. Jefferson wrote that "I view cities as pestilential to the morals, the health, the liberties of man," and Rousseau claimed that "cities are the abyss of the human species." There are various theories that explain the connection between cities and crime. The earliest ideas on this topic were written by Emile Durkheim, George Simmel, and Max Weber. They claimed that urban crime rates reflect the more anonymous and unstable nature of urban life. They also focused on high numbers of temporary urban residents who fear neither recognition nor community reprisals. And claimed that cities create an informational overload that leads bystanders to avoid involvement in crime against their neighbours. They also argued that cities abet crime when urban neighborhoods lose (as they often do) their traditional social structures. Owing to a higher population density and close proximity of residences, the urban areas are generally recognized as breeding grounds for the criminals and prone to higher incidence of crimes compared to rural areas. The unchecked and unplanned growth of cities as well as a large floating population are major contributing factors for urban crimes. Criminal propensity is also known to be higher in urban industrial areas. Migration of poor, illiterate people having roots in a simple environment consequent to migration suffer emotional instability in the urban setting. They are easily lured into theft, robbery, smuggling, and other illegal activities. The increasing trend in urban crimes tends to disturb peace and tranquility of the cities and make them unsafe to live in, particularly for the women.

It is becoming more complicated in current situations because criminals often get shelter from politicians, bureaucrats, and leaders of the urban society. Urbanization represents complex socio-economic and political forces as well as a distinct local culture and history which also portray the inferior status of women. It apparently raises the status of women as compared to the rural areas, however, the gender-based discrimination only changes its forms. Urban crimes like rape, murder, kidnapping, dacoity, robbery, sexual harassment, molestation, etc., are getting more pronounced in present days. The geography of urban crime can add a dimension, hitherto largely neglected, to the broad field of criminology. However, the geographical perspectives have no special claims in relation to problem solving but rather offer a limited and partial role which must ultimately be fused with the roles offered by other disciplines in the attempt to understand the phenomena being investigated.

Rationale for Selecting Patna City as Study Area

In north India, Gangetic and son plain is one of the very diverse cultural regions. This is the region where the socio-economic and cultural transformations are taking place at very rapid rates. In all these transformations, the urbanism is on the verge of transition and resultant to that many urban centers grown up. In this region, Bihar has significant status because of its centric position and having a very diverse cultural heritage city, Patna.

Patna has been a desired dwelling of cross-cultural and cross-regional city since the ancient era due to its religious value and academic values. With the passing of time, onset of modernization, and the growth of rapid urbanization, Patna has also been affected and is currently in the phase of high social transition. Hence, all the components of development via urbanization takes place in this region. Women empowerment and gender equality is one of the important components of such transformation. Although being a traditional society, the level of acceptance of such an indicator that challenges the patriarchy superiority is a subject to indigestible at all. Hence, the chance of crime against women increases as a reaction of such identical challenges. Therefore, to study Patna City for the crime against women with exploratory nature of study design in this particular perspective is much justifiable, soundly more applicable for generalization to the northern Indian cultural values along with the emergence of new urban centers and cultural of urbanism. Conceptual framework of crime against women in Patna city have shown in Figure 8.1.

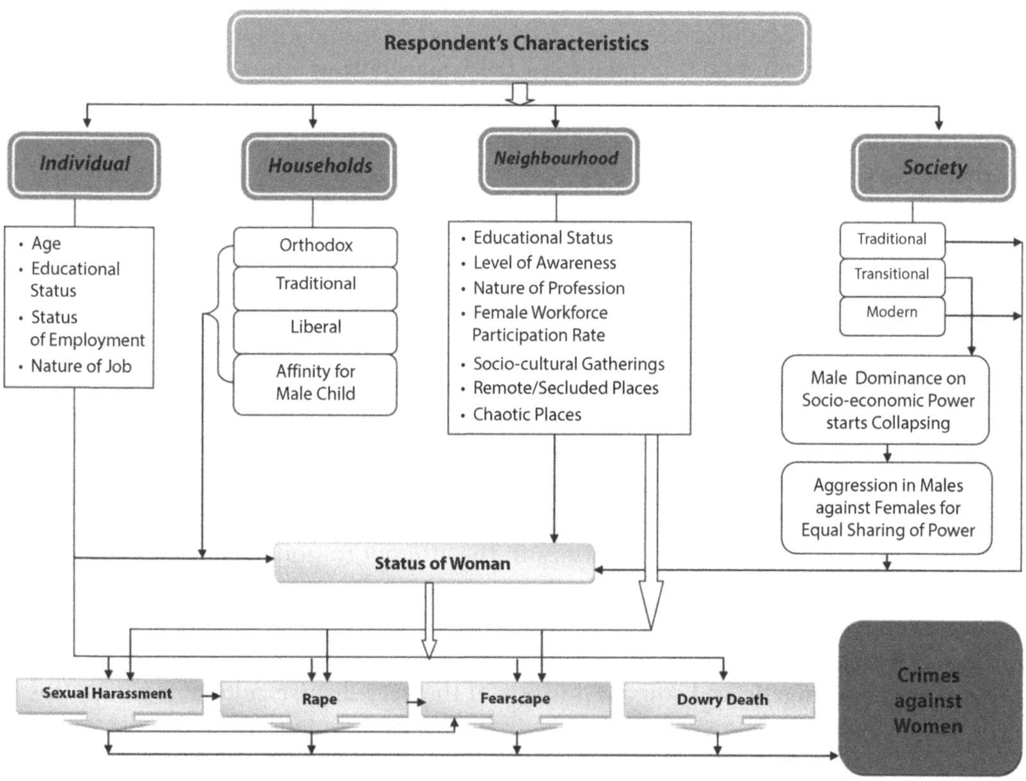

Figure 8.1 Conceptual framework of crime against women in Patna city.

Study Area

Patna town lies between the 25° 12'N to 25° 44' North latitude and 84° 42'E to 86° 04' East longitude. Here, the River Ganga flows East to West, with the world famous Ghats on the left bank of the river. Patna, also known as Patliputra, is the Capital of Bihar. It is one of the oldest continuously inhabited cities in the world and the oldest in India. It is a city on the Banks of the Ganges in Bihar. Patna is the microcosm of Hinduism, a city of traditional classical culture, glorified by myth and legend, and sanctified by religion; it has always attracted a large number of pilgrims and worshippers from time immemorial. Patna is also renowned for its rich tapestry of music, arts, crafts, and education. Some of the world-renowned exponents India has produced in these fields were schooled in Patna's cultural ethos. Geologically, it is situated in the fertile alluvial Gangetic plains and is underlaid with sediments deposited in successive stages. Patna experiences a humid subtropical climate with large variations between summer and winter temperatures. Summers are long, from early April to October, with intervening monsoon seasons, and are also extremely hot, even by South Asian standards. The temperature ranges between 22°C and 46°C (72°F and 115°F) in the summers. Winters in Patna see very large diurnal variations, with warm days and downright cold nights. Cold waves from the Himalayan region cause temperatures to dip across the city in the winter from December to February, and temperatures below 5°C are not uncommon. The average annual rainfall is 1110 mm. Location map of study area have shown in Figure 8.2.

Population Growth

The present area under Municipal Corporation of Patna (MCP) jurisdiction is 32.2 km^2 with a population of 1,684,222 in census 2011. Owing to its rich tourism potential, the estimated daily flow of tourists and pilgrims to the city is about 0.25 million (District Statistical Handbook, 2011). Patna City shows a constant increase in population with a varying rate of increase from decade to decade.

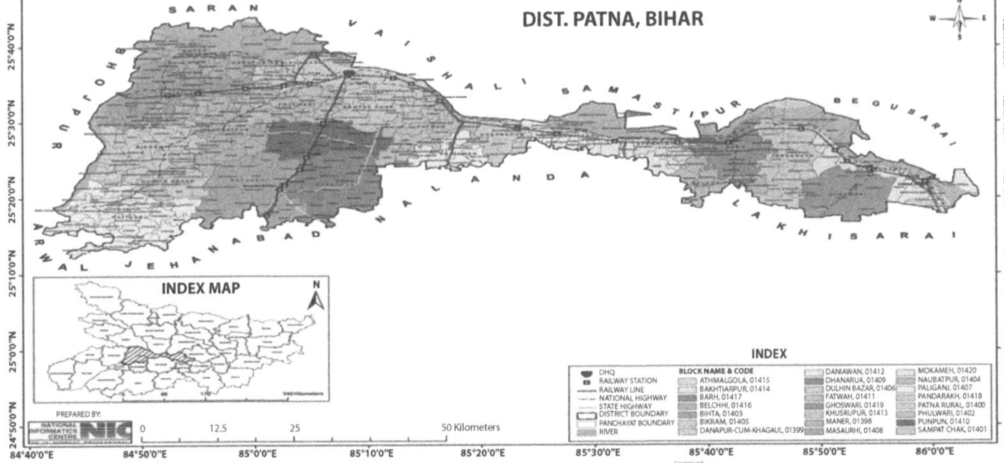

Figure 8.2 Study area map.

Population Density

The total area of Patna City is 25.0 km^2. The area under jurisdiction of the Municipal Corporation is a function not of its population growth or densities, but of political and administrative changes related to development/urbanization of the urban fringe areas. Population density of the town is 1823 persons per km^2.

Literacy

Across districts, Patna district has moderate literacy rates (70.68%, as per 2011 census). The current total literacy rate within the MCP area is high (83.37%) as compared to the state urban average of 56.3% and national urban average of 70.1%, but low as compared to the prevalent literacy rates in other Danapur cities. Patna City also has a low female literacy rate (District Statistical Handbook, 2011). This is a cause of concern considering the fact that the city is renowned for its education and philosophy.

Economy

Approximately 29% of Patna's population is employed, of those, approximately 40% work in manufacturing, 26% work in trade and commerce, 19% work in other services, 8% work in transport and communication, 4% work in agriculture, 2% work in construction, and 2% are marginal workers (working for less than half of the year).Among manufacturing workers, 51% work in spinning and weaving, 15% work in metal, 6% work in printing and publishing, 5% work in electrical machinery, and the rest work in a wide variety of industry sectors. Patna's manufacturing industry is not well developed and is dominated by small-scale industries and household production.

Transport and Communication

Patna is well connected by air, rail, and road. One of the major factors in Patna's sustained existence as an inhabited city is its role as an established transportation hub between different cities.

Research Questions

The research questions for this study are:

1. What are the nature, composition, trend, and pattern of crimes against women in Patna City?
2. What is the prevalence, nature, and extent of sexual harassment in the traditional societies like Patna?
3. What is the understanding of women's perception of feared places and locations in Patna City from the point of women-safety?
4. Do prior victimization experiences and vulnerability of women to sexual crimes affect the fear level among women?
5. Is there any relationship between the educational status of women and fear level among women?

Research Objectives

The primary aim of the present study is to make a comprehensive study of crime against women in the holy city of Patna which is least addressed as suggested by existing published literature. It would be accomplished through considering the following specific objectives with reference to Patna City and its environs.

1. To analyze the nature and composition of crime against women.
2. To attempt a thorough analysis of trend and pattern of crimes against women in the city.
3. To know the perception of women about their vulnerability and victimization experience of sexual crimes against them.
4. To study the geography of women-specific fearscapes of the city.

Sources of Data and Methodology

With the above fundamental research questions in mind, Patna City has been taken as the field for empirical exploration. In the empiric exercise, the focus will be on women-specific fearscapes of the city and four particular crime heads, namely, rape, dowry deaths, sexual harassment, and molestation. This is purely an exploratory kind of research. The research used primary as well as secondary sources of data sets. While primary data was collected through the research tools, the secondary data was taken from different governmental and non-governmental sources. The secondary sources are national crime bureau reports (NCRB), district crime bureau reports (DCRB), police records of crime and crime against women, census data for Patna city, and statistical handbooks for the Patna district.

As the present research is a detailed study of crime against women, the type of interaction held with respondents goes beyond the restrictions of formal interviews. The work was carried out with the help of secondary data as well. There are two types of data in secondary sources. One set gives detailed information regarding population and various socio-economic variables that have bearing upon crime pattern. The second set pertains to details regarding various aspects of crime. For the first set, the data was taken from the census of India publication. For the second set, the data was collected from 'Crime in India,' published by the NCRB. The secondary data was also collected from various literatures available on crime against women.

8.3 Conclusions and Suggestions

Women with higher qualification levels are more scared and feel more unsafe than those who hardly have any qualification or who are at a lower qualification level. With more knowledge and information, one is easily aware of the cases of women violence and crime against women. Which in turn increases the sense of fear among women and they start feeling insecure to go out or travel alone. The findings revealed the negative outcomes of fear among women which can adversely affect women's overall quality of life. The preceding discussion attempted to unravel the interlinkages between space. Gender and fear of crime. Female body and sexuality create different cognitive understanding about space. Fear of crime shapes women's access to and mobility in public spaces in Patna city. The results presented here suggest that there is not one simple explanation for why women are afraid of crime rather the explanations seem to vary.

References

Banerjee P. (2013). Dowry in 21st century India: the sociocultural face of exploitation. *Trauma Violence Abuse*, Vol. 15(1):34–40.

Daigle, L.E., Fisher, B.S. & Cullen, F.T. (2008). The violent and sexual victimization of college women: Is repeat victimization a problem? *J. Interpers. Violence*, Vol.23(2):1296-1313.

Koskela, Hille, & Pain, Rachel. (2000). Revisiting fear and place: Women's fear of attack and the built environment. *Geoforum*, Vol.31(2), 269–280.

Owens, J.G. (2015). A gender-biased definition: Unintended impacts of fear requirement in stalking victimization. *Crime Delinquency*, Vol. 63(11):1339-1362.

Pain, R. (2001). Gender, Race, Age and Fear in the City. *Urban Stud.*, Vol.36(5):899-913.

Phadke, S. (2005). You can be lonely in a crowd: The production of safety in Mumbai. Indian *J. Gend. Stud.*, Vol.21(1):41–62.

Phadke, S. (2007). Dangerous liaisons- Women and men: Risk and reputation in Mumbai, *Econ. Polit. Wkly.*, Vol.42(15):1510–1518.

Ranade, S. (2007). The way she moves: Mapping the everyday production of gendered-space. *Econ. Polit. Wkly.*, Vol.42(15):1519–1526.

Sharma, D.D., & Choudhary, S. (2012). Geography of crime in Shimla City, Himachal Pradesh, *National Geographical Society of India*, Vol.58(1):30- 41.

Singh, P. (2012). Impact of development and quality of life on crime profile of India. *National Geographical Journal of India*, Vol.58(1):45-56. among the buildings are not adequate. In urban planning some liberties are given to the state government. But liberty turns to rustic administrations,

9

Expansion of Irrigation Facilities and Its Impact on Cropping Intensity: A Spatio-Temporal Analysis With Reference to Haryana

Neeraj

Department of Geography, Maharani Kishori Jat Kanya Mahavidyalaya, Rohtak, India

Abstract

Cropping intensity is an efficient indicator for measuring agricultural development in a particular region. In the quest for a strategy for maximizing agricultural production and for providing more work and income to farmers, multiple cropping, *i.e.*, raising more than one crop in a sequence during an agricultural year on a unit of cropland, holds vast potential in most agricultural regions in India. The scope for expanding the net area sown, having already reached a saturation level, and the potential for raising the yield nearly exhausted in many crops and regions. Stepping up the incidence of multiple cropping will be necessary to augment agricultural production. The extension of irrigation facilities has caused a shift in cultivation from single cropping to double cropping regime in many parts of India. Hence, an attempt has been made in this chapter to analyze the impact of irrigation on cropping intensity using the secondary data from 1991–92 to 2017–18 in Haryana. For meeting the gap between increasing human population and food production, multiple cropping, *i.e.*, growing more than one crop on a unit of cropland in an agriculture year, is the only option left behind. In fact, the higher degree of cropping intensity indicates a higher level of agricultural development.

***Keywords*:** Irrigation facilities, cropping intensity, green revolution, food crops

9.1 Introduction

Haryana is predominantly an agricultural economy. About 70% of its residents earn their bread and butter from agriculture directly or indirectly. Haryana has occupied a pioneering position in agriculture in a very short period and it is widely regarded as the granary of India. Adoption of green revolution technology can be seen as the biggest push in the agricultural performance of the state. The spread of green revolution technology from core areas of green revolution the non-core areas has lead to the steady increase in area under irrigation, use of pesticides and fertilizer in the state. Regarding cropping patterns, wheat and rice are the major crops grown in the state. Haryana is self-sufficient in food production and the second largest contributor to India's central pool of food grains. The state has made rapid progress in the field of agriculture since 1966–67 and holds a leading position among

Email: nirajsehrawat0@gmail.com

Sanjay Kumar, Suraj Kumar Singh, Shruti Kanga, Gowhar Meraj, Majid Farooq and M. S. Nathawat (eds.) Disaster Management and Environmental Sustainability, (121–130) © 2024 Scrivener Publishing LLC

other agriculturally advanced regions of India (Debnath *et al.*, 2023a,b; Kanga *et al.*, 2022a). The state which constitutes only 1.44% of the geographical area and 2.4% of the total population of the country now contributes as much as about 7.2% towards the country's total food grains production. The cropping intensity of the state of Haryana has increased regularly. The present study is an attempt to study spatio-temporal variations of cropping intensity in Haryana and tries to analyze the role of irrigation in cropping intensity (Debnath *et al.*, 2023c; Farooq *et al.*, 2022a).

In multiple cropping systems, water is one of the predominant elements. Besides this, the annual rainfall of most of Haryana is insufficient for raising more than one crop in a year. Therefore, irrigation becomes essential for raising two or three crops in a year (Lahon *et al.*, 2023b; Lavane *et al.*, 2023 Nath *et al.*, 2023; Rafiq *et al.*, 2022). The cropping intensity has a direct correlation with assured irrigation which enables farmers to go for multiple cropping and use a higher dose of fertilizers and HYV seeds. A cropping system in irrigated areas can be developed to make the best use of all the resources available in a particular situation (Gulati *et al.*, 2023; Meraj *et al.*, 2022).

Studies found a positive relationship between the intensity of cropping and irrigation (Farooq *et al.*, 2022b; Kanga *et al.*, 2022b). An area irrigated more than once during the agricultural year grew much faster, which is a measure of the intensity of cropping. Studies also examined the trends in cropping intensity in Karnataka and found that expansion of irrigation in the North Maidan and Malnad has led to faster growth of multiple cropping. Results shows that deficiency of irrigation appears to be the main cause for low cropping intensity in the state. (Kotecha *et al.*, 2023; Farooq *et al.*, 2022c; Meraj *et al.*, 2021a, b; Singh *et al.*, 2023b)) examined the impact of irrigation on cropping intensity in Tamilnadu at the state level. In the state, dug-well irrigation showed a significantly positive impact on the cropping intensity up to 1979–80. After that, tube well irrigation had more impact on cropping intensity (Singh *et al.*, 2022).

(Lahon *et al.*, 2023a; Singh *et al.*, 2023a; Sud *et al.*, 2023) examined the spatio-temporal pattern of agricultural development emphasizing crop intensity, irrigation status at meso and micro level in Kandi Block. He found that on the one hand, the massive development of agriculture has been doubling or tripling the production and productivity level, but on the other hand, there is depletion of groundwater as well as environmental quality (Nath *et al.*, 2023; Upadhyay *et al.*, 2023).

9.2 Objectives

The main objectives of the study are:

(i) To analyze the spatio-temporal variations in cropping intensity in Haryana for a period of 17 y, *i.e.*, from 1991–92 to 2017–18.

(ii) To evaluate the role of irrigation in cropping intensity in Haryana.

9.3 Study Area

Haryana lies between 27°39' N and 30°55'5" N latitudes and 74°27'8" E and 77°36'5" E longitudes, covering an area of 44,212 km^2. It constitutes 1.44% of the total geographical area of the country. It is situated in the northwestern part of India and is a part of the Indo-Gangetic plain. It is bounded by Himachal Pradesh in the north, Uttar Pradesh and Delhi in the east, Punjab and Chandigarh on the northwest, and Rajasthan in the southwest. According to the 2011 census, the population of the state is 25,353,081. The average density of the population of the state is 573 persons per square kilometer. Starting originally with seven districts and one division, at present, the state has 22 districts and six divisions, *i.e.*, Gurgaon, Ambala, Rohtak, Karnal, Faridabad, and Hisar divisions. It must be noted that as the study pertains to comparative analysis of the periods 1991–94 and 2015–18, therefore, for analysis purposes, only 21 districts were taken.

The northern part of the state is bounded by Siwaliks, which provides a slope towards the south and southwest (Singh *et al.*, 2022b). Likewise, the Aravalli in the south and southeast provides a slope towards the north. It is the only one in the eastern part of Gurgaon that land in general slopes towards the south and southeast (Singh *et al.*, 2022a). The state is characterized by arid and semi-arid type of climatic conditions. The state is normally deficient in rainfall over its greater part. It is located in the transition zone between the Thar Desert of Rajasthan and the moderately humid upper Ganga plain.

9.4 Database and Methodology

The present study is based on secondary data drawn from published as well as unpublished sources. The time-series data relating to various aspects of agriculture such as net area sown, total cropped area, and gross irrigated area at the state and district levels have been obtained from the Statistical Abstract of Haryana published by the Department of Economic and Statistical Analysis, Haryana, Government of Haryana for the period 1991–92 to 2017–18. As agriculture is a very dynamic activity and it tends to change under given geo-climatic conditions, to minimize the impact of fluctuations due to weather and climatic factors, the study is based on triennial averages of the variables for the year 1991–94 and 2015–18. Cropping intensity is a statistical measure for expressing the total cropped area as a proportion of net sown area in percentage:

$$\text{Cropping intensity} = \frac{\text{Gross cropped area}}{\text{Net area sown}} \times 100$$

To understand the association of cropping intensity with the irrigated area was established with the help of Karl Pearson's coefficient of correlation, and a t-test was done to test the significance of correlation.

9.5 Results and Discussion

9.5.1 Trends of Cropping Intensity in Haryana

The trend line drawn for time series data shows the rapid increase in cropping intensity with time. Figure 9.1 shows that the cropping intensity fluctuates highly from year to year.

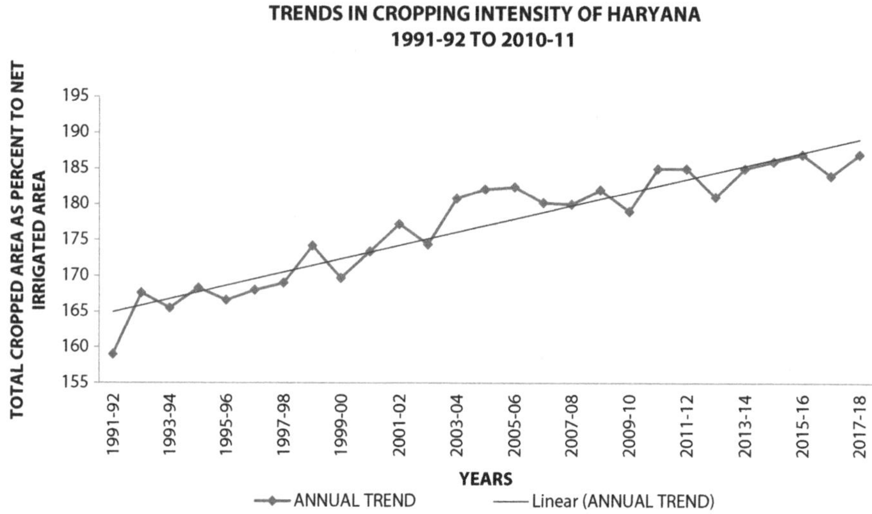

Figure 9.1 Trends in cropping intensity of Haryana from 1991–92 to 2010–11.

It is revealed that fluctuations in cropping intensity do not occur with any definable regularity. The cropping intensity was fairly high up with a few fluctuations.

In 1991–92, cropping intensity was 159%, which increased to 187% in 2017–18. A glance at the line graph reveals that cropping intensity followed the trend line which shows the ideal increase in cropping intensity. The consistent increase in cropping intensity in the state may be attributed to well develop irrigation facilities.

9.5.2 Spatial Pattern of Cropping Intensity

Haryana is one of the most intensively cultivated states in India. More than 87% of the net sown area that sown in the state is cultivated more than once (2017–18). But within the state, there have been wide spatial variations in the cropping intensity.

Cropping Intensity Regions (1991–94):

i) **Very low cropping intensity regions (<140):**
In years 1991–94, the districts of Rohtak and Jhajjar comprised the very low cropping intensity region. These districts are flood-prone. The soil and water quality of these districts are saline. Apart from this, the irrigated area was also very low.

ii) **Low cropping intensity regions (141–160):**
The Rewari, Gurugram, Nuh, Faridabad, Palwal, Sonipat, Bhiwani, and Yamunanagar districts were in the low cropping intensity region. In most of these districts, there is less area under irrigation (Sajan *et al.*, 2022).

iii) **High cropping intensity regions (161–180):**
Ambala, Panchkula, Mahendragarh, Fatehabad, Hisar, Sirsa, Kaithal, and Panipat constituted the high cropping intensity region.

iv) **Very high cropping intensity regions (>180):**
In years 1991–94, cropping intensity of Karnal district was highest in the state, *i.e.*, 194% followed by Jind and Kurushetra (183% and 182%, respectively). In these districts, the larger proportion of the net sown area is under irrigation.

Cropping intensity regions (2015–18):

i) **Low cropping intensity regions (141–160):**
The Rewari district constituted the low cropping intensity region.

ii) **High cropping intensity regions (161–180):**
Jhajjar, Gurugram, Rohtak, and Nuh constituted the high cropping intensity region.

iii) **Very high cropping intensity regions (>180):**
Yamunanagar, Palwal, Bhiwani, Sonipat, Faridabad Mahendragarh, Panchkula, Ambala, Fatehabad, Hisar, Sirsa, Panipat, Kaithal, Jind, Kurukshetra, and Karnal were in the high cropping intensity region. During 2015–18, the cropping intensity of the Sonipat district was highest in the state, *i.e.*, 198%, followed by Karnal and Panchkula (197% and 196%, respectively). In 1991–94, the Sonipat district maintained the low status, but in 2015–18, the Sonipat district was raised to very high cropping intensity region. The main reason for the highest concentration in this district was because of the freshwater zone, which has been harnessed for irrigation. During both periods, the Jind, Karnal, and Kurukshetra districts were identified as very high cropping intensity region. Fertile soil and adequate irrigation facilities offer greater opportunities to the farmers to produce more than one crop during the same year (Figures 9.2 and 9.3).

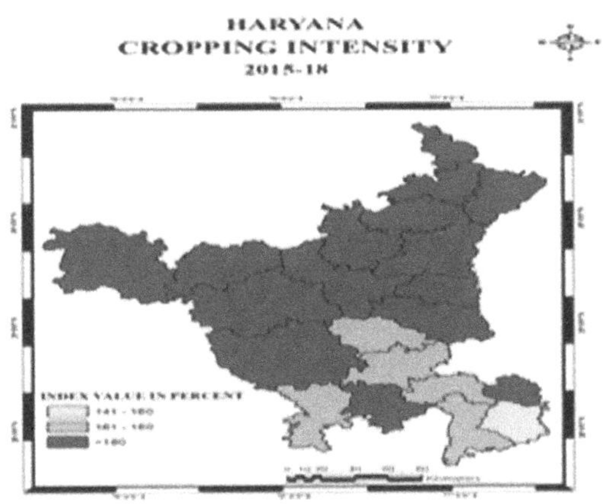

Figure 9.2 Cropping intensity index of the state of Haryana (2015-18).

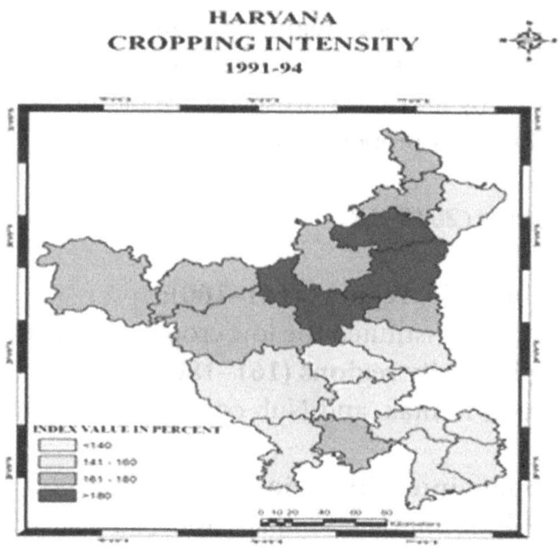

Figure 9.3 Cropping intensity index of the state of Haryana (1991-94).

9.6 Relationship Between the Extent of Irrigation and Cropping Intensity

Table 9.1 reveals that the cropping intensity increased all over the state. An increase in irrigation facilities in the state either by canals or tube wells also played a significant role in

Table 9.1 Haryana district-wise cropping intensity and percentage net irrigated area out of net area sown [1991–1994 and 2015–18].

Districts	1991–94		2015–18	
	Cropping intensity	% Net irrigated area out of net area sown	Cropping intensity	% Net irrigated area out of net area sown
Jhajjar	131	69.29	162	95
Rohtak	131	69.29	169	96.33
Rewari	148	77.03	156	96.3
Gurugram	152	44.29	162	100
Nuh	152	44.29	171	64.6
Faridabad	153	77.88	193	97.93
Palwal	153	62.5	191	98.67
Sonipat	153	97	198	99.77
Bhiwani	156	47.18	186	65.2
Yamunanagar	158	76.53	185	96.93
Ambala	163	60.7	192	96.27

(Continued)

Table 9.1 Haryana district-wise cropping intensity and percentage net irrigated area out of net area sown [1991–1994 and 2015–18]. (*Continued*)

Districts	1991–94		2015–18	
	Cropping intensity	% Net irrigated area out of net area sown	Cropping intensity	% Net irrigated area out of net area sown
Panchkula	163	60.7	196	82.53
Mahendragarh	164	55.69	189	52
Fatehabad	167	62.5	193	96.87
Hisar	167	77.88	195	97.9
Sirsa	170	75.67	188	96
Kaithal	176	97.44	191	99.1
Panipat	180	97.06	188	100
Kurukshetra	182	99.23	189	98.1
Jind	183	86	188	98.13
Karnal	194	98.52	197	99.83
HARYANA	**164**	**74.59**	**186**	**89.8**

Source: Calculated by the author/primary data.

Table 9.2 Haryana correlation between cropping intensity and percentage of net irrigated area out of net area sown [2015–18].

District	r value	District	r value
Jind	0.91*	Rohtak	0.54*
Karnal	0.85*	Bhiwani	0.52*
Sirsa	0.84*	Kaithal	0.50*
Hisar	0.81*	Panchkula	0.31
Ambala	0.79*	Gurgaon	0.22
Faridabad	0.76*	Palwal	0.2
Kurukshetra	0.73*	Jhajjar	0.08
Sonipat	0.69*	Nuh	0.05
Yumunanager	0.60*	Rewari	0.05
Panipat	0.55*	Fatehabad	0.02
Mahendragarh	0.54*	Haryana	0.93

*Coefficients are significant at 1% level of significance.

the increase in cropping intensity. The net irrigated area as a percent to net area sown was 74.59% from 1991–92 to 1993–94, whereas it increased by 89.8% from 2015–16 to 2017–18. But it should also be kept in mind that cropping intensity increased even in the western parts of the state where there were limited water resources. The increasing population pressure on the cultivated land also led to the vertical expansion of the cultivated area.

Table 9.2 shows the correlation coefficient between the extent of irrigation and cropping intensity. The correlation coefficient was computed for districts on the time series data. For the state as a whole, the correlation between cropping intensity and irrigated proportion of net area sown shows nearly perfect positive (r = 0.93). It was found that in all 21 districts, there is a significant positive correlation between these two variables. Among all the districts, Jind shows the highest positive relationship (r = 0.91). Out of these districts in Jind, Karnal Sirsa, Hisar, Ambala, Faridabad, Kurukshetra, Sonipat, Yamunanagar, Panipat, Mahendragarh, Rohtak, Bhiwani, and Kaithal districts, correlation is significant at 1% level of significance.

9.7 Conclusion

Cropping intensity is the most essential role of play for agriculture, agricultural practices, and agricultural production. Rainfall is deficient in the state; there is an acute need for irrigation that has been identified for agricultural development in the state. It is assumed that the increase in irrigation facilities might enhance the probability of more cropping intensity. The study marked variations in cropping intensity in Haryana during the post-liberalization. The interplay of several economic and cultural factors with varied agro-ecological conditions determines the magnitudes of crop diversification in Haryana. In general, cropping intensity increased as a result of the increase in irrigation intensity. But exceptions were also observed. Yamunanagar, Palwal, Mewat, Bhiwani, Sonipat, Ambala, Mahendragarh, Hisar, Jind, Karnal, Sirsa, Fatehabad, Rohtak, Panipat, and Kurukshetra followed the general trend while the Panchkula, Rewari, Jhajjar, Gurgaon, Faridabad, and Kaithal districts were exceptions. Due to favorable physical agro conditions and the substantial use of modern technology like chemical fertilizers, high yield verity seeds played a significant role in increasing cropping intensity in addition to the use of irrigated water. The correlation between cropping intensity and irrigation shows positive in all the districts. The introduction of a modern irrigation system through tube wells and canals changed the agricultural landscape in many ways. The percentage of arable lands increased enormously. With the development of irrigation, Haryana state showed a sharp fall of the area under dry crops while rice and double cropping gained increasing popularity. Early rice is more important in the tube well-irrigated area while at present, rice in the canal is irrigated. Due to expansion irrigation, this natural resource is facing tremendous stress. In-state freshwater zone, the larger part of irrigation, is being sustained by groundwater irrigation through tube wells. This has resulted in the continuous lowering of the water tables in the state. On the one hand, the water table is rising due to the poor quality of groundwater and inadequate drainage facilities leading to the problems of waterlogging. The unfavorable geo-hydrological conditions in the state coupled with changing patterns in favor of water-intensive crops have been major reasons for the degradation of water resources. If suitable measures not taken on time, the situation will become unmanageable.

References

Debnath, J., Sahariah, D., Lahon, D., Nath, N., Chand, K., Meraj, G., Kumar, P., Singh, S. K., Kanga, S., & Singh, S. K. (2022a). Geospatial modeling to assess the past and future land use-land cover changes in the Brahmaputra Valley, NE India, for sustainable land resource management. *Environ. Sci. Pollut. Res.*, 1-24.

Debnath, J., Sahariah, D., Lahon, D., Nath, N., Chand, K., Meraj, G., Kumar, P., Singh, S. K., Kanga, S., & Farooq, M. (2023b). Assessing the impacts of current and future changes of the planforms of River Brahmaputra on its land use-land cover. *Geosci. Front.*, https://doi.org/10.1016/j.gsf.2023.101557.

Debnath, J., Sahariah, D., Saikia, A., Meraj, G., Nath, N., Lahon, D., Annayat, W., Kumar, P., Chand, K., Singh, S. K., & Kanga, S. (2023c). Shifting Sands: Assessing Bankline Shift Using an Automated Approach in the Jia Bharali River, India. *Land*, 12(3), 703.

Farooq, M., Meraj, G., Rishabh, Kanga, S., Nathawat, R., Singh, S. K., & Ranga, V. (2022a). Slum Categorization for Efficient Development Plan—A Case Study of Udhampur City, Jammu and Kashmir Using Remote Sensing and GIS. In *Geospatial Technology for Landscape and Environmental Management: Sustainable Assessment and Planning* (pp. 283-299). Cham: Springer International Publishing.

Farooq, M., Mushtaq, F., Meraj, G., Singh, S. K., Kanga, S., Gupta, A., Kumar, P., Singh, D., & Avtar, R. (2022b). Strategic Slum Upgrading and Redevelopment Action Plan for Jammu City. *Resources*, 11(12), 120.

Farooq, M., Rashid, H., Meraj, G., Kanga, S., & Singh, S. K. (2022c). Assessing the Microclimatic Environmental Indicators of Climate Change of a Temperate Valley in the Western Himalayan region. In A. R. Siddiqui and Avijit Sahay (Eds.), *Climate Change, Disaster and Adaptations - Contextualizing human responses to ecological change* (pp. 47–61). https://link.springer.com/chapter/10.1007/978-3-030-91010-5_5

Gulati, B., Sharma, R., Kanga, S., Singh, S. K., Sajan, B., Meraj, G., Kumar, P., & Ramanathan, A. L. (2023). Unraveling the Relationship Between Stubble Burning and Air Quality Degradation in Punjab: A Temporal and Spatial Analysis (2019-2022). *J. Clim. Change*, 9(2), 43-53.

Kanga, S., Meraj, G., Farooq, M., Singh, S. K., & Nathawat, M. S. (Eds.). (2022a). Disaster Management in the Complex Himalayan Terrains. *Geography of the Physical Environment*. Springer, Cham.

Kanga, S., Singh, S. K., Meraj, G., Kumar, A., Parveen, R., Kranjčić, N., & Đurin, B. (2022b). Assessment of the Impact of Urbanization on Geoenvironmental Settings Using Geospatial Techniques: A Study of Panchkula District, Haryana. *Geographies*, 2(1), 1-10. ISSN No. 2673-7086.

Kotecha, M. J., Kanga, S., Pankhaniya, S. K., Agrawal, S., Meraj, G., & Singh, S. K. (2023). Urban Heat Island (UHI) Resilience Plan in Varying Climatic Conditions Using Geospatial Approach: A Case Study of Rajkot City. In *Climate Change, Agriculture and Society: Approaches Toward Sustainability* (pp. 307-338). Cham: Springer International Publishing.

Lahon, D., Sahariah, D., Debnath, J., Nath, N., Meraj, G., Farooq, M., Kanga, S., Singh, S. K., & Chand, K. (2023a). Growth of water hyacinth biomass and its impact on the floristic composition of aquatic plants in a wetland ecosystem of the Brahmaputra floodplain of Assam, India. *PeerJ*, 1-32. http://dx.doi.org/10.7717/peerj.14811

Lahon, D., Sahariah, D., Debnath, J., Nath, N., Meraj, G., Kumar, P., Hashimoto, S. and Farooq, M. (2023b). Assessment of Ecosystem Service Value in Response to LULC Changes Using Geospatial Techniques: A Case Study in the Merbil Wetland of the Brahmaputra Valley, Assam, India. *ISPRS Int. J. Geo-Inf.*, 12(4), p.165.

Lavane, K., Kumar, P., Meraj, G., Han, T. G., Ngan, L. H. B., Lien, B. T. B., Ty, T. V., Thanh, N. T., Downes, N. K., Nam, N. D. G., Minh, H. V. T., Singh, S. K., & Kanga, S. (2023a). Assessing the Effects of Drought on Rice Yields in the Mekong Delta. *Climate*, 11(1), 13.

Meraj, G., Farooq, M., Singh, S. K., Islam, M. N., & Kanga, S. (2022). Modeling the sediment retention and ecosystem provisioning services in the Kashmir valley, India, Western Himalayas. *Model. Earth Syst. Environ.*, 8(3), 3859-3884.

Meraj, G., Kanga, S., Kranjčić, N., Đurin, B., & Singh, S. K. (2021a). Role of natural capital economics for sustainable management of earth resources. *Earth*, 2(3), 622-634.

Meraj, G., Singh, S. K., Kanga, S., Islam, M. N., & Singh, S. K. (2021b). Modeling on comparison of ecosystem services concepts, tools, methods and their ecological-economic implications: A review. *Model. Earth Syst. Environ.*, 1-20.

Nath, N., Sahariah, D., Meraj, G., Debnath, J., Kumar, P., Lahon, D., Chand, K., Farooq, M., Chandan, P., Singh, S. K., & Kanga, S. (2023). Land Use and Land Cover Change Monitoring and Prediction of a UNESCO World Heritage Site: Kaziranga Eco-Sensitive Zone Using Cellular Automata-Markov Model. *Land*, 12(1), 151.

Rafiq, M., Meraj, G., Kesarkar, A. P., Farooq, M., Singh, S. K., Kanga, S., & Bhardwaj, V. (2022). Hazard Mitigation and Climate Change in the Himalayas–Policy and Decision Making. In S. Kanga, G. Meraj, M. Farooq, S. K. Singh, & M. S. Nathawat (Eds.), Disaster Management in the Complex Himalayan Terrains. *Geography of the Physical Environment*. Springer, Cham. https://doi.org/10.1007/978-3-030-89308-8_12.

Sajan, B., Mishra, V. N., Kanga, S., Meraj, G., Singh, S. K., & Kumar, P. (2022). Cellular automata-based artificial neural network model for spatial-temporal dynamics and land use/land cover change prediction. *Agronomy*, 12, 2772.

Singh, A., Singh, S. K., Meraj, G., Kanga, S., Farooq, M., Kranjcic, N., **Đurin**, B., & Sudhanshu. (2022b). Designing Geographic Information System Based Property Tax Assessment in India. *Smart Cities*, 5, 364–381.

Singh, H., Meraj, G., Singh, S., Shrivastava, V., Sharma, V., Farooq, M., Kanga, S., & Kumar, P. (2022a). Status of Air Pollution during COVID-19-Induced Lockdown in Delhi, India. *Atmosphere*, 13(12), 2090.

Singh, S. K., Gupta, S. K., Meraj, G., Kumar, P., & Kanga, S. (2023b). Uncovering the hydrometeorological drivers responsible for forest fires utilizing geospatial techniques. *Theor. Appl. Climatol.*. https://doi.org/10.1007/s00704-023-04497-y

Singh, S. K., Meraj, G., Mondal, N., Bera, A., Verma, M. K., Tomar, J. S., & Kanga, S. (2022). *Assessment of seasonal vegetation dynamics over parts of thar desert using geospatial techniques*. ANGRAU, 105.

Singh, S., Singh, S. K., Prajapat, D. K., Pandey, V., Kanga, S., Kumar, P., & Meraj, G. (2023a). Assessing the Impact of the 2004 Indian Ocean Tsunami on South Andaman's Coastal Shoreline: A Geospatial Analysis of Erosion and Accretion Patterns. *J. Mar. Sci. Eng.*, 11(6), 1134.

Sud, A., Kanga, R., Singh, S. K., Meraj, G., Kanga, S., Kumar, P., Ramanathan, A. L., & Bhardwaj, V. (2023). Simulating Groundwater Potential Zones in Mountainous Indian Himalayas—A Case Study of Himachal Pradesh. *J. Geophys. Res.: Earth Surf.*, 10(3), 65.

Upadhyay, R. K., Tripathi, G., Đurin, B., Šamanović, S., Cetl, V., Kishore, N., Sharma, M., Singh, S. K., Kanga, S., Wasim, M., Rai, P. K., & Bhardwaj, V. (2023). Groundwater Potential Zone Mapping in the Ghaggar River Basin, North-West India, Using Integrated Remote Sensing and GIS Techniques. *Water*, 15(5), 961.

10

Dwindling Forest Cover and Environmental Degradation: A Case Study of Bihar

Vinita Prasad

Dept. of Geography, A. N. College (P.P.U.), Patna, Bihar, India

Abstract

The forest is an important resource and is essential for maintaining ecological balance. It is a store house of numerous resources and has many direct and indirect uses. In recent years, the forest cover has been removed to acquire land for various economic pursuits, consequently, Bihar has been left with just 7.75% percent of forest cover. The present chapter intends to explore the problems and prospect of forest resource in Bihar. The chapter discusses the distribution of forest cover in the state; it identifies the causes behind the disappearance of forest cover; and suggests measures to improve and enhance the forest cover to make it sustainable. It is an empirical study based on secondary data, detailed literature survey, and self-observation.

Keywords: Forest, deforestation, ecological balance, restoration, wildlife

10.1 Introduction

The forest is an important biotic resource. It maintains ecological balance and also provides a wide range of forest products. At present, Bihar has a forest cover that accounts for about **7.75% of the state's geographical area**. The forest cover in Bihar is much below the scientific recommendation, which calls for 33% of the total geographical area be under forests (Kanga *et al.*, 2022a). There is wide variation in the spatial distribution of forests in the state and only 11 districts out of a total 38 districts contain forest cover. The declining forest cover has a direct bearing on the environment, causing environmental degradation (Singh *et al.*, 2022b; Debnath *et al.*, 2022a; Kotecha *et al.*, 2023). The forest cover protects the environment in various ways (Debnath *et al.*, 2023b; Gulati *et al.*, 2023). It protects soil cover, regulates temperature and rainfall, helps in underground water percolation and evo-transpiration, and provides shelter to a wide range of flora and fauna preserving biodiversity (Farooq *et al.*, 2022c; Debnath *et al.*, 2023c; Farooq *et al.*, 2022a; Farooq *et al.*, 2022b). It is obvious that forest cover is essential to maintain ecological balance.

Email: wnprsd@yahoo.co.in

Sanjay Kumar, Suraj Kumar Singh, Shruti Kanga, Gowhar Meraj, Majid Farooq and M. S. Nathawat (eds.) Disaster Management and Environmental Sustainability, (131–138) © 2024 Scrivener Publishing LLC

10.2 Objectives

In the above backdrop, this study proceeds with following objectives:

- The chapter examines the distribution of forest cover in the state,
- It further identifies the causes behind the disappearance of forest cover,
- Again, it explores the constrains in the way of restoring forest cover in the state, and
- Finally, suggests measures to improve and enhance the forest cover to make it sustainable.

10.3 Research Design

It is an empirical study based on secondary data, satellite imagery, personal observation, and extensive library work. The secondary data and satellite imagery were obtained from various authentic government and non-government agencies. Appropriate cartographic and statistical techniques were adopted to represent the data. Satellite image of study area have shown in Figure 10.1.

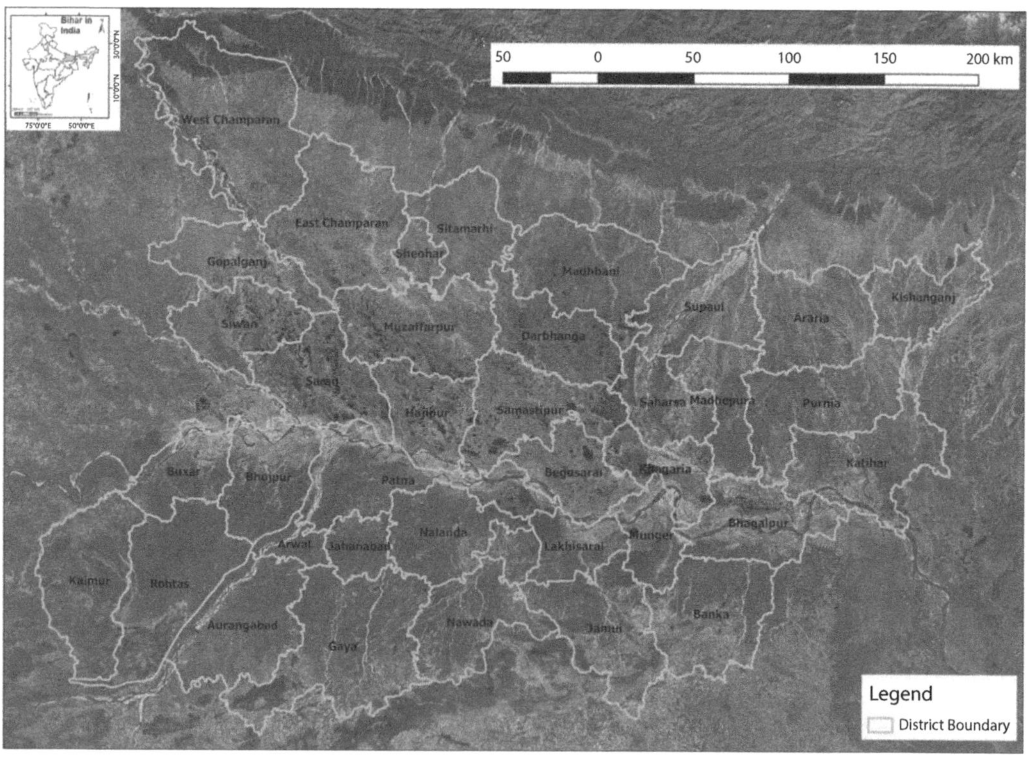

Figure 10.1 Source: Satellite Imagery, 2020.

10.4 Study Area

The study area Bihar lies between 24°21′10″ to 27°31′15″ latitudes and 85°19′30″ to 88°17′4″ longitudes, having a total geographical area of about 94,613 km^2. It occupies 2.16% of the country's total geographical area. Bihar has a forest cover of **7,299 km^2** which is about **7.75% of the state's geographical area** (*Satellite data,* 2015).

10.5 Results and Discussion

Bihar: Distribution of Forests

At present, Bihar's 7.75% of geographical area (**7,299** km^2) is under forest cover, which is further decreasing at an alarming rate. The forest cover also includes trees along roadsides, canals, and railway tracks (Kanga *et al.,* 2022b). Bihar lies in the tropical to sub-tropical monsoon region, receiving average annual rainfall of 1200 mm, which is conducive for the growth of moist deciduous forests. Bihar's forest cover is confined to the districts of Kaimur, Rohtas, Aurangabad, Gaya, Jehanabad, Nawada, Nalanda, Munger, Jamui, Banka, and West Champaran. The northern part of the state comprised of West Champaran, Saharasa, and Purnia districts, contain moist deciduous sal forest, whereas the districts of south Bihar like Kaimur, Rohtas, Aurangabad, Gaya, Jamui, Munger, and Banka have dry deciduous sal forests (http://forest.bih.nic.in). The forest canopy of Bihar is dominated by trees such as sal, shisham, teak, gamahar, kadamb, semal, neem, peepal, bargad, arjun, asan, haldu, mahua, and kendo.

Table 10.1 shows the intensity of forests in Bihar. It is clear from the table that the thick dense forest occupies only 4.5% (332 km^2) of the forested area. The dense forest covers 44.6% (3,260 km^2) of the forest area. The partially dense forest occupies 50.7% (3707 km^2) of forest cover, the largest forest in terms of area. Pie diagram showing distribution of forests is shown in Figure 10.2.

Apart from this, Bihar contains 12 wildlife sanctuaries and one national park. The national park is an area which is strictly reserved for the betterment of the wildlife and biodiversity, and where activities like developmental, forestry, poaching, hunting, and grazing on cultivation are not permitted (http://natureconservation.in). The Wildlife sanctuary is a

Table 10.1 Forest cover in Bihar (2017).

Sl. no.	Forest type	Area (in km^2)	Area (in %)
1	Thick dense forest	332	4.5%
2	Moderately dense forest	3260	44.6%
3	Open forest	3707	50.7%
	Total	**7299**	**100%**

Source: Van Vibhag, Bihar.

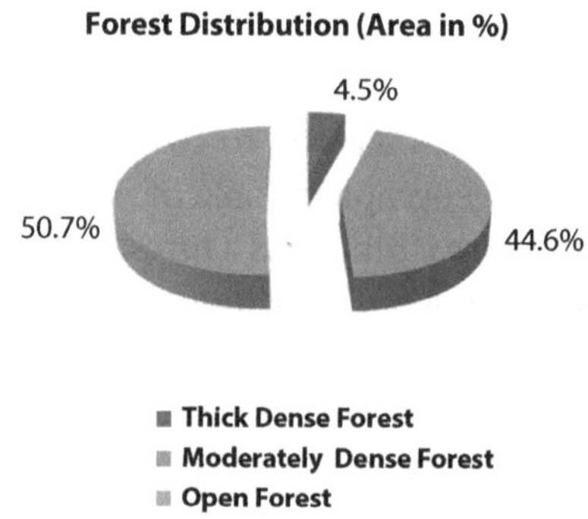

Figure 10.2 Distribution of forests.

protected geographic area within which wildlife is protected. The national parks and wildlife sanctuaries are meant to protect and preserve wildlife. The 12 wildlife sanctuaries in the state occupy a total area of 2904.68 km^2 (Figure 10.3).

There is just one national park, Valmiki National Park, which is a Tiger reserve, located in West Champaran district, occupying an area of 335.6 km^2, and coincides with the thick dense forest area of the state (Figure 10.4).

Disappearance of Forest Cover in Bihar: Causes

Bihar consists of highly fertile alluvial plains suitable for agriculture. Consequently, since time immemorial, the forest cover has been removed to reclaim every possible patch of land for cultivation, which has dwindled the forest cover. In the year 2000, the bifurcation of the state in Bihar and Jharkhand further reduced the forest area in the state and Bihar was left with about 14% of the forest cover, which keep declining and is down to a meager 7.75%. The expansion of anthropogenic activities has been responsible for this plight (Lahon *et al.*,

Figure 10.3 Forest cover map of Bihar.

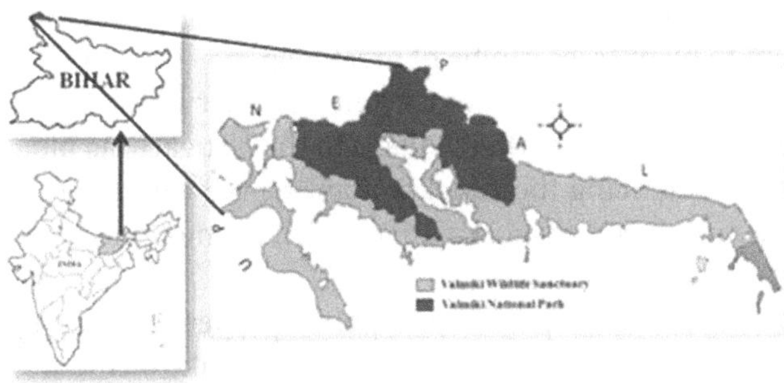

Figure 10.4 Location of the Valmiki National Park.

2023b). The continuous rise in population has accelerated demand for accommodation and resulted in clearing of trees for the expansion of settlement in both rural and urban areas (Lahon *et al.*, 2023a). A number of orchards were removed from both urban and rural areas for the construction of houses. The ignorance of the stakeholder regarding the benefits of green cover also led to clearing of tress in large numbers. The construction of highways, railway tracks, and widening of roads have also led to merciless cutting of trees (Sajan *et al.*, 2022).

In the last few years, a number of new roads have been constructed and highways were widened and led to blind cutting of trees across the state. Similarly, the rapid construction of flyovers in most of the cities in the state has led to the mass removal of tree cover. The lack of strict implementation of law is also responsible for this.

Restoration of Forest Cover: Problems and Constrains

A host of factors come in the way of deforestation in the state. The anthropogenic activities have resulted in rapid deforestation and degradation. The expansion of rural and urban areas has engulfed a huge chunk of green cover. Apart from this, there is no replacement of trees cut for construction or widening of roads, highways, railway tracks, dams, houses, etc. There is also a legal lacuna as the construction or widening of roads, highways, railway tracks, dams falls under the departments of roadways, railways, and irrigation, respectively, whereas tree plantation falls under the jurisdiction of the forest department, and the lack of coordination between the various departments makes the situation worse.

High demand for trees for firewood in rural areas further aggravates the situation. There is hardly any alternative source of fuel in rural areas and the masses largely depend on firewood. Inadequate laws, inefficient law enforcement agencies, lack of awareness and community participation in plantation and forest conservation, use of firewood for cremation purposes, and indiscriminate grazing are collectively responsible for the decline of forest cover (Singh *et al.*, 2022a).

Measures to Make Forest Cover Sustainable

There is an urgent need to protect and conserve the forests in Bihar. It is essential to check overuse, misuse, and abuse of this precious biotic resource. Generating mass awareness regarding the benefits of forests and the consequences of deforestation among the people and mobilizing them to plant trees can be of great help.

There is a need to highlight the initiative taken on either individual or community basis to augment green cover, such as Dharhara village in the Naugachhia block of Bhagalpur district, where people are practicing the ritual of planting ten fruit trees on the birth of every girl child (gulfnews.com). It is a novel initiative which promotes forest conservation on the one hand and gender equality on the other.

Large scale plantation, on both private and public properties, is an effective way to augment green cover. Plantation of trees on special occasions like birthdays, anniversaries, and festivals is a unique way of plantation. Firewood is largely used for cremation purposes, which consumes large amounts of wood, thus, the alternative arrangement of cremation especially such as the promotion of electric cremation, or use of cow dung for the purpose, can save forests to a great extent. The arrangement of proper grazing ground is also an important step to conserve forests. The strict implementation of law regarding forest conservation is a must to control further degradation of forests. There is a need to have severe penalties for the violation of laws related to protection and conservation of forests.

The government of Bihar has initiated a number of programs to preserve and conserve the forest cover. The forest department has asked government officials to plant one sapling each to increase the state's green cover. The Bihar government observes Earth Day on August 9th every year, and the state government is committed to plant 2.51 million trees across the state (http://forest.bih.nic.in). The government is keen to convert government premises into green zones by doing massive planting. Agro-forestry is also being promoted to increase green cover.

Under the ongoing scheme "Mukhya Mantri Chhatra Vriksharopan Yojna," more than 10 lakh students have planted trees in different parts of the state since 2006. Government is also promoting social forestry, which involves raising fast growing trees to meet the community demand for fuel and fodder, thus, reducing the stress on forests. Quick-growing trees are being planted along the sides of the roads, canals, and railway lines. Trees are also being planted under MNREGA.

The "Jal-Jeevan-Hariyali Abhiyan" launched in Bihar aims at water conservation and afforestation to counter climate change (liveyojana.com). Besides water conservation, this program also focuses on plantation of trees including water absorbing plants along the road and government and private lands on a large scale.

10.6 Conclusion

It is clear from the above discussion that forest resource is passing through a critical phase. There is a lot to do to conserve the green cover and make it sustainable. It needs sincere effort on the part of all stakeholders, and public–private partnership can do a lot in this direction.

References

Debnath, J., Sahariah, D., Lahon, D., Nath, N., Chand, K., Meraj, G., Kumar, P., Singh, S. K., Kanga, S., & Singh, S. K. (2022a). Geospatial modeling to assess the past and future land use-land cover changes in the Brahmaputra Valley, NE India, for sustainable land resource management. *Environ. Sci. Pollut. Res.*, 1-24.

Debnath, J., Sahariah, D., Lahon, D., Nath, N., Chand, K., Meraj, G., Kumar, P., Singh, S. K., Kanga, S., & Farooq, M. (2023b). Assessing the impacts of current and future changes of the planforms of River Brahmaputra on its land use-land cover. *Geosci. Front.*. https://doi.org/10.1016/j.gsf.2023.101557.

Debnath, J., Sahariah, D., Saikia, A., Meraj, G., Nath, N., Lahon, D., Annayat, W., Kumar, P., Chand, K., Singh, S. K., & Kanga, S. (2023c). Shifting Sands: Assessing Bankline Shift Using an Automated Approach in the Jia Bharali River, India. *Land*, 12(3), 703.

Farooq, M., Meraj, G., Rishabh, Kanga, S., Nathawat, R., Singh, S. K., & Ranga, V. (2022a). Slum Categorization for Efficient Development Plan—A Case Study of Udhampur City, Jammu and Kashmir Using Remote Sensing and GIS. In *Geospatial Technology for Landscape and Environmental Management: Sustainable Assessment and Planning* (pp. 283-299). Cham: Springer International Publishing.

Farooq, M., Mushtaq, F., Meraj, G., Singh, S. K., Kanga, S., Gupta, A., Kumar, P., Singh, D., & Avtar, R. (2022b). Strategic Slum Upgrading and Redevelopment Action Plan for Jammu City. *Resources*, 11(12), 120.

Farooq, M., Rashid, H., Meraj, G., Kanga, S., & Singh, S. K. (2022c). Assessing the Microclimatic Environmental Indicators of Climate Change of a Temperate Valley in the Western Himalayan region. In A. R. Siddiqui and Avijit Sahay (Eds.), *Climate Change, Disaster and Adaptations - Contextualizing human responses to ecological change* (pp. 47–61). https://link.springer.com/chapter/10.1007/978-3-030-91010-5_5

Gulati, B., Sharma, R., Kanga, S., Singh, S. K., Sajan, B., Meraj, G., Kumar, P., & Ramanathan, A. L. (2023). Unraveling the Relationship Between Stubble Burning and Air Quality Degradation in Punjab: A Temporal and Spatial Analysis (2019-2022). *J. Clim. Change*, 9(2), 43-53.

Kanga, S., Meraj, G., Farooq, M., Singh, S. K., & Nathawat, M. S. (Eds.). (2022a). Disaster Management in the Complex Himalayan Terrains. *Geography of the Physical Environment*. Springer, Cham.

Kanga, S., Singh, S. K., Meraj, G., Kumar, A., Parveen, R., Kranjčić, N., & Đurin, B. (2022b). Assessment of the Impact of Urbanization on Geoenvironmental Settings Using Geospatial Techniques: A Study of Panchkula District, Haryana. *Geographies*, 2(1), 1-10. ISSN No. 2673-7086.

Kotecha, M. J., Kanga, S., Pankhaniya, S. K., Agrawal, S., Meraj, G., & Singh, S. K. (2023). Urban Heat Island (UHI) Resilience Plan in Varying Climatic Conditions Using Geospatial Approach: A Case Study of Rajkot City. In *Climate Change, Agriculture and Society: Approaches Toward Sustainability* (pp. 307-338). Cham: Springer International Publishing.

Lahon, D., Sahariah, D., Debnath, J., Nath, N., Meraj, G., Farooq, M., Kanga, S., Singh, S. K., & Chand, K. (2023a). Growth of water hyacinth biomass and its impact on the floristic composition of aquatic plants in a wetland ecosystem of the Brahmaputra floodplain of Assam, India. *PeerJ*, 1-32. http://dx.doi.org/10.7717/peerj.14811

Lahon, D., Sahariah, D., Debnath, J., Nath, N., Meraj, G., Kumar, P., Hashimoto, S. and Farooq, M. (2023b). Assessment of Ecosystem Service Value in Response to LULC Changes Using Geospatial Techniques: A Case Study in the Merbil Wetland of the Brahmaputra Valley, Assam, India. *ISPRS Int. J. Geo-Inf.*, 12(4), p.165.

Sajan, B., Mishra, V. N., Kanga, S., Meraj, G., Singh, S. K., & Kumar, P. (2022). Cellular automata-based artificial neural network model for spatial-temporal dynamics and land use/land cover change prediction. *Agronomy*, 12, 2772.

Singh, A., Singh, S. K., Meraj, G., Kanga, S., Farooq, M., Kranjcic, N., Đurin, B., & Sudhanshu. (2022b). Designing Geographic Information System Based Property Tax Assessment in India. *Smart Cities*, 5, 364–381.

Singh, H., Meraj, G., Singh, S., Shrivastava, V., Sharma, V., Farooq, M., Kanga, S., & Kumar, P. (2022a). Status of Air Pollution during COVID-19-Induced Lockdown in Delhi, India. *Atmosphere*, 13(12), 2090.

11

Environment and Sustainable Development: Issues and Challenges

Anshu Veenapani

Department of Commerce, T.M. Bhagalpur University, Bhagalpur, India

Abstract
Environment refers to all external conditions and factors that affect all living organisms on this Earth. The entire lithosphere, atmosphere, hydrosphere, and biosphere come within the ambit of environment. Development means economic and social growth like income, investment, production, health, and education. In the modern century, we face the grave challenge of environmental degradation. The population explosion and excessive exploitation of natural resources have brought disastrous effects on the environment and the survival of human beings. Sustainable development is the only panacea to combat the effects of environment. There should be a rational and equitable balance between the two, only then, may we be able to save our Mother Earth for the future generations.

Keywords: Environment, sustainable, excessive, exploitation, equitable

11.1 Introduction

The overwhelming preoccupation of materialistic development is mostly with the questions "growth," "equity," and "sustainability." The issues of long-term ecological and environmental pollution, public health hazards, and waste disposal have long been neglected (Upadhyay *et al.*, 2023). During the last few decades, the brute exploitation of natural resources and industrial growth, unplanned and unregulated with the intense pursuit of economic growth at any cost, have inflicted upon the environment mercilessly and the consequence is what is called environmental pollution (G.T. Miller, 2000; Singh *et al.*, 2022a). This has emerged as a major problem all over the world. The conflict between the accelerating pace of economic development and the need for conservation of degrading natural environment have prompted thinking about the major environmental issues involved and the challenges to overcome them (Singh *et al.*, 2022b). Our aim is to make the planet Earth a better, safer, healthier, and wealthier place to live for the posterity. Our purpose is to raise awareness among the masses towards environmental conservation and management for improving the quality of life of the people and safeguarding the planet (Singh *et al.*, 2022; Singh *et al.*, 2023a; Sud *et al.*, 2023).

Email: anshuveenapani@gmail.com

Sanjay Kumar, Suraj Kumar Singh, Shruti Kanga, Gowhar Meraj, Majid Farooq and M. S. Nathawat (eds.)
Disaster Management and Environmental Sustainability, (139–144) © 2024 Scrivener Publishing LLC

The term 'environment' refers to all external conditions and factors that affect living organismsplant, animal, and man. In fact, all of lithosphere, hydrosphere, atmosphere, and biosphere come within the ambit of environment. Development means economic growth which is usually measured by an increase in the country's GNP (Gross National Product) and GDP (Gross Domestic Product) (Pandey *et al.*, 2010; Rafiq *et al.*, 2022; Sajan *et al.*, 2022). The other indicators of development are the status of education, health, and infrastructure. While economic indicators such as investment, production, and trade are consistently positive, the key environmental indicators are increasingly negative (Meraj *et al.*, 2021a). Forests are shrinking, water tables are going down, soils are eroding, wetlands are disappearing, rivers are drying, fisheries are collapsing, temperatures are rising, and plant and animal species are disappearing (Meraj *et al.*, 2021b; Nath *et al.*, 2023). The major environmental problems we face today are exponential growth, depletion of resources, destruction and degradation of wildlife habitat, loss of biodiversity, poverty, and pollution. World population has more than doubled in the last 50 y. Unless birth rates are controlled or death rates rise sharply, it may reach more than 1400 crore by 2150. Thus, the challenge is to develop new models for dealing with environmental problems more sustainably (Singh *et al.*, 2023b).

11.2 Environment and Development

At the dawn of civilization, there was no conflict between environment and development. We may divide the cultural as well as environmental changes that happened in the world into the following subheads:

1. Early man as hunters and gatherers
2. Agricultural revolution
3. Industrial revolution
4. Information revolution
5. Knowledge revolution

We are in the midst of the information revolution in which new technologies such as telephones, radios, television, internet, and computers are enabling people to deal with more information more rapidly. What is needed is a knowledge revolution in order to revise the education system at all levels to train people how to evaluate the environmental impact of human activities on the environment (Debnath *et al.*, 2022a).

It is undoubtedly accepted that we have altered the composition of the air, fouled the water, and over exploited the fossil fuel stored in the womb of the earth for millennia (Farooq *et al.*, 2022a). We have made the earth less fertile and in many places, barren, and we have diminished the diversity of the living biotic communities (Debnath *et al.*, 2023c). Thus, the problems we have made for ourselves are many, such as air, water and land pollution, acid rain, ozone depletion, desertification, global warming, deforestation, loss of species, and scarcity of fresh water (Debnath *et al.*, 2023b). The pollution of air and water are serious problems. Air pollution leads to many lung diseases. The burning of fossil fuel is increasing carbon dioxide content in the air, resulting in global warming. CFC (Chloro Fluoro Carbon) and HCFCs (Hydro Chloro Fluoro Carbon) will do more damage to the ozone layer. The scarcity of water in many areas is threatening human survival. The pollution of surface water is another problem which affects our health. Most of the rivers of the world are polluted and so is our great Ganga. Soil degradation is a serious problem; overgrazing, unscientific cultivation, and deforestation are leading to soil creation and soil fertility depletion (Lahon *et al.*, 2023b; Lavane *et al.*, 2023; Meraj *et al.*, 2022).

The greatest problem facing the world is exponential growth of population especially in the developing countries or the third world. Excessive population growth leads to poverty and poverty breeds population growth, and poverty and environment are inextricably linked in a chain of cause and effect (Farooq *et al.*, 2022b). Environmental problems cannot be tackled in isolation from those national and global economic factors that perpetuate poverty. In fact, there are two nations in one state—the privileged (rich) and the people (poor) (Gulati *et al.*, 2023). In India or in Bihar, the economy is growing at the rate of 8.9% per annum, but at the same time, the number of poor people is also growing. At least 40% of people in Bihar live below the poverty line. The new shadow of fear is cast far less by the possibility of a nuclear holocaust than by the prospect of environmental disaster (Kotecha *et al.*, 2023; Lahon *et al.*, 2023a). The grave question is the survival of the human race and the saving of civilization. In the words of His Holiness Dalai Lama, "Mother planet is showing us the red warning light, she is telling us to take care of our house the planet;" we can manage the Earth so as to create environments that are ecologically stable, economically profitable, aesthetically rewarding, and favorable to the continued evolution of civilization. We have to become more and more conscious about the environmental issues and challenges (Kanga *et al.*, 2022b).

11.3 Environmental Ethical Issues and Development

The ethics of human survival requires us to recognize how we have ravaged the planet and endangered its life forms, including our species. It encourages us to rise to the challenge that faces us, accepting that all must care for the Earth if its future is to be secured (Kanga *et al.*, 2022a). To secure that future requires a unique and united act of statesmanship that reflects human values and human spirit at their highest and the best.

The probable answer to make a secure world is to attain a development model which is ecologically viable. Economic growth fostered by a supportive international environment and guided by the right domestic policies is an important part of the answer to poverty alleviation. But it must be economic growth of a kind, compatible with ecological health. The Bruntland Report propounded the concept of sustainable development in 1980. By sustainable development we mean that the world must meet its present needs in ways that do not diminish the ability of future generations to meet their needs, it must ensure that human progress is sustainable.

11.4 Measures and Suggestions

To conserve and safeguard the environment, the World Conservation Union, UNEP (UN Environment Programme), and WWF (World Wide Fund for Nature) have suggested two things: one is to secure a widespread and deeply held commitment to the new ethics of sustainable living, and the other is to integrate conservation and development (Farooq *et al.*, 2022c). Conservation is needed to keep our actions within the Earth's capacity, and development with a view to enable people everywhere to enjoy long, healthy, and fulfilling lives. Some of the measures are suggested below:

- Energy consumption should be reduced 1–2% per year.
- Forest reduction should be stopped and new forests on marginal barren wastelands should be developed.
- Reduction of pollution by reducing the emissions of sulphur dioxide, nitrogen oxide, carbon dioxide, and CFC.
- Poverty reduction by increasing annual income by 2–3%.
- Population control by bringing fertility rates down to 2.1 births per woman as in China and Kerala.
- Legal regimes must be enforced to conserve biodiversity.

Lastly, for living a sustainable life on Earth, the Gandhian way of simple living, respect of nature, and reduction of our wants and desires are pertinent.

References

G.T. Miller (2000): *Living in the Environment* 11th Edition, Cole Publishing Co., Newyork.

Debnath, J., Sahariah, D., Lahon, D., Nath, N., Chand, K., Meraj, G., Kumar, P., Singh, S. K., Kanga, S., & Singh, S. K. (2022a). Geospatial modeling to assess the past and future land use-land cover changes in the Brahmaputra Valley, NE India, for sustainable land resource management. *Environ. Sci. Pollut. Res.,* 1-24.

Debnath, J., Sahariah, D., Lahon, D., Nath, N., Chand, K., Meraj, G., Kumar, P., Singh, S. K., Kanga, S., & Farooq, M. (2023b). Assessing the impacts of current and future changes of the planforms of River Brahmaputra on its land use-land cover. *Geosci. Front.*, https://doi.org/10.1016/j.gsf.2023.101557.

Debnath, J., Sahariah, D., Saikia, A., Meraj, G., Nath, N., Lahon, D., Annayat, W., Kumar, P., Chand, K., Singh, S. K., & Kanga, S. (2023c). Shifting Sands: Assessing Bankline Shift Using an Automated Approach in the Jia Bharali River, India. *Land*, 12(3), 703.

Farooq, M., Meraj, G., Rishabh, Kanga, S., Nathawat, R., Singh, S. K., & Ranga, V. (2022a). Slum Categorization for Efficient Development Plan—A Case Study of Udhampur City, Jammu and Kashmir Using Remote Sensing and GIS. In *Geospatial Technology for Landscape and Environmental Management: Sustainable Assessment and Planning*, (pp. 283-299). Cham: Springer International Publishing.

Farooq, M., Mushtaq, F., Meraj, G., Singh, S. K., Kanga, S., Gupta, A., Kumar, P., Singh, D., & Avtar, R. (2022b). Strategic Slum Upgrading and Redevelopment Action Plan for Jammu City. *Resources*, 11(12), 120.

Farooq, M., Rashid, H., Meraj, G., Kanga, S., & Singh, S. K. (2022c). Assessing the Microclimatic Environmental Indicators of Climate Change of a Temperate Valley in the Western Himalayan region. In A. R. Siddiqui and Avijit Sahay (Eds.), *Climate Change, Disaster and Adaptations - Contextualizing human responses to ecological change*, (pp. 47–61). https://link.springer.com/chapter/10.1007/978-3-030-91010-5_5

Gulati, B., Sharma, R., Kanga, S., Singh, S. K., Sajan, B., Meraj, G., Kumar, P., & Ramanathan, A. L. (2023). Unraveling the Relationship Between Stubble Burning and Air Quality Degradation in Punjab: A Temporal and Spatial Analysis (2019-2022). *J. Clim. Change*, 9(2), 43-53.

Kanga, S., Meraj, G., Farooq, M., Singh, S. K., & Nathawat, M. S. (Eds.). (2022a). Disaster Management in the Complex Himalayan Terrains. *Geography of the Physical Environment*. Springer, Cham.

Kanga, S., Singh, S. K., Meraj, G., Kumar, A., Parveen, R., Kranjˇ, N., & Đurin, B. (2022b). Assessment of the Impact of Urbanization on Geoenvironmental Settings Using Geospatial Techniques: A Study of Panchkula District, Haryana. *Geographies*, 2(1), 1-10. ISSN No. 2673-7086.

Kotecha, M. J., Kanga, S., Pankhaniya, S. K., Agrawal, S., Meraj, G., & Singh, S. K. (2023). Urban Heat Island (UHI) Resilience Plan in Varying Climatic Conditions Using Geospatial Approach: A Case Study of Rajkot City. In *Climate Change, Agriculture and Society: Approaches Toward Sustainability*, (pp. 307-338). Cham: Springer International Publishing.

Lahon, D., Sahariah, D., Debnath, J., Nath, N., Meraj, G., Farooq, M., Kanga, S., Singh, S. K., & Chand, K. (2023a). Growth of water hyacinth biomass and its impact on the floristic composition of aquatic plants in a wetland ecosystem of the Brahmaputra floodplain of Assam, India. *PeerJ*, 1-32. http://dx.doi.org/10.7717/peerj.14811

Lahon, D., Sahariah, D., Debnath, J., Nath, N., Meraj, G., Kumar, P., Hashimoto, S. and Farooq, M. (2023b). Assessment of Ecosystem Service Value in Response to LULC Changes Using Geospatial Techniques: A Case Study in the Merbil Wetland of the Brahmaputra Valley, Assam, India. *ISPRS Int. J. Geo-Inf.*, 12(4), p. 165.

Lavane, K., Kumar, P., Meraj, G., Han, T. G., Ngan, L. H. B., Lien, B. T. B., Ty, T. V., Thanh, N. T., Downes, N. K., Nam, N. D. G., Minh, H. V. T., Singh, S. K., & Kanga, S. (2023a). Assessing the Effects of Drought on Rice Yields in the Mekong Delta. *Climate*, 11(1), 13.

Meraj, G., Farooq, M., Singh, S. K., Islam, M. N., & Kanga, S. (2022). Modeling the sediment retention and ecosystem provisioning services in the Kashmir valley, India, Western Himalayas. *Model. Earth Syst. Environ.*, 8(3), 3859-3884.

Meraj, G., Kanga, S., Kranjčić, N., Đurin, B., & Singh, S. K. (2021a). Role of natural capital economics for sustainable management of earth resources. *Earth*, 2(3), 622-634.

Meraj, G., Singh, S. K., Kanga, S., Islam, M. N., & Singh, S. K. (2021b). Modeling on comparison of ecosystem services concepts, tools, methods and their ecological-economic implications: A review. *Model. Earth Syst. Environ.*, 1-20.

Nath, N., Sahariah, D., Meraj, G., Debnath, J., Kumar, P., Lahon, D., Chand, K., Farooq, M., Chandan, P., Singh, S. K., & Kanga, S. (2023). Land Use and Land Cover Change Monitoring and Prediction of a UNESCO World Heritage Site: Kaziranga Eco-Sensitive Zone Using Cellular Automata-Markov Model. *Land*, 12(1), 151.

Rafiq, M., Meraj, G., Kesarkar, A. P., Farooq, M., Singh, S. K., Kanga, S., & Bhardwaj, V. (2022). Hazard Mitigation and Climate Change in the Himalayas–Policy and Decision Making. In S. Kanga, G. Meraj, M. Farooq, S. K. Singh, & M. S. Nathawat (Eds.), *Disaster Management in the Complex Himalayan Terrains. Geography of the Physical Environment*. Springer, Cham. https://doi.org/10.1007/978-3-030-89308-8_12.

Sajan, B., Mishra, V. N., Kanga, S., Meraj, G., Singh, S. K., & Kumar, P. (2022). Cellular automata-based artificial neural network model for spatial-temporal dynamics and land use/land cover change prediction. *Agronomy*, 12, 2772.

Singh, A., Singh, S. K., Meraj, G., Kanga, S., Farooq, M., Kranjcic, N., Đurin, B., & Sudhanshu. (2022b). Designing Geographic Information System Based Property Tax Assessment in India. *Smart Cities*, 5, 364–381.

Singh, H., Meraj, G., Singh, S., Shrivastava, V., Sharma, V., Farooq, M., Kanga, S., & Kumar, P. (2022a). Status of Air Pollution during COVID-19-Induced Lockdown in Delhi, India. *Atmosphere*, 13(12), 2090.

Singh, S. K., Gupta, S. K., Meraj, G., Kumar, P., & Kanga, S. (2023b). Uncovering the hydro-meteorological drivers responsible for forest fires utilizing geospatial techniques. *Theor. Appl. Climatol.*, https://doi.org/10.1007/s00704-023-04497-y

Singh, S. K., Meraj, G., Mondal, N., Bera, A., Verma, M. K., Tomar, J. S., & Kanga, S. (2022). Assessment of seasonal vegetation dynamics over parts of thar desert using geospatial techniques. *ANGRAU*, 105.

Singh, S., Singh, S. K., Prajapat, D. K., Pandey, V., Kanga, S., Kumar, P., & Meraj, G. (2023a). Assessing the Impact of the 2004 Indian Ocean Tsunami on South Andaman's Coastal Shoreline: A Geospatial Analysis of Erosion and Accretion Patterns. *J. Mar. Sci. Eng.*, 11(6), 1134.

Sud, A., Kanga, R., Singh, S. K., Meraj, G., Kanga, S., Kumar, P., Ramanathan, A. L., & Bhardwaj, V. (2023). Simulating Groundwater Potential Zones in Mountainous Indian Himalayas—A Case Study of Himachal Pradesh. *J. Geophys. Res.: Earth Surf.*, 10(3), 65.

Upadhyay, R. K., Tripathi, G., Đurin, B., Šamanović, S., Cetl, V., Kishore, N., Sharma, M., Singh, S. K., Kanga, S., Wasim, M., Rai, P. K., & Bhardwaj, V. (2023). Groundwater Potential Zone Mapping in the Ghaggar River Basin, North-West India, Using Integrated Remote Sensing and GIS Techniques. *Water*, 15(5), 961.

12

Government Programs and Policies Towards Green Environment

Suraj Dev

Department of Geography, Delhi School of Economics, New Delhi, India

Abstract

During colonial India, only one law related to the environment existed: the Indian Forest Act (1927). After independence, India expressed its concerns about the environment much earlier as compared to other developed countries. The United Nations Conference on the Human Environment (also known as the Stockholm Conference) was held in 1972 in Stockholm. In February 1972, a National Committee on Environmental Planning and Coordination (NCEPC) was set up, and just after 4 y, The 42nd Amendment to the Indian Constitution of 1976 integrated environmental concerns into its text in which article 48A to the Directive Principles of State Policy was incorporated. It declares "the State shall endeavor to protect and improve the environment and to safeguard the forests and wildlife of the country". A similar responsibility is imposed upon every citizen in the form of Fundamental Duty (Article 51(A)(g)) "to protect and improve the natural environment including forest, lakes, rivers, and wildlife, and to have compassion for living creatures." Environmental laws on water (1974), air (1981), and forest conservation (1981) were passed, as well as the umbrella act of Environment Protection (1986). Under the Water (Prevention and Control of Pollution) Act, 1974, The Central Pollution Control Board (CPCB) was constituted in September 1974. In November 1980, the Tiwari Committee constituted a separate Department of Environment (DoE), and finally, the ministry of Environment and Forest was set up in 1985. Further several environmental laws were enacted.

***Keywords*:** Environment protection, environmental laws, green environment, environmental policy, legal laws

> "God forbid that India should ever take to industrialization after the manner of the West. The economic imperialism of a single tiny island kingdom (England) is today keeping the world in chains. If an entire nation of 300 million took to similar economic exploitation, it would strip the world bare like locusts."
>
> **– Mahatma Gandhi**

Email: devs6827@gmail.com

Sanjay Kumar, Suraj Kumar Singh, Shruti Kanga, Gowhar Meraj, Majid Farooq and M. S. Nathawat (eds.) Disaster Management and Environmental Sustainability, (145–154) © 2024 Scrivener Publishing LLC

12.1 Environmental Laws

Environmental law is the collection of laws, regulations, agreements, and common laws that govern how humans interact with the environment. Environmental law has two objectives: protecting the environment and guiding the use of natural resources. In addition to protecting the environment from harm, environmental laws determine who can and cannot use natural resources. It may be possible to regulate pollution, natural resource use, forest protection, mineral extraction, and the numbers of animals and fish.

12.1.1 Importance of Legal Laws Related to Environment

The past (recent times) itself has seen a vast array of environmental problems that threaten life and prosperity for individuals. Environmental problems are not restricted to the area where they are generated; they affect communities far from their source. Environmental degradation can be prevented by enacting effective laws. Effective laws are needed to prevent rogue people, the jungle mafia, poaching, pollution, and overuse of natural resources. Environmental pollution is an independent force without regard to political or legal walls [1–3].

12.2 Major Programs and Policies of India Towards Green Environment

12.2.1 Wildlife Protection Act (1972) of India

In 1972, Parliament passed the Wildlife Protection Act, 1972, which protects wild animals, birds, and plants by economic, ecological, and environmental security concerns. There is a National Board for Wildlife assembled by the act, chaired by the Prime Minister. The National Board takes all measures necessary to protect and develop wildlife and forests. The act also provides for the creation of a Wildlife Management Board to advise the state on matters related to wildlife management. The act regulates the hunting of wild animals and protects specified plants from being hunted. In addition, the State Government is authorized to notify sanctuaries, national parks, and game reserves. As part of its wildlife policy, it has also created a National Tiger Conservation Authority and regulates commerce in wild animals, animal products, and trophies. In violation of the act, one can be imprisoned for 1–7 y, and/or fined up to Rs. 50 lakhs.

12.2.2 Water (Prevention and Control of Pollution) Act, 1974

Water (Prevention and Control of Pollution) Act was enacted in 1974 to prevent and control water pollution as well as restore or maintain water wholesomeness. Amendments to the Act were made in 1988. The Water (Prevention and Control of Pollution) Cess Act of 1977 provided for the imposition and collection of a levy on water consumed by persons operating and conducting certain types of industrial activities. It is collected to enhance the resources of the Central Board and the State Boards for the prevention and control of water pollution established by the Water (Prevention and Control of Pollution) Act, 1974. Amendments to the Act were last made in 2003.

12.2.3 Tiwary Committee (1980)

The Tiwary committee was formed in February 1980 to suggest administrative and legislative measures to protect the environment. It made recommendations in itsreport, which was submitted in September 1980. During the review, the committee pointed out that laws related to the environment were insufficient or ineffective for protecting the environment.

After the publication of this report, an environmental department was established at both the state and central levels, and an independent ministry and department were established in 1981.

In addition to the Tiwari committee survey, a survey of the environmental law enforcers at that time would find some gaps to fill; some areas of the environment were uncovered by air pollution. Further, it was suggested that tackling the environment requires a cooperative, coordinated, and creative approach from both the government and non-government organizations (NGOs). The law committee, based on their study and analysis, suggests many recommendations.

It has also been criticized that in its conclusion on the Tiwary committee, it concludes that only one symmetric approach was applied to arrive at its conclusion on the matter of environment protection.

12.2.4 Air (Prevention and Control of Pollution) Act, 1981

According to the definition of air pollution, any liquid or gaseous substance present in the atmosphere at levels that can be harmful to humans, animals, plants, or the environment. The Air Act was passed under Article 253 of the Constitution of India and in compliance with the decision of the Stockholm Conference.

Purpose: to prevent, control, and mitigate air pollution to provide a healthy atmosphere for living creatures and ecosystems.
Defining terms were "air pollution," "air pollutants," "vehicle exhausts," and "industrial plants". The Act also includes automobiles, diesel vehicles, transport, railways, and domestic fuels.
Demarcate high pollution areas: the Act defines certain heavily polluted as areas as air pollution control areas, and no industrial plant shall be operated in these areas without the consent of the State Pollution Control Board.

The Central and State Water Boards are responsible for controlling and preventing air pollution. For the prevention and control of air pollution, the State Boards have to enact and enforce standards.

12.2.5 Forests (Conservation) Act, 1980

The Forest (Conservation) Act was enacted to increase the protection provided to forests and to regulate their diversion for non-forestry purposes. Under the Act, "forest land" refers to forest land so far as it has been recorded in official government records as reserved forest, protected forest, or any other forest area. As defined by the Indian Forest Act of 1927, the term "tree" will refer to the same thing. By the Forest (Conservation) Act, 1980, the permission of the Central Government for the de-reservation/diversion of forest land must be obtained before being used for non-forestry purposes. Deforestation has been checked and afforestation of non-forest areas has been encouraged by this legislation. Under this law, the user(s) must deposit the required amount to undertake compensatory afforestation to mitigate the negative effects of forest land conversion.

Forest land may not be leased out to anybody other than the government under the amended Act (1988). It enhances the conservation and plantation of trees while increasing the forest cover by 30%.

12.2.6 Amended Forest Act, 1992

Under the Act, activities other than forest cutting were allowed in forests with permission from the Central Government. These activities include setting transmission lines, conducting seismic surveys and drilling, and constructing hydroelectric plants. National parks, wildlife sanctuaries, etc., are prohibited from exploration or survey without prior approval of the Central Government, even if no trees are cut. Tea, coffee, spices, rubber, mulberry for raising silkworms, and cash crops are not allowed in reserve forests. Mineral exploration is a non-forestry activity, and prior approval from the Central Government is required. To maintain the balance in the ecology of the forest, fruit-bearing trees, oil-producing plants, or medicinal plants need to be approved first.

12.2.7 Environment (Protection) Act, 1986

The Environment (Protection) Act (EPA) of 1986 was passed to protect and improve the environment. It empowers the government to create authorities that have the responsibility of preventing environmental pollution. Furthermore, the act addresses specific environmental problems specific to different regions of the country. It was the United Nations Conference on the Human Environment held in Stockholm in June 1972 (Stockholm Conference), in which India took part, which laid the groundwork for the enactment of the EPA. The Act implements the decisions made at the Stockholm Conference. The EPA Act was enacted under Article 253 of the Indian Constitution, which allows legislation to give effect to international agreements. According to the Constitution, Article 48A, the state shall endeavor to protect and improve the environment and to safeguard the country's forests and wildlife. According to Article 51A, every citizen must protect the environment.

12.2.7.1 Salient Features

The Environment (Protection) Act, 1986 allows the government to control pollution, improve environmental quality, prohibit or restrict the setting up, and/or operation of any industrial facility for environmental reasons. In 1986, The Environment (Protection) Act was passed to protect and improve the environment. It empowers the Central Government to establish authorities charged with the mandate of preventing environmental pollution in all its forms and to tackle specific environmental problems that are peculiar to different parts of the country. The Act was last amended in 1991.

12.2.8 Indian Environmental Policy (1992)

The Ministry of Environment and Forests, Government of India, in the same year, 1992, announced a policy statement for the abatement of pollution. The Preamble not only spells out the objectives but also shifts the emphasis from defining the problem area towards actual implementation for achieving the objectives.

The policy stresses:

- the need for a positive attitude on the part of everyone in the society topre-vent pollution.
- the need for a comprehensive approach to integrating environmental andeco-nomic aspects in development planning.
- preventive aspects of pollution abatement and promotion of technological inputs to reduce industrial pollutants and reliance upon public cooperation in securing a clean environment to respond to the coming challenges.
- the integration of environmental considerations into decision-making at all levels.
- The policy statement adopts the following fundamental guiding principles:
- prevention of pollution at source
- the adoption of the best available technology
- the polluter pays principle, and
- public participation in decision making.

The policy favors a combination of regulatory and market-based instruments to prevent and control pollution. There has also been considerable activity in the states.

12.2.9 National Environmental Policy 2006

As the main objective of this policy is to secure livelihoods and wellbeing for all, conservation of the environment is the best means of protecting it. It is to ensure that people who depend on the resource benefit from conservation, rather than from degradation of the resource. The policy also aims to generate partnerships between the public sector, local communities, academic, scientific, and investment communities, to share resources and harness abilities to manage the environment. The plan also seeks to reconsider coastal zoning for environmental sustainability, to ensure the protection of ecological systems, waters, and potential sea-level rise vulnerability along with some coastal areas.

Environmental impact assessments will continue to be the primary method for evaluating and reviewing new projects. Panchayati Raj institutions and metropolitan local authorities were included. Environment protection is an integral component of sustainable development, and cannot be viewed as an isolated entity.

12.2.9.1 Objectives of National Environment Policy, 2006

- The protection and conservation of critical ecosystems and resources.

- To promote equitable access to environmental resources and materials for all sections of society, particularly for poor communities that are most dependent on environmental resources for their livelihoods.
- To ensure that environmental resources are used judiciously to meet the expectations of the present and future generations.
- To integrate environmental issues into economic and social policies, plans, and programs.
- It is important to minimize adverse environmental effects by ensuring efficient use of environmental resources per unit economy.
- Aim for good governance (transparency, rationality, accountability, and reduction of time and cost, participation, and regulatory independence) when managing and regulating the use of environmental resources.
- The objective of this group is to provide higher funding and technology to support environmental conservation through multi-stakeholder partnerships involving local and public agencies, academia, research institutes, investors, and multilateral and bilateral development partners.

12.2.10 National Action Plan on Climate Change

The National Action Plan on Climate Change was officially launched on June 30, 2008. The NAPCC identifies measures that contribute to development objectives and help address climate change effectively at the same time. In the National Action Plan, eight "National Missions" constitute the core. Their objectives include promoting awareness of climate change, adaptation and mitigation, energy efficiency, and natural resource conservation [4, 5].

The eight missions are:

1. **National Solar Mission:** the Jawaharlal Nehru National Solar Mission (JNNSM), or the National Solar Mission, is a government initiative to promote solar power in India. The JNNSM, which was launched in January 2010, was revised twice and now boasts a target of 100 GW of solar energy by 2022.

   To make India a leader in solar energy, the JNNSM aims to create the policy conditions that will lead to its widespread deployment.
2. **National Mission for Enhanced Energy Efficiency:** the National Mission for Enhanced Energy Efficiency (NMEEE) consists of four initiatives to boost energy efficiency in energy-intensive industries, including:
   (i) The Perform, Achieve, and Trade (PAT) scheme was launched by the Bureau of Energy Efficiency (BEE) to promote energy efficiency and reduce energy consumption in certain industries.
3. **National Mission on Sustainable Habitat:** the prime minister has approved a national mission on sustainable habitat. The project aims to make cities sustainable through better energy efficiency in buildings, better solid waste management, and public transportation.
4. **National Water Mission:** the main objective of the National Water Mission is to conserve water, minimize water wastage, and ensure a more equitable distribution of water throughout the states by integrating water resources development and management.

5. **The National Mission for Sustaining the Himalayan Ecosystem:** it is a multi-pronged, cross-cutting initiative involving a range of sectors. A significant part of India's population depends for sustenance on the Himalayas, a region not only affected by climate change but also requires adaptation measures. This project contributes to the sustainable development of India by promoting knowledge surrounding climate change, its impacts, and adaptation actions for this region.
6. **National Mission for a Green India:** It works toward conserving, restoring, and enhancing India's forest cover and responding to climate change through adaptive measures and mitigation. The plan outlines a holistic vision of greening that goes beyond tree planting. It focuses on biodiversity, water, biomass, the preservation of mangroves, wetlands, critical habitats, etc., along with carbon sequestration.
7. **National Mission for Sustainable Agriculture:** the mission aims to promote sustainable agriculture by enhancing agriculture productivity especially in rainfed areas through integrated farming, soil health management, and synergizing resource conservation.
8. **National Mission on Strategic Knowledge for Climate Change:** the mission has been formulated to establish a good network with existing knowledge institutions to facilitate capacity building and help to understand climate processes and risks better.

12.2.11 Recent Environmental Programs in India

1. **Namami Gange Programme:** this is an Integrated Conservation Mission, approved by the Government of India in June 2014. This plan aims to achieve two objectives: effective abatement of pollution and conservation and rejuvenation of the National River Ganga. It is operated by the Ministry of Jal Shakti.

   Implementation:

   National Mission for Clean Ganga is implementing the program, along with its state counterpart organizations, State Program Management Groups. The National Ganga Council (NGC) was established in October 2016 under the River Ganga Authorities Order, 2016. It is headed by the prime minister.

   The Namami Gange Programme has the following key pillars: Sewerage treatment, infrastructure, river surface cleaning, afforestation, industrial effluent monitoring, river front development, biodiversity, public awareness, and Ganga Gram.
2. **Green Skill Development Programme (GSDP):** the Green Skill Development Program, initiated by the Ministry of Environment, Forest and Climate Change, aims to enable young persons in India to find gainful employment and/or self-employment.
3. **Swachha Bharat Abhiyan:** on October 2, 2014, India's prime minister launched the Swachha Bharat Mission as a way to push the nation towards achieving universal sanitation coverage. During the campaign, over 100 million toilets would be built in rural India by the 150th birth anniversary of

Mahatma Gandhi to declare all villages, Gram Panchayats, Districts, States, and Union Territories open defecation-free (ODF). With the mission moving on to the next phase of SBMG (Swachh Bharat Mission Gramin), ODF-Plus, it will be able to sustain open defecation-free behavior, ensure that no one is left behind, and provide access to solid and liquid waste management facilities. During Phase II of Swachh Bharat Mission (Grameen), ODF-Plus activities will reinforce ODF behaviors and provide interventions for the safe management of solid and liquid waste in villages.

4. **Nagar Van Scheme:** the Nagar Van (Urban Forest Policy) aims to develop 200 urban forests across the country within 5 y. Warje Urban Forest in Pune (Maharashtra) will be considered as a role model for the scheme. The scheme enforces the participation of citizens, and collaboration between the Forest Department, municipalities, NGOs, and corporations. Local bodies will be offering vacant land from their respective parks or forests in place of these urban forests.
5. **Atal Bhujal Yojna (ABHY):** Atal Bhujal Yojana (or Atal Jal) is an initiative to manage groundwater launched by Prime Minister Narendra Modi during the 95th birth anniversary of former Prime Minister Atal Bihari Vajpayee, on December 25, 2019. The scheme aims to improve groundwater management in seven states of India.
6. **Jal Jeevan Mission:** with Jal Jeevan Mission, it is envisioned that all rural households in India will have access to safe and adequate drinking water by 2024. The program will also include mandatory elements regarding source sustainability, such as gray water management, water conservation, and rainwater harvesting. The Jal Jeevan Mission will be a community-based initiative and will include extensive information, education, and communication efforts. With this approach, JJM will make water a priority for everyone.
7. **National Clean Air Program:** under the National Clean Air Programme (NCAP), the government has committed to tackling the issue of air pollution at a national level on a time-bound basis with a target of reducing the particle matter concentration by 20% to 30% by 2024 keeping 2017 as the reference year. The NCAP has identified 122 non-attainment cities based on air quality data from 2014–2018.

12.3 Conclusion

For the last 50 years, the Indian government is doing so much for the environment. Currently, the government is promoting electronic vehicles to get rid of pollution. Nowadays, India is much aware of the environmental issues and changes its policies according to the current problem and situation. However, it is not very effective due to some factors like overpopulation, illiteracy, and unemployment. Transition in livelihood can make a path to a green environment. India must explore new methods. These options are available in agriculture, forests, and other ecosystems.

References

1. Guha, Ramachandra (2013, March 27). The past & present of Indian Environmentalism, The Hindu.
2. Padhi, Debi (2020, June 6). World Environment: India's Concern And Contribution. Odisha Bytes.
3. Ghosh, A. K. Environmental governance: India's changing scenario (2018,January, 22). Down to Earth
4. पर्यावरणीय वधिान (कानून) (2018, सतिंबर 30). India Water Portal
5. Ministry Of Science and Technology. Climate Change Programme
6. Centre for Science and Environment. National Mission on Sustainable Habitat https://www.cseindia.org/national-mission-on-sustainable- habitat-1977

13

Use of Geospatial Technique in Urban Flood Hazard Management

Pallavi Tomar[1], Suraj Kumar Singh[1*], Shruti Kanga[2], Amitanshu Pattanaik[3] and Gowhar Meraj[4]

[1]Centre for Climate Change and Water Research, Suresh Gyan Vihar University, Jaipur, India
[2]Department of Geography, School of Environment and Earth Sciences, Central University of Punjab, Bathinda, Punjab, India
[3]Defence Geoinformatics Research Establishment (DGRE), Defence Research & Development Organisation, Ministry of Defence, India
[4]Graduate School of Agricultural and Life Sciences, The University of Tokyo, 1-1-1 Yayoi, Tokyo, Japan

Abstract
Natural disasters such as flooding affect cities and their residents in a variety of ways. They can destroy personal property and ruin lives, as well as damage the environment. Social and environmental factors may cause a worsening of these effects. As it is understood now, flooding is a significant urban problem in the world and India. Urban flood occurs due to the disparate rainfall distribution, uncontrolled urban expansion, and lack of suitable drainage systems, encroachments on storm drains, dwindling wetlands, and other factors. Remote sensing and geographic information systems (GIS) can be used to limit flood damage. Flood monitoring and management have increasingly been dependent on remote sensing technology that integrates GIS. Geographic information system technology is beneficial to flood control since it allows mapping flood risks in sensitive areas and defines flood zones. The field of remote sensing has progressed from optical to microwave, which provides flood mapping in all-weather circumstances. To create and maintain flood hazard maps, the most critical parameter is flood depth, and determining flood depth from remotely sensed or hydrological data depends on the accuracy of the digital elevation model. This research examines the remote sensing and GIS techniques, methodologies, and models for urban flood management in cities worldwide. By offering a scientific method of flood management, these strategies have helped city planners and policymakers keep hazards under control.

***Keywords*:** Remote sensing, GIS, digital elevation model, urban floods, hazards, risk reduction

13.1 Introduction

Over the previous decade, flooding has affected more than 800,000 individuals and caused economic damages totalling US$300 billion. An increased risk of flooding is anticipated in the future because of global climate change and growing urbanisation. This is important for

**Corresponding author*: suraj.kumar@mygyanvihar.com

Sanjay Kumar, Suraj Kumar Singh, Shruti Kanga, Gowhar Meraj, Majid Farooq and M. S. Nathawat (eds.) Disaster Management and Environmental Sustainability, (155–168) © 2024 Scrivener Publishing LLC

disaster management and mitigation efforts, such as providing search and rescue services and drainage of floodwater [13]. Flooding is the most common and destructive kind of natural disasters in the world involving: (a) loss of lives of people and animals; (b) destruction of agricultural crops and loss of soil; (c) damage to infrastructure and communication networks; and (d) transport of sediment loads and pollutants. Approximately 40% of the losses that occur in natural catastrophes come from flooding. For many types of floods, like catastrophic, flash, extreme, river, coastal, pluvial, and big floods, definitions and classifications vary. Due to the geomorphology of the region, which is characterized by steep slopes and permeable soil and rock formations, flash floods are the most prevalent type of flood in the Mediterranean, which occur because of the narrow river basins and short flow concentration time. Because of this, flood mapping and monitoring, along with flood event effect identification and compensation services, are critical components of any flood risk reduction plan, catastrophe detection, and recompense service. Because inundated areas are temporary, changing quickly, and cover enormous areas, and because they, by nature, restrict accessibility, this information is hard to come by using typical procedures [6]. The timely and accurate information supplied by Earth observation (EO) data from space about floods and other natural disasters might prove invaluable in developing efficient mitigation and response strategies. Satellite imagery lets us gather data from hard-to-reach places and allows us to see the full extent of flooding, which is important for damage assessment and risk management.

Using radar and optical data is currently the most prevalent method. Landsat-7, Landsat-8, and Sentinel-2 (S2) satellites take optical sensor data with visible through shortwave infrared (SWIR) bands and deliver spectral resolution and a huge archive for hydrological data. Optical sensors identify naturally reflected or emitted energy in the visible and infrared spectrums that is reflected off the Earth's surface [33]. Visible/infrared sensors provide adequate delineation of inundated areas when water surfaces are unobstructed by clouds, trees, and floating vegetation. In studies employing Landsat imagery, scientists have relied on a number of different water indices to classify flooded areas. A modified version of the original NDWI, referred to as the Modified Normalized Difference Water Index (MNDWI), was later created by swapping the middle infrared band with the NIR (Near Infra Red) band [5]. The MNDWI can increase open water characteristics, as well as efficiently reduce and eliminate both built-up land noise and the noise produced by plant and soil. Landsat, the longest-running programme in global satellite surface observation, has provided us with some of the most impressive and extensive records in the world.

The application potential of optical sensors is limited in the case of flood occurrences that are accompanied by clouds because these sensors are unable to capture clean images during inclement weather. SAR images are helpful in such situations because they offer information in low-visibility and bad weather conditions and are less susceptible to the environment (*i.e.*, daylight and weather). Monitoring flood events during extended rainfall is crucial due to the fact that cloud penetration capacity is usually high [27]. The image texture algorithm, histogram thresholding, and various multi-temporal change detection methods are some of the most widely utilized SAR flood extent mapping approaches. The use of horizontal polarization (H = horizontal) is more efficient in providing a good separation of flooded zones compared to vertical (V = vertical) or horizontal and vertical (HV = horizontal and vertical) polarizations. Vertical and vertical (VV) polarized data had a strong dependence on surface roughness, although in some circumstances, the results were enough for delineating flooded areas. A primary disadvantage of SAR data is that the backscattered signal it returns is the result of a complicated collection of effects, all of which are influenced by things like incidence angle, vegetation density, and wind [25]. Further details are required to unravel the opposite backscatter reactions for similar situations.

Floods are a vital component of the ecology and occur every year. Floods are a widespread natural danger that affects all regions. The economy, lifestyle, and culture of this deltaic country were all influenced by it. Flooding in metropolitan areas is common both throughout the world and in India. The increase in rainfall and the lack of drainage infrastructure, the lack of natural buffers (such as wetlands), and urban expansion contribute to urban flooding [35]. The work of urban flood management is both complex and challenging for decision-makers, planners, and designers. The proper management of urban flooding should protect human lives and property while maximizing the use of natural resources for the benefit of the community. It may be possible to accomplish this by lowering flood risks to a tolerable level or by keeping, sharing, or transferring flood risks through appropriate methods [28].

India is known as one of the world's most flood-prone countries, with 113 million people susceptible to floods, and urban flooding has been witnessed across the globe. The average yearly economic loss due to disasters in India is estimated to be $9.8 billion, and over half of that amount is attributed to floods. Heavy rainfall in cities with undeveloped landscapes, unable to absorb water, creates urban flood. Floods are an annual natural occurrence and an integral aspect of the environment. Floods are among the most frequent and widely prevalent hazards seen in all forms of natural disasters [15]. Urban flooding has become commonplace throughout the world, including in India. The increase in rainfall and the lack of drainage infrastructure, natural buffers (such as wetlands), and urban expansion contribute to urban flooding. Due to the current flood crisis in Indian cities, drains overflow after 1 h of steady rain. Flooding in urban areas is a global problem, and it has become an almost common occurrence in Indian cities.

As climate change progresses, urban floods are predicted to become more common (especially in tropical locations), the length of the flooding season is likely to increase, and new places are at risk of flooding. Water bodies in cities like Kolkata, Delhi, and Hyderabad, are frequently illegally filled. The number of water bodies in Delhi (approximately 800) has gone down to 600, and the rest have disappeared. This has resulted in planning and infrastructure being discarded and natural drainage becoming restricted, leaving the city vulnerable to urban flooding [12]. The NCT (National Capital Territory) of Delhi is particularly susceptible to floods from the Yamuna River and its surrounding areas in Haryana and Rajasthan due to Sahibi River runoff. The Khadar floodplains are likewise at risk of repeated flooding. Heavy rain, flash floods, urbanization, illegal settlements, storm water drain trespassing, storm water drains' siltation, and water bodies' siltation.

Urbanization of the Indian metropolis fills up natural drainage channels and urban lakes with artificial structures and industries, reducing the amount of high-value land available for development and industry. State authorities should be made accountable for flood control measures that they implement within a time-bound framework because nothing is more expensive than the loss of life in the world. Flood monitoring and management have lately been helped out by remote sensing technology and geographic information systems (GIS).

Geographic information systems use in flood control allows flood risks to be mapped and flood zones to be identified as critical focus areas [19]. The use of microwave remote sensing has improved flood mapping in all weather conditions. A robust method for locating and measuring water spread areas is satellite remote sensing data analysis. Sensing of remote from satellites benefits a broad overview of the flood situation, quick access to data, and convenience of data acquisition. Remote sensing satellites help discover and retrieve information on flooded and waterlogged areas using various spectral and spatial resolutions. A significant drawback to optical sensors is that they are not reliable during the monsoon season because of limited cloud-free data. Microwave data obtained from ERS and RADARSAT satellites have cloud penetration capacity and can aid flood mapping and management in situations such as these [31].

13.2 A GIS-Based Method for Identifying Flood Risk

Damaging flooding can result in damage to buildings, people, and the natural environment. A flood disaster has been the cause of 39.26% of worldwide natural disasters since 2000, and it has cost $397.3 billion in damage. There are many ways to get flooded: quick snowmelt, strong storm surge, or dam collapse. As well, a variety of catastrophic situations or occurrences can produce flooding, including earthquake-triggered landslides. A large number of urban researchers found that while it is impossible to minimize flood danger, it can be reduced to an acceptable level that can be achieved through the systematic inclusion of community-based, non-structural events with structural events [14]. The real-time tracking of maximum flood extent is one of the most critical aspects of disaster management. It allows for immediate action and both short- and long-term recovery and future mitigation. Due to its synoptic view and its coupling with GIS techniques, remote sensing data has proven particularly successful in real-time flood inundation mapping and monitoring. As many types of active and passive sensors have shown promise in the demarcation of flood boundaries and the actual assessment of the inundation area in a cost-effective manner, the floodplain and the river bed have been better explored. Varied kinds of remote sensing satellites with different spatial and spectral resolutions can be used to find and gather vital information on places that have been flooded or waterlogged [16]. The data concerning the afflicted region can be supplied using LISS-III and PAN sensors. Compared to IRS-1D and P6, which are currently in orbit, the WiFS (Wide Field Sensor Images) sensor has triple the good receptivity over 5 d. The most significant downside of optical sensors is the inability to collect data when it is cloudy. Microwave data from ERS and RADARSAT can be utilized to assist with flood mapping and management, especially in instances of heavy cloud cover.

Urban floods can be caused by insufficient drainage systems, while flooding in rural areas usually results from high river discharge. Two primary factors affecting urban flooding risk are imperviousness and various rainfall intensities. It is simple to quantify both factors using multiple remote sensing and GIS techniques. Climate change and urbanization have exacerbated flood risk, and the process of mitigating flood damage has grown more complex and drawn out [4]. Because floods are the most common disaster type, and also that over the years, they have only gotten more severe, more focus should be made on flood forecasting, flood emergency preparation and response, and post-flood recovery.

Because data is collected at the micro-level and put into the hands of the government and the public, they will be able to prepare appropriately and respond to flood disasters.

A flood risk analysis that considers how land is used is critical for flood management, design, and planning, as well as for response. Calculations of water balance models using GIS for urban surfaces might assist in predicting the likely regions where flooding or critical storage may occur in a watershed area [30]. The GIS software and indications were used to determine physical damage, helping to define a hierarchy among the various structures and infrastructures so that victims could be aided and rescue operations launched. City governments can discover and enhance the strategies to prevent urban flooding. Through the remote sensing images, census details, and field studies, a vulnerability map was developed by the multi-criteria evaluation approach for the flood's analysis and assessment using a GIS environment. This research aims to locate numerous risk-prone sites and places and can be utilized to prevent or lessen post-disaster damage. The overlay method combines many GIS layers, including land use information and the flood risk map helps prioritize spatial flood risk areas and identify flood management successes [22].

Satellite data, which is now more available, has been essential in assisting with the examination of various natural and man-made phenomena, as well as other aspects of research. One of the many advantages of the Copernicus program is the Sentinel satellite constellation, which offers SAR and multispectral data with global coverage, high-frequency pass, and high spatial resolution [21]. The Landsat program has given remote sensing data since 1972, while the MODIS daily satellites offer multispectral images. Atmospheric and radiometric calibration usually provides these data, enabling them to be used by people with a variety of expertise levels, including remote sensing processing experts. Worldview 8, a user-friendly data portal, and G-Pod from the European Space Agency (ESA) that processes ENVISAT and Sentinel-1 SAR data are also examples of space agency data portals that are user-friendly. As well, free GIS plugins provide access to free multispectral satellite pictures. It is helpful for the management of natural hazard impacts to have the availability of these resources. Each year, the floods cause significant economic damage and injury. Flood mapping and modeling are critical for flood hazard assessment, damage estimation, and planning for sustainable urban management to appropriately manage flood risk. Satellites offer a low-cost solution for flood mapping in this context, using extracted flood areas to be identified through multispectral satellite data and its derived indexes [32]. Extraction of flooded areas can also be performed with SAR images or a combination of these data. Satellite data of other kinds can be beneficial in improving flood mapping. For instance, satellite data, such as SRTM and ASTER, was utilized to create digital elevation models (DEMs) that were then used to help estimate flood-prone areas or to develop a SAR/multispectral data-based map. Additionally, the GRACE satellite and ASCAT soil moisture datasets were used to develop flood indicators. The techniques used for flood mapping each have their own benefits and limitations and need to be examined on a case-by-case basis.

Satellite passes and instant processing chains enabled the deployment of satellite-based emergency management services, like the Copernicus EMS operated by the European Union, the Global Flood Detection System, and the NASA Global Flood Mapping System. Yet, because these services offer fast mapping products that might be influenced by uncertainty and are not always validated, they are not ideal for certain purposes. The accuracy of the maps of flooded areas created by official agencies, which are based on customized aerial pictures and field surveys, is improved. However, they are slow and expensive to construct. Further, it should be noted that flood mapping might be influenced by ambiguity in hydraulic models or DEM resolution.

Models for Flood

Because of human meddling in the natural environment and the effects of global climate change, floods are occurring more frequently nowadays. The number of flash floods and surface runoff can be estimated using several hydrological models. These hydrological models can build on the GIS platform by integrating land-use land cover information, particularly alluvial and moral parameters collected from the DEM. The WetSpa hydrologic model is used to model flood generation and propagation in the upper sub-basins. The research team here focused on calculating the effects of various types of precipitation, water movement, and water distribution on flooding. This model predicts water level and flood flow accurately, which makes it reliable. A regression tree-based loss model, RT3, includes three predictors (water depth, inundation time, and floor space) recommended for flood risk assessments and building loss estimation. To manage floods, multiple computational processes should be employed. A web-based USISM (Urban Strom-Inundation Simulation Model) model was developed to assess rainfall inundation and storm inundation for quick decision-making and practical interventions. The primary causes of floods are either flowing water or rainfall-induced runoff. Higher-return-period floods are lasting longer and covering more areas than they did in the past. Flooding depths show some fluctuation, but the lengths of the floods are relatively similar. To work on large-scale flood problems more quickly, a two-dimensional cellular automata-based model was developed. Overtopping of a river or stream is also another prevalent cause of flooding due to sudden rainfall. One of the crucial factors for flood hazard mapping is flood frequency and flood water-surface elevations [31]. It takes both hydraulic and hydrologic studies to map flood plains. An effective rainfall-runoff model can be created with HEC-HMS (hydrological model – Hydrologic Engineering Center) in watersheds that are already involved. The change of watershed spatial and temporal conditions made GIS an essential tool for floodplain modeling. HEC-Geo RAS (Hydrologic Engineering Center's River Analysis System) is an attractive option for mapping floodplain areas and floodwater-surface elevations for various design storms, ranging from short- to long-term return periods.

Climate change and global warming are behind the rise in floods, increasing frequency and intensity due to their effects on weather and the environment. Weather records reveal that precipitation has grown in both amount and variability. There has also been a gradual increase in excessive rains, resulting in heavy losses of life and property in flooded areas. Because of their growing frequency and intensity, floods are anticipated to constitute a global menace in most cities. Flood warning systems are currently being employed in monitoring and forecasting water resource disasters and floods. However, gauge station data, which are only one-dimensional measurements of water levels and streamflow, are insufficient for reflecting the complexity of interactions between runoff and land [24]. Even more troubling, flood risk models currently used are poor in specificity and inadequate for analyzing smaller areas. In addition, those responsible for disaster prevention and danger reduction have insufficient visual field information to make decisions. Over half of the world's population lives in urban areas today due to large-scale urbanization. However, because of the concentration of urban inhabitants and economic activity, the metropolitan regions also suffer from more significant evacuating and sheltering during floods. Urban flood events, a type of calamity, can be tracked in real-time using city cameras. For this study, we proposed using network cameras for hydrological management to realize intelligent surveillance and warning of river overflows based on the demands of decision-makers for urban flood control [17]. This instrument delivers flood alerts and real-time information to disaster relief sectors for disaster reduction measures as a sensing system for urban intelligent flood control. This study was used to develop a visual-based water level sensing technology and depict the results. A flood event can be identified by scanning waterbody surfaces for abnormalities and comparing water levels to historical records, thereby enabling the automated detection and alerting of flood events. Right now, flooding alerts are primarily dependent on information gathered from in-place measurements and remote sensing imagery. A collective and practical application of runoff field visual data is lacking. It is thus insufficient for small urban regions to have timely decision-making to lessen tragedies when it comes to dynamic field information. Therefore, we used a visual sensing technology that utilizes "smart" cameras that can monitor flooding in a small urban area using passive monitoring cameras. Based on the findings of this real-world flood instance, the method provided in the experiment was shown to produce accurate photos and time-series water level fluctuation data that appropriately portrays the flood condition. Using this technique, the water management and disaster management sectors may easily grasp the local hydrological situation and swiftly put disaster mitigation measures into practice [23].

The flood modelling process is filled with many uncertainties. Historical flood inventory maps can be used to confidently solve these issues with machine learning, provided they are available. A machine learning model can deliver less expensive and less time-consuming solutions for flood hazard and risk assessment than the traditional method of field surveys [10]. It offers direction to authorities on what extra data is needed to create more accurate flood maps for preventing future harm. Because flood susceptibility maps are crucial for future investigations and hazard and risk disaster management and mapping, they are essential. One of the complex and confusing aspects of flood modeling is the high level of uncertainty. Still, we have proven that machine learning techniques may help create better flood susceptibility maps. This machine learning model can improve flood hazard mapping and increase land-use planning efficiency, which is helpful in disaster management [9]. Our approach offers a reasonable balance between immediate, real-time flood mapping for emergencies and the typical flood mapping that uses on-demand data. A more accurate but slower method of mapping flood areas is real-time automatic flood mapping. However, classical flood mapping is a proper method, but it is resource-intensive. We also suggest a user-friendly way to processing satellite data that people with little remote sensing experience may use and, in addition, be used to studies of hydrology, urban planning, or flood damage assessment [18]. It is crucial to flooding mapping and hazard/risk management in floodplain areas and in disadvantaged communities worldwide to have a network of operating satellites that provide free- and open-access data with continuity. Flood extent mapping and damage assessment are being progressively performed using remotely sensed data. You can find free optical sensor data with easy-to-use, well-established data processing algorithms.

Flood Management

The effects of urban floods are more severe than those of flooding in other regions that are crossed by a river. This is because of the population and number of items that are exposed to or at risk of damage or loss of function. Urban regions are, in fact, complicated networks of homes, community health centers, and schools linked by streets. Not only can floods be capable of causing direct harm by completely ruining some system or damaging many other parts of an area, but they can also result in an indirect loss in functionality due to suspension and deaths. In practice, markers of susceptibility tend to be excessively narrow in scope. Even now, we have not produced all the conceptual vulnerability indicators or parameters and adequate operational instruments that we need to measure weaknesses and strengths in dealing with environmental dangers. To overcome drainage problems, the newly developed algorithms are built on the foundation of modern geographic information technologies used to process and interpret satellite images and topographic map boards, carry out spatial analysis, and find solutions to the functional tasks of the drainage network [3]. A flash flood is characterized by rapid onset, a brief duration, and a rapid rise in the water level. It happens almost immediately after rain, sometimes even in the same hour, and it may be followed by landslides, mudflows, bridge collapses, damage to structures, and even fatalities. The elements that impact the quantity of runoff in a watershed and thus the strength of flooding are numerous. Because of that, the hydraulic rating of soil, impervious surfaces, and water flow accumulation were incorporated with demographic data to develop a composite flood danger index. The approach put out utilizes GIS and MCDA together MCDA (Multi-Criteria Decision Analysis). This methodology evaluates six elements that affect the risk of flooding in metropolitan areas: sewage network elevation, slope, distance, water surface distance, water table height, and land use. The AHP (Analytic Hierarchy Process) weight criterion comparison process was carried out in three different methods: interval rough numbers, fuzzy numbers, and crisp values. Maps representing flood risk areas were created, based on various assumptions, using a weighted linear combination (WLC). The results were validated by comparing data from the region's history of devastating floods with data from the new maps, which delineated risk zones [1].

This research explores the possibilities of using GIS in flood modeling and forecast of water flow. Several instances show how to develop flood danger indices or their zones using hydrologic-type models utilizing simple GIS approaches. The HMS model (hydrological modeling system) can also be informally connected to a GIS, such as those used for flood prediction. Furthermore, the SCS (soil conservation service) model, for example, is fully incorporated into a GIS by embedding it. Installation of flood forecasting systems in watersheds with missing hydrometric data may help prevent flooding-related damage [8]. Geographical information systems were utilized in this study to keep up-to-date watershed data and SCS model parameters for real-time flood forecasting. The primary objective of this work is to assess the capacity of GIS to link with a comprehensive hydrologic model, in particular HMS. The output of the hydrological model should be more accurate if the collected rainfall and runoff data (from the stations in the area) can be compared to the results. The results from the model simulation during rainfall-runoff processes in this research demonstrated that the submergence of the primary watershed does not independently depend on each one's outlet discharge rate. Nevertheless, it is connected to how the two main river watershed exit hydrographs are merged. In each cross-section of the channel network, the model shows flood characteristics temporally and geographically.

The increased frequency of floods throughout the world, as well as in Nigeria, is a severe issue. This study aims to locate the contributing elements to flooding and map out the flood-prone areas using GIS. The DEM and land use/land cover map was built in the ArcGIS environment using IKONOS imagery and the 1:50,000 toposheets. Findings indicate that the bulk of Surulere's land is vulnerable to flooding. Adequate flood warnings and swift emergency action go hand in hand with all types of intervention. The most prepared for flood recovery is the one who is most aware of the community's preparedness and their own and knows that the clearer the warnings, the better. Prevention of hazards developing into disasters requires public education campaigns, and they are vital to the success of these warnings [34].

Evacuation routes are an essential part of any disaster plan, and evacuees may flee to a flood refuge at a higher elevation or away from the flood itself, depending on the local conditions. Large water depths necessitate outward evacuations when high flood velocities and when buildings are likely to be compromised. Successful flood evacuations demand that people know what to do and have a plan of action in place [2]. Involving community members and carrying out regular drills guarantee the evacuation process is efficient. Essential utilities such as water supply, sanitation, and security, particularly vital in developing a workable evacuation mechanism, should be available in refugee gathering places. Blocking channels or making channels worse by human factors like inappropriate land use and deforestation in headwater regions. The devastation done by floods is of a deadly and destructive nature.

Increased population will lead to more urbanization, a higher percentage of impervious surface area, a decrease in infiltration and runoff, and a higher flood peak. Urban environments contain a substantial amount of impervious surfaces that inhibit the ability of rainfall to permeate the soil. These create massive runoff and high flood levels, causing widespread flooding and catastrophic devastation [20]. The vast majority of urban flooding occurs because of the presence of rubbish in the drainage systems. Drainage facilities must be maintained and cleaned for optimal reliability. Stormwater retention techniques are both necessary for mitigating urban floods and protecting downstream communities from flooding. The risk of urban flooding is a combination of hazard and susceptibility. Municipalities can achieve many purposes, such as flood control, water supply, recreational space, groundwater recharge, and improvement of the urban environment, using multiple purpose measures [26]. Tracking and assessment of implemented measures assist in identifying best practices in flood risk management plans, helping to enhance these plans under specific circumstances constantly. This research's findings include recommendations on storm water management, and the results draw conclusions based on the urban resilience theory. First of all, the rate of rainwater use is relatively low.

Moreover, the water system network is ruined. Finally, the revetment vegetation is excessively simplistic. The combination of urban park landscape design and storm water management could be incorporated to ensure proper function. The original monolithic structure and its flood concerns will never be solved, and a thorough resolution needs to be sought throughout the landscape process of the heritage region. By creating an ecological infrastructure, locating the primary point of control for these processes, creating a suitable landscape security pattern, and coordinating numerous relationships, the ecology and heritage can be secured. The VIKOR (VIekriterijumsko KOmpromisno Rangiranje) approach is a viable disaster response program, as demonstrated by the results of a study on how to mitigate losses from sudden flooding disasters quickly. First, build the index system to evaluate flood disasters on an evaluation of the disaster level. To complete the fuzzy VIKOR model, you should include the weights for utility. That will help reflect the preferences of decision-makers. At the end of the process, do some empirical testing to pick the most workable solution, which you can then persuade the decision-makers to adopt. The rapid assessment model may be used to scientifically and speedily analyze the flood disaster. It can be employed to choose a feasible emergency response program from the existing scheme based on evaluation results, allowing the flood disaster's severity to be reduced quickly [11].

The article employs to speed up assessments of different types of emergencies, like the severity of earthquakes, the degree of fires, and the prevalence of infectious diseases. Additionally, it will help the urban flood catastrophe response theory, however, to a limited extent. Large urban centers are increasingly worried about flooding. Urban flooding and precipitation are growing in urban areas due to global warming. Reducing the effects of flooding, mitigating climate change, and adapting to future flooding are standard urban flood resilience priorities [29]. The study found that area has no flood-control agency or flood-protection policies, which has resulted in extensive growth in flood-risk areas and increased flood vulnerability. Interview findings show that flood management techniques rely more on structural flood response measures and that infrastructure should be used less. Research findings urge effective policy creation, collaborative planning, education, and awareness programs to recognize the gravitational effects of floods and promote the behavioral adjustments essential to improve pre-disaster and pre-disaster preparedness. These include preparing home emergency kits, developing partnerships among organizations to share expertise, and forming networks among community members to assist with disaster response.

Conclusion

The authors believe that GIS and RS tools should be used at all stages of flood management. An overview of the GIS and RS-based flood management case studies is presented. We may extrapolate that RS methods offer a more affordable and faster solution, which allows us to gather spatial data even in locations that are not easily accessible. On the other hand, GIS tools enable data gathering, analysis, querying, and presenting information in a simplified format to carry out hydrological models. The flood management process comprises three stages: pre-flood, during a flood, and post-flood operations. The flood prediction, flood prevention, and flood mitigation stages can be further broken into several steps, including flood risk identification and mapping and flood damage assessment. We can conclude from this that flood management is complex and necessitates multidisciplinary collaboration.

A flood prediction example requires the skills of hydrology, soil science, and geography. Hydrological engineering was the primary means of flood control and water management in early times. Despite the recent dramatic climate fluctuations, resources have not generated successful outcomes in these programs. And in addition to flood control measures like hydrological engineering, new ideas like flood risk avoidance and disaster mitigation of floods have been added to flood management. The flood forecasting system, which uses a limited investment, intends to deliver early warnings to the sectors responsible for disaster prevention so that they can take quick action to minimize the damage from disasters [7]. To avoid flooding, the city opens a reservoir early, closes a water gate, and evacuates citizens from low-lying areas before disaster strikes. Water level and precipitation monitoring stations, remote sensing pictures, and numerical prediction models are the primary means of implementing warning systems. Systems like these can provide prior warnings of massive floods with hours or even a day or two to spare; therefore, they remain the most commonly utilized method.

On the other hand, since hazard reduction in small urban locations is not ideal for large-scale forecasting, it has low specificity and timeliness. Comparing flood projections to actual flood conditions is complicated. As a result, this method can't give policymakers sufficient data to manage crises and determine catastrophe mitigation solutions quickly and adequately. To meet this goal, we presented a real-time flood monitoring system in this study, which uses the current river monitoring cameras and can deliver time-series data about water levels. Through this approach, managers can have a better understanding of the situation and may better predict when evacuations are needed before a flood strikes. The several flood warning and prediction systems each have their unique advantages. Additionally, they can implement effective strategies for flood mitigation afterward, thereby lessening the impact of the flood on metropolitan areas. However, it cannot respond to a field flooding or overflow incident because it is not intelligent. Vulnerable locations can acquire only relevant data from the information systems for large-scale flood forecasts, as the designs are incapable of providing adequate field information to decision-making agencies. To paraphrase, we may see a lack of flexibility in combining catastrophe prevention knowledge and the joint implementation of mitigation efforts. Thus, the future of catastrophe reduction methods may include integrating several systems. Managers need to be well-informed on disaster details to plan for long-term recovery and execute instant disaster mitigation plans. To facilitate future study, we will need to figure out how to combine measurements from different sources, such as the field, satellites, and cameras, into a viable information system.

References

1. Abanish, K., Shruti, K., Ajay, K.T., Suraj, K.S., Bojan, Đ. 2021. Surface runoff estimation of Sind river basin using integrated SCS-CN and GIS techniques. *HydroResearch*, 4, 61-74.
2. Anusha, N., Bharathi, B. 2020. Flood detection and flood mapping using multi-temporal synthetic aperture radar and optical data. *Egypt. J. Remote Sens. Space Sci.*, 23(2), 207-219.
3. Aparna, B., Ajay, K.T., Gowhar, M., Shruti, K., Suraj, K.S., Bojan, Đ., Subhash, A. 2021. Climate vulnerability and economic determinants: Linkages and risk reduction in Sagar Island, India; A geospatial approach. *Quat. Sci. Adv.*, 4, 100038.

4. Bera, A., Singh, S.K. 2021. Comparative Assessment of Livelihood Vulnerability of Climate Induced Migrants: A Micro Level Study on Sagar Island, India. *Sustain. Agri. Food Environ. Res.*, 9 (2), 216-230.
5. Brenden, J., Jurjen, W., Beatriz, R.R., Erin, C.P. 2015. Early Flood Detection for Rapid Humanitarian Response: Harnessing Near Real-Time Satellite and Twitter Signals. *ISPRS Int. J. Geo-Inf.*, 4, 2246-2266.
6. Chao, Y., Futao, W., Shixin, W., Yi, Z. 2019. Accuracy evaluation of flood monitoring based on multiscale remote sensing for different landscapes. Geomatics, *Nat. Hazards Risk*, 10 (1), 1389-1411.
7. Fakhera, N., Muhammad, M.R., Abdul, G., Fahim Arif. 2014. Flood Detection/Monitoring Using Adjustable Histogram Equalization Technique. *Multidimens. Signal Process Appl.*, Volume 2014, Article ID 809636, https://doi.org/10.1155/2014/809636
8. Frazier, P., Page, K. 2000. Water body detection and delineation with Landsat TM data. *Photogramm. Eng. Remote Sens.*, 66 (12), 1461-1467.
9. Gonçalves, P., Marafuz, I., Gomes, A. 2015. Flood hazard, Santa Cruz do Bispo Sector, Leça River, Portugal: a methodological contribution to improve land use planning. *J. Maps*, 11(5), 760–771.
10. Jain, M.K, Kothyari, U.C. 2000. Estimation of soil erosion and sediment yield using GIS. *Hydrol. Sci. J.*, 45(5), 771-786.
11. Jain, S.K., Saraf, A.K., Goswami, A., Tanvear, A. 2006. Flood inundation mapping using NOAA AVHRR data. *Water Resour. Manage.*, 20, 949-959.
12. Joy J., Kanga, S., Singh, S.K., Sudhanshu. 2021. Cadastral level Soil and Water conservation Priority Zonation using Geospatial technology. *Int. J. Agric. Syst.*, 9 (1), 10-26.
13. Kikuko, S., Qinglin, C., Makoto, H., Hiroaki, S., Yuichiro, U. 2021. Emergency flood detection using multiple information sources: Integrated analysis of natural hazard monitoring and social media data. *Sci. Total Environ.*, 767, 144371.
14. Kundzewicz, Z. W., Pińskwar, I., Brakenridge, G.R. 2013. Large floods in Europe, 1985–2009. *Hydrol. Sci.*, 58(1), 1-7.
15. Liliana, Z., Romulus, C., Remus, P., Gabriela, I. 2017. Mapping flood and flooding potential indices: a methodological approach to identifying areas susceptible to flood and flooding risk. Case study: The Prahova catchment (Romania). *Front. Earth Sci.*, 11, 229–247.
16. Majid, F., Singh, S.K., Kanga, S. 2021a. Inherent vulnerability profiles of agriculture sector in temperate Himalayan Region: A preliminary assessment. *Indian J. Ecol.*, 48 (2), 434-441.
17. Majid, F., Singh, S.K., Kanga, S. 2021b. Mainstreaming Adaptation Strategies in Relevant Flagship Schemes to Overcome Vulnerabilities of Climate Change to Agriculture Sector, *Res. J. Agric. Sci.: An International Journal*, 12(2), 637-646.
18. McFeeters, S. K. 1996. The use of Normalised Difference Water Index (NDWI) in the delineation of open water features. *Int. J. Remote Sens.*, 17(7), 1425–1432.
19. Meraj, G., Singh, S.K., Kanga, S., Islam, M.N. 2021. Modeling on comparison of ecosystem services concepts, tools, methods and their ecological-economic implications: a review. *Model. Earth Syst. Environ.*, 1-20.
20. Mohapatra, P.K., Singh, R. D. 2003. Flood management in India. *Nat. Hazard.*, 28, 131-143.
21. Notti, D., Giordan, D., Caló, F., Pepe, A., Zucca, F., Galve, J.P. 2018. Potential and Limitations of Open Satellite Data for Flood Mapping. *Remote Sens.*, 2018, 10, 1673.
22. Pandey, A.C., Singh, S. K., Nathawat, M.S. 2010. Waterlogging and flood hazards vulnerability and risk assessment in Indo Gangetic plain. *Nat. Hazard.*, 55, 273-289.
23. Pandey, A.C., Singh, S. K., Nathawat, M.S. 2012. Analysing the impact of anthropogenic activities on waterlogging dynamics in Indo-Gangetic plains, Northern Bihar, India. *Int. J. Remote Sens.*, 33 (1), 135-149.

24. Pandey, A.C., Singh, S. K., Nathawat, M.S., Saha, D. 2013. Assessment of surface and subsurface waterlogging, water level fluctuations and lithological variations for evaluating ground water prospects in Ganga Plains. *Int. J. Digital Earth*, 6 (3), 276 - 296.
25. Pierdicca, N., Chini, M., Pulvirenti, L., Macina, F. 2008. Integrating physical and topographic information into a fuzzy scheme to map flooded area by SAR. *Sensors*, 8, 4151–4164.
26. Qiu, J., Cao, B., Park, E., Yang, X., Zhang, W., Tarolli, P. 2021. Flood Monitoring in Rural Areas of the Pearl River Basin (China) Using Sentinel-1 SAR. *Remote Sens.*, *13*, 1384.
27. Rosser, J.F., Leibovici, D.G., Jackson, M.J. 2017. Rapid flood inundation mapping using social media, remote sensing and topographic data. *Nat. Hazard.*, 87, 103–120.
28. Schnebele, E., Cervone, G. 2013. Improving remote sensing flood assessment using volunteered geographical data. *Nat. Hazards Earth Syst. Sci.*, 13, 669–677.
29. Shahabi, H., Shirzadi, A., Ghaderi, K., Omidvar, E., Al-Ansari, N., Clague, J.J., Geertsema, M., Khosravi, K., Amini, A., Bahrami, S., Rahmati, O., Habibi, K., Mohammadi, A., Nguyen, H., Melesse, A.M., Ahmad, B.B., Ahmad, A. 2020. Flood Detection and Susceptibility Mapping Using Sentinel-1 Remote Sensing Data and a Machine Learning Approach: Hybrid Intelligence of Bagging Ensemble Based on K-Nearest Neighbor Classifier. *Remote Sens.*, 12, 266.
30. Singh, S. K., Pandey, A.C. 2014. Geomorphology and the controls of Geohydrology on water-logging in Gangetic Plains, North Bihar, India. *Environ. Earth Sci.*, 71 (4), 1561- 1579.
31. Singh, S. K., Pandey, A.C., Nathawat, M.S. 2011. Rainfall variability and Spatio temporal dynamics of flood inundation during the 18th August 2008 Kosi Flood in Bihar, India. *Asian J. Earth Sci.*, 4, 9-19.
32. Singh, S. K., Pandey, A.C., Rathore, V.S., Nathawat, M.S. 2014. Evaluating factors responsible for contrastic signature of wasteland development in northern & southern Ganga plains (Bihar State, India) with focus on waterlogging. *Arabian J. Geosci.*, 7 (10), 4175-4190.
33. Supattra, P., Paramate, H. 2020. Internetworking flood disaster mitigation system based on remote sensing and mobile GIS. Geomatics, *Nat. Hazards Risk*, 11 (1), 1886-1911.
34. Wang, Y., Colby, J. D., Mulcahy, K. A. 2002. An efficient method for mapping flood extent in a coastal flood plain using Landsat TM and DEM data. *Int. J. Remote Sens.*, 23 (18), 3681-3696.
35. Xu, H. 2006. Modification of normalised difference water index (NDWI) to enhance open water features in remotely sensed imagery. *Int. J. Remote Sens.*, 27, 3025–3033.

14

Impact of Urbanization on Infrastructure and Environment of the Cities in Bihar

Anshu Veenapani[1]* and Sharat Chandra[2]

[1]*Department of Commerce, T.M. Bhagalpur University, Bhagalpur, India*
[2]*Department of Geography, T.M. Bhagalpur University, Bhagalpur, India*

Abstract
The level of urbanization is the proportion of urban population to total population. Though it is low in Bihar, but in absolute terms, the urban population is very high. Most of the urban population is concentrated in the capital city of Patna and in some district headquarters like Gaya, Bhagalpur, Muzaffarpur, and Darbhanga. Rapid urbanization in Bihar can be attributed to mass scale rural–urban migration. The quality of life is under threat particularly in these urban centers. Environmental degradation has been taking place very rapidly and causing excessive air and water pollution. The social environment of the cities is also under threat owing to vital problems like unchecked migration, illegal settlements, socio-cultural disparities, uneven distribution of income, lack of law and order, and shortage of energy. The unintended cost of increasing urbanization results in the deterioration of the environmental quality and pressure on the existing infrastructure reinforcing a further degradation of the environment.

***Keywords*:** Urbanization, infrastructure, degradation, sprawl, migration

14.1 Introduction

India is the second largest populated country in the world next to China. Out of the total population, about 30% reside in urban areas. The level of urbanization in terms of population is low in Bihar, but urban population is mostly concentrated in major cities like Patna, Muzaffarpur, Gaya, Bhagalpur, Darbhanga and other district headquarters. The areal extents of these towns are growing rapidly and the numbers of inhabitants are also increasing fast. The population of these towns accounts for 11% of Bihar's total population in 2011. The population in these towns grew by 29% (2011), which is higher than the total population of Bihar.

Rapid urbanization in Bihar can be largely attributed to the massive exodus of population from rural to urban areas (Lahon *et al.*, 2023a). Due to the uncontrolled urbanization, the quality of life is under threat, particularly in the cities. Environmental degradation has

**Corresponding author*: anshuveenapani@gmail.com

Sanjay Kumar, Suraj Kumar Singh, Shruti Kanga, Gowhar Meraj, Majid Farooq and M. S. Nathawat (eds.)
Disaster Management and Environmental Sustainability, (169–176) © 2024 Scrivener Publishing LLC

been taking place very rapidly and causing excessive air and water pollution, waterlogging, noise pollution, disposal of waste materials, and mushrooming of slums in the towns and cities of Bihar (Debnath *et al.*, 2023a, b, c; Farooq *et al.*, 2022a, b, c; Kotecha *et al.*, 2023; Gulati *et al.*, 2023; Kanga *et al.*, 2022a, b).

Objectives of Study

The major objectives of the study are as follows:

- to analyze the trends of urbanization
- to assess the impact of urbanization on environment and infrastructure
- to suggest policies for urban development
- to pinpoint some measures for making our urban system sustainable

Trends of Urbanization (1901–2011)

According to the census of 2011, the total population of Bihar is 10, 38, 04,637 which is nearly 8.58% of the total population of India. In 2021, its population increased to the extent of more than 12 crore. It is the third major populous state of India.

The population share of the state is higher than the land share (2.86%). In terms of population density, the state ranks first. Bihar has only 11.73 million urban populations, which is 11.30% of the total population of Bihar, while 29% population lives in urban centers of India.

The Table 14.1 below shows the growth of urbanization in Bihar, which is due to poor development of education, poverty, lack of non–agricultural activities, negligible industrial development, etc.

In 1901, only 47% of the people of the state lived in urban centers. Before 1951, the growth rate was only 5% to 6%, but after independence, it increased gradually and became 11.30% in 2011. Still, the pace of urban development is very slow in comparison to states like Punjab, Haryana, Gujarat, Maharashtra, Andhra Pradesh, Tamilnadu, and Karnataka. Even states like Jharkhand, M.P., and Chhattisgarh have greater shares than Bihar. Table 14.2

Table 14.1 Bihar trends of urbanization [1901–2011].

Year	Bihar	India
	(In percent)	
1901	4.70	10.80
1911	4.72	10.30
1921	4.60	11.20
1931	4.80	12.00
1941	5.40	13.90
1951	6.80	17.30
1961	8.40	18.00
1971	10.00	19.90
1981	12.72	23.30
1991	13.17	25.70
2001	10.47	28.20
2011	11.30	29.00

Table 14.2 Bihar decadal growth of population in some selected districts [2001–2011].

Sl. No.	District	Total	Rural	Urban
1	Motihari	29.01	26.98	58.93
2	Araria	30.00	30.17	27.35
3	Purnea	29.66	26.32	53.15
4	Darbhanga	19.00	16.95	42.18
5	Siwan	21.725	22.26	21.98
6	Khagaria	29.46	30.43	14.15
7	Munger	19.45	18.84	21.01
8	Patna	22.34	18.34	27.97
9	Bhojpur	21.27	20.75	24.42
10	Gaya	26.08	26.90	20.925
11	Jamui	25.54	25.54	40.10
12	Banka	26.14	26.13	26.46
13	Bihar	25.07	23.90	32.11

Source: Census Hand Book, 2011.

shows the growth of urban population in some districts of Bihar. In 2001, Aurangabad had the highest growth of population (66.90%), whereas Jehanabad got the second highest growth rate (55.77%). The third highest growth rate is found in Patna (42.87%). The districts of Gaya, Bhagalpur, Muzaffarpur, Darbhanga, Biharsarif, and Ara also recorded significant growth of urban population.

Purnia, East Champaran, Nawada, and Kaimur districts recorded more than 50% growth in urban population. Above 40% growth rate was registered by districts like Darbhanga, Begusarai, Jamui, and Aurangabad. Most of the other districts showed 20–30% growth rate in urban population. Among the cities, Gaya and Bhagalpur showed insignificant growth, while Patna, Ara, Muzaffarpur, and Darbhanga cities recorded higher growth rates during the last decade.

Urban Problems

Air Pollution

The quality of air in urban areas has been deteriorating rapidly over the past few decades due to vehicular, thermal, generator, and domestic fuel emissions. With the fast increase in the number of vehicles, there has been a 30% increase in air pollution.

Moreover, congestion leads to road degradation, decrease in fuel efficiency and fossil fuels, particularly coal, which is the major source of energy, and contributes to air pollution in the form of suspended particles. The urban households generally use LPG while the household of lower income groups use wood. Coal is a biomass fuel, highly polluting and due to poor ventilation inside their houses, it deteriorates indoor air quality and consequently affects individual growth.

The most common air pollutants in urban areas are sulfur dioxide (SO_2), nitrogen dioxide (NO_2), and suspended particulate matter (SPM). Vehicular emissions contribute carbon dioxide (CO_2) in the air.

Quality of Water

Water pollution is also one of the major problems in the urban areas of Bihar, resulting from increased human activities and interference with the natural ecosystems. The major water resources of Bihar, including both surface and underground resources, are not only diminishing but are increasingly getting polluted due to human interventions with rapid urbanization, huge quantities of waste water enter rivers and often unscientific disposal of solid wastes also contaminates groundwater, thereby reducing the freshwater resources. Groundwater quality is also degraded due to pollution from pesticides leaching and toxic metal leaching from the industrial effluents that have serious health effects on pollution (Meraj *et al.*, 2022; Singh *et al.*, 2022b; Sajan *et al.*, 2022).

Solid Waste Generation

Urban growth with inadequate infrastructural support is turning India's major cities into waste dumps. The daily per capita generation of municipal solid waste (MSW) of Indian cities is above 100 g in small towns and 5400 g in major cities. The annual solid waste generation in the towns of Bihar is increasing without proper care and treatment. Urban municipal waste is a mix of paper, plastic, metal, glass, and organic matter (Nath *et al.*, 2023; Pandey *et al.*, 2010; Rafiq *et al.*, 2022). The plastic materials are not recyclable. So, they create a major problem in drainage of the cities. Ninety percent of municipal waste is collected by the civic authorities and dumped into low-lying areas outside the towns and cities. These waste materials are creating an alarming situation in the environment and as well as in underground water resources (Singh *et al.*, 2023a; Sud *et al.*, 2023; Meraj *et al.*, 2021a, b).

Noise Pollution

The main causes of noise pollution are vehicular traffic, industrial activities, and sources of entertainment, such as musical systems like TVs and sound boxes. There is more noise during the daytime. The pollution level is more than the recommended level of 55 dB, which can be controlled and checked by the conscience of people, tight administration, and civic law.

Social Environment

Slums are the areas of unhygienic living conditions with degraded physical and social living, and economic environment (Singh *et al.*, 2022a). The skilled, semi-skilled, and low income group people are attracted to the slum areas and reside there (Upadhyay *et al.*, 2023). Such areas are marked by a deplorable living environment and often have a lack of civic amenities and facilities, leading to environmental degradation and higher incidences of mortality (Singh *et al.*, 2023b).

Crime in Cities

Due to increasing disparities in the urban areas, particularly in major cities, crime has been rampantly increasing. In recent years, high-tech of cybercrimes have also increased in cities.

Determination of Quality of Life

Due to the growing population, demand of basic services and infrastructure in urban centers are resulting in a serious deterioration of service quality across housing, transport, health care, power, water sanitation, and education. The neglect of urban sprawl by civic authorities has led cities to be vulnerable to natural disasters and diseases. The houses in urban centers are very much overcrowded. More than 25% are residing in a single room without proper light and ventilation. With regards to sources of safe drinking water, about 20% of people are dependent on municipal-supplied water, 60% are dependent on personal water boring, and 20% are dependent on the source of hand pump water.

Shortage of Energy

The process of urbanization leads to the substitution of commercial energy for non-commercial energy in industry, business, and domestic sectors. Thus, one can relate urban development with increased commercial energy use. Urban centers travel by energy intensive transportation model. Building infrastructure in urban areas to support the high population density requires significant quantity of energy. The changes in lifestyle of urban people in recent years have contributed significantly to deficiency in energy utilization levels.

In the residential sector, more people are using LPG and some are using kerosene stores. The propensity to consume more and more electricity is marked by an increase in the uses of modern facilities. Due to increased availability and affordability, LPG and other electric and electronic goods are generally in use in the urban areas.

Urban Environmental Management

The aforesaid discussion shows that creating sustainable urban systems has become a necessity for making the cities habitable for the future generation. The Ministry of Urban Development and the Ministry of Urban Employment and Poverty Alleviation form the apex authority of the government at the national level to formulate policies, addressing all issues of urban development. Some of the key interventions of the GOI (Government of India) are the National Urban Transport Policy, guidelines for sector reform and public–private partnership in the urban sanitation sector. Centrally sponsored schemes (CSS) such as integrated development of small and medium towns and megacities are also some of the policy decisions.

Conclusion

The unplanned growth of urbanization is responsible for the deterioration in environmental quality. It puts a pressure on the existing infrastructure, reinforcing a further degradation of the environment. However, environmental degradation is not only a manifestation of urban population pressure but also of changes in their lifestyles. Widening disparities among urban social groups have resulted in the deterioration of the urban social environment and an increase in crime rates in major cities. Awareness about sustainable development, which includes environmental concerns, is gaining ground and various tools such as ecological footprinting and green accounting are being developed to ensure the quality of the environment that we leave for our next generation.

References

Debnath, J., Sahariah, D., Lahon, D., Nath, N., Chand, K., Meraj, G., Kumar, P., Singh, S. K., Kanga, S., & Singh, S. K. (2022a). Geospatial modeling to assess the past and future land use-land cover changes in the Brahmaputra Valley, NE India, for sustainable land resource management. *Environ. Sci. Pollut. Res.*, 1-24.

Debnath, J., Sahariah, D., Lahon, D., Nath, N., Chand, K., Meraj, G., Kumar, P., Singh, S. K., Kanga, S., & Farooq, M. (2023b). Assessing the impacts of current and future changes of the planforms of River Brahmaputra on its land use-land cover. *Geosci. Front.*, https://doi.org/10.1016/j.gsf.2023.101557.

Debnath, J., Sahariah, D., Saikia, A., Meraj, G., Nath, N., Lahon, D., Annayat, W., Kumar, P., Chand, K., Singh, S. K., & Kanga, S. (2023c). Shifting Sands: Assessing Bankline Shift Using an Automated Approach in the Jia Bharali River, India. *Land*, 12(3), 703.

Farooq, M., Meraj, G., Rishabh, Kanga, S., Nathawat, R., Singh, S. K., & Ranga, V. (2022a). Slum Categorization for Efficient Development Plan—A Case Study of Udhampur City, Jammu and Kashmir Using Remote Sensing and GIS. In *Geospatial Technology for Landscape and Environmental Management: Sustainable Assessment and Planning*, (pp. 283-299). Cham: Springer International Publishing.

Farooq, M., Mushtaq, F., Meraj, G., Singh, S. K., Kanga, S., Gupta, A., Kumar, P., Singh, D., & Avtar, R. (2022b). Strategic Slum Upgrading and Redevelopment Action Plan for Jammu City. *Resources*, 11(12), 120.

Farooq, M., Rashid, H., Meraj, G., Kanga, S., & Singh, S. K. (2022c). Assessing the Microclimatic Environmental Indicators of Climate Change of a Temperate Valley in the Western Himalayan region. In A. R. Siddiqui and Avijit Sahay (Eds.), *Climate Change, Disaster and Adaptations - Contextualizing human responses to ecological change*, (pp. 47–61). https://link.springer.com/chapter/10.1007/978-3-030-91010-5_5

Gulati, B., Sharma, R., Kanga, S., Singh, S. K., Sajan, B., Meraj, G., Kumar, P., & Ramanathan, A. L. (2023). Unraveling the Relationship Between Stubble Burning and Air Quality Degradation in Punjab: A Temporal and Spatial Analysis (2019-2022). *J. Clim. Change*, 9(2), 43-53.

Kanga, S., Meraj, G., Farooq, M., Singh, S. K., & Nathawat, M. S. (Eds.). (2022a). *Disaster Management in the Complex Himalayan Terrains*. Geography of the Physical Environment. Springer, Cham.

Kanga, S., Singh, S. K., Meraj, G., Kumar, A., Parveen, R., Kranjˇ, N., & Đurin, B. (2022b). Assessment of the Impact of Urbanization on Geoenvironmental Settings Using Geospatial Techniques: A Study of Panchkula District, Haryana. *Geographies*, 2(1), 1-10. ISSN No. 2673-7086.

Kotecha, M. J., Kanga, S., Pankhaniya, S. K., Agrawal, S., Meraj, G., & Singh, S. K. (2023). Urban Heat Island (UHI) Resilience Plan in Varying Climatic Conditions Using Geospatial Approach: A Case Study of Rajkot City. In *Climate Change, Agriculture and Society: Approaches Toward Sustainability* (pp. 307-338). Cham: Springer International Publishing.

Lahon, D., Sahariah, D., Debnath, J., Nath, N., Meraj, G., Farooq, M., Kanga, S., Singh, S. K., & Chand, K. (2023a). Growth of water hyacinth biomass and its impact on the floristic composition of aquatic plants in a wetland ecosystem of the Brahmaputra floodplain of Assam, India. *PeerJ*, 1-32. http://dx.doi.org/10.7717/peerj.14811

Meraj, G., Farooq, M., Singh, S. K., Islam, M. N., & Kanga, S. (2022). Modeling the sediment retention and ecosystem provisioning services in the Kashmir valley, India, Western Himalayas. *Model. Earth Syst. Environ.*, 8(3), 3859-3884.

Meraj, G., Kanga, S., Kranjčić, N., Đurin, B., & Singh, S. K. (2021a). Role of natural capital economics for sustainable management of earth resources. *Earth*, 2(3), 622-634.

Meraj, G., Singh, S. K., Kanga, S., Islam, M. N., & Singh, S. K. (2021b). Modeling on comparison of ecosystem services concepts, tools, methods and their ecological-economic implications: A review. *Model. Earth Syst. Environ.*, 1-20.

Nath, N., Sahariah, D., Meraj, G., Debnath, J., Kumar, P., Lahon, D., Chand, K., Farooq, M., Chandan, P., Singh, S. K., & Kanga, S. (2023). Land Use and Land Cover Change Monitoring and Prediction of a UNESCO World Heritage Site: Kaziranga Eco-Sensitive Zone Using Cellular Automata-Markov Model. *Land*, 12(1), 151.

Rafiq, M., Meraj, G., Kesarkar, A. P., Farooq, M., Singh, S. K., Kanga, S., & Bhardwaj, V. (2022). Hazard Mitigation and Climate Change in the Himalayas–Policy and Decision Making. In S. Kanga, G. Meraj, M. Farooq, S. K. Singh, & M. S. Nathawat (Eds.), *Disaster Management in the Complex Himalayan Terrains*. Geography of the Physical Environment. Springer, Cham. https://doi.org/10.1007/978-3-030-89308-8_12.

Sajan, B., Mishra, V. N., Kanga, S., Meraj, G., Singh, S. K., & Kumar, P. (2022). Cellular automata-based artificial neural network model for spatial-temporal dynamics and land use/land cover change prediction. *Agronomy*, 12, 2772.

Singh, A., Singh, S. K., Meraj, G., Kanga, S., Farooq, M., Kranjcic, N., Đurin, B., & Sudhanshu. (2022b). Designing Geographic Information System Based Property Tax Assessment in India. *Smart Cities*, 5, 364–381.

Singh, H., Meraj, G., Singh, S., Shrivastava, V., Sharma, V., Farooq, M., Kanga, S., & Kumar, P. (2022a). Status of Air Pollution during COVID-19-Induced Lockdown in Delhi, India. *Atmosphere*, 13(12), 2090.

Singh, S. K., Gupta, S. K., Meraj, G., Kumar, P., & Kanga, S. (2023b). Uncovering the hydro-meteorological drivers responsible for forest fires utilizing geospatial techniques. *Theor. Appl. Climatol.*, https://doi.org/10.1007/s00704-023-04497-y

Singh, S. K., Meraj, G., Mondal, N., Bera, A., Verma, M. K., Tomar, J. S., & Kanga, S. (2022). Assessment of seasonal vegetation dynamics over parts of thar desert using geospatial techniques. *ANGRAU*, 105.

Singh, S., Singh, S. K., Prajapat, D. K., Pandey, V., Kanga, S., Kumar, P., & Meraj, G. (2023a). Assessing the Impact of the 2004 Indian Ocean Tsunami on South Andaman's Coastal Shoreline: A Geospatial Analysis of Erosion and Accretion Patterns. *J. Mar. Sci. Eng.*, 11(6), 1134.

Sud, A., Kanga, R., Singh, S. K., Meraj, G., Kanga, S., Kumar, P., Ramanathan, A. L., & Bhardwaj, V. (2023). Simulating Groundwater Potential Zones in Mountainous Indian Himalayas—A Case Study of Himachal Pradesh. *J. Geophys. Res.: Earth Surf.*, 10(3), 65.

Upadhyay, R. K., Tripathi, G., Đurin, B., Šamanović, S., Cetl, V., Kishore, N., Sharma, M., Singh, S. K., Kanga, S., Wasim, M., Rai, P. K., & Bhardwaj, V. (2023). Groundwater Potential Zone Mapping in the Ghaggar River Basin, North-West India, Using Integrated Remote Sensing and GIS Techniques. *Water*, 15(5), 961.

15

Uses of Wi-Fi Technology and Its Effects on Social Life

Vishal Kumar

Voltech HR Services, Bihar Wi-Fi CAPEX Project Arrah, Bihar, India

Abstract
The chapter presents a detailed analysis of the Network, its usage and effect on our social life. At present, the uses of the Internet is increasing day by day. Almost all things became online, whether you are buying or selling products. We want to know new things related to tech or non-tech as so many things are now online. The Internet connects people, enables knowledge sharing, and provides pieces of information related to each and everything. After the pandemic, all things became online. Whether it is an online coaching institute, classes, or shopping.

***Keywords*:** Uses of Wi-Fi, Wi-Fi working, tools and communication, equipment, effect on society

15.1 Introduction

It is no longer debatable that technologies have revolutionized our society and everyday lives. By using devices and tools, we are able to bring everything at our fingertips. Wi-Fi is wireless local area network(LAN)—it is one of the important technologies of computer networking. It allows the users to connect to the internet technology via wireless devices, it also allows the router to be cordless, the network is connected through an access point for the internet, and it is connected to a Universal Serial Bus (USB) port. There are many ports available and each and every port is assigned for different access point. It is necessary that all devices, which are used to communicate with each other, must be from same company. We can never use other companies' devices in a single wireless LAN connectivity area. Wi-Fi is used for free at restaurants, coffee houses, schools, or colleges. It sends out data to the LAN and WAN networks and these networks allow any connection without using cords; you can take your laptop wherever you want, with the option of having a wireless connection with the help of the router. Hotspots are connected through wi-fi, the cell phone companies are supported by GSM (Global System for Mobile Communications) and wi-fi facilities, which is shown on the slide bar of your mobile and task bar on your laptop; it allows people to always stay connected through Wi-Fi. Thus, it can also be used in emergencies. Wi-Fi is already connecting the Internet to every application that consumers want

Email: Kumarvishal2k11@gmail.com

Sanjay Kumar, Suraj Kumar Singh, Shruti Kanga, Gowhar Meraj, Majid Farooq and M. S. Nathawat (eds.)
Disaster Management and Environmental Sustainability, (177–190) © 2024 Scrivener Publishing LLC

today, it is enabled for light bulbs, home security, monitoring and control systems, door bells, motors, electronics appliances, appliances, and automotive products. Wi-Fi is a single technology that can help make all connectivity seamless, Wi-Fi applications with practical uses will have success, it has industry-standard security protections, and consumers can easily rely on. Wi-Fi networks using Wi-Fi Protected Access 2 (WPA2)-personal security is available on our cellphones, which we use while using Wi-Fi services. In this security, you can control who connects to or use your Wi-Fi. It provides privacy that the transmission cannot be read by others and for communications as they travel across your network, and for maximum security, your network should include only devices with the latest in security technology.

15.2 Origin of Wi-Fi

When we talk about modern Wi-Fi, we consider IEE802.11 protocol. It is part of IEE802 and set of LAN and the set of media access control (MAC) and physical layers (PHY) protocols for implementing the wireless local area network (WLAN) of computer communication. IEEE 802.11 is used in most home and office networks to allow laptops, printers, smartphones, and other devices to communicate with each other and access the Internet via wireless devices. The term "Wi-Fi" was created by the Wireless Ethernet Compatibility Alliance, which later became the Wi-Fi Alliance. In April 2000, the group announced the first set of Wi-Fi certified products, starting with IEEE 802.11b. More than 20 years later, there are more than 15 billion Wi-Fi products that are in use around the world, according to the alliance.

15.3 Uses of Wi-Fi

By enabling Wi-Fi on your cellphones and laptop, you can easily access online sites like YouTube, Google, and Facebook.

It is a very useful tool because by communicating with the router directly, many processes of networking can be simplified. To use Wi-Fi, we must have an ISP (Internet service provider). In Wi-Fi shown in Figure 15.1, an ISP works like a backbone, without an ISP we cannot access Wi-Fi. In Wi-Fi, an ISP is directly connected with the router via POE (power over ethernet). It is built with a highly secured protocol where a person can register their user id and create his/her own user id. For security, there is a person who has access to the firewall, where all the data is placed securely. If anyone want to create, edit, and delete their data, then that particular person will do the required things asked by the consumers. Communication module have shown Figure 15.1a.

Figure 15.1 Symbol of Wi-Fi.

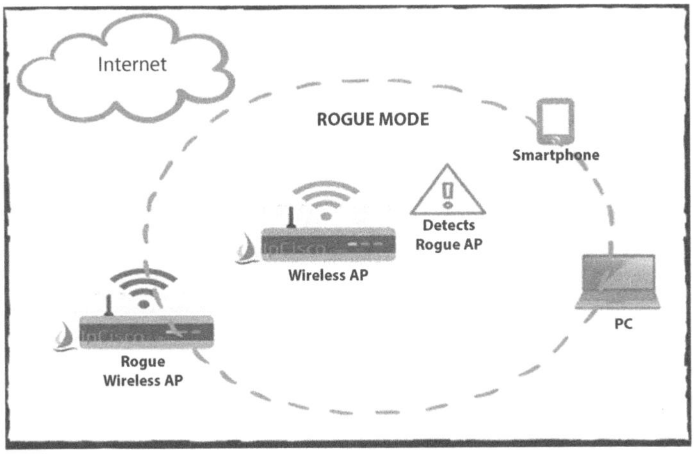

Figure 15.1(a) Communication module.

15.4 How Wi-Fi Works

Wi-Fi includes flexible user interfaces, vast security, built-in network protocols, and support for networked and offline applications that can be downloaded dynamically. So, this platform is what connects the user and his device to the server onto another laptop connected to the same network. For connecting to the device of any person X, a user needs to ensure that person X is connected to the network and registered on the server. The user will request the server to send the IP of person X, which upon receiving a request, will reply with an Internet Protocol (IP) (if available). Upon receiving the IP, the device will create a socket and communication will begin over that particular socket. If person X moves out of the range of the network, the call will automatically be terminated and the entry of person X will be removed from the server if person X is not registered. If person X is already registered in server, then Person X does not need to register again.

A Sequence of Steps-

- When a device connects to the network, it receives an IP from the wireless router which is then added to the server's routing table.
- Each device is identified by a user name. So, the routing table updates its IP with a corresponding user name.
- If any activity will be done by any user in the routing table, that person is identified by a user name and this information is made available to all users logged into the network.

Devices are connected to each other by a specific wire, which is known as an ethernet cable. Ethernet cabling has many types but for our convenience, mostly use rolled cable. These are probably the easiest cables to make because you just cut the end of one side of a straight-through cable, turn it over, and put it back on with a new connector. Types of cabling have been shown in Figure 15.2.

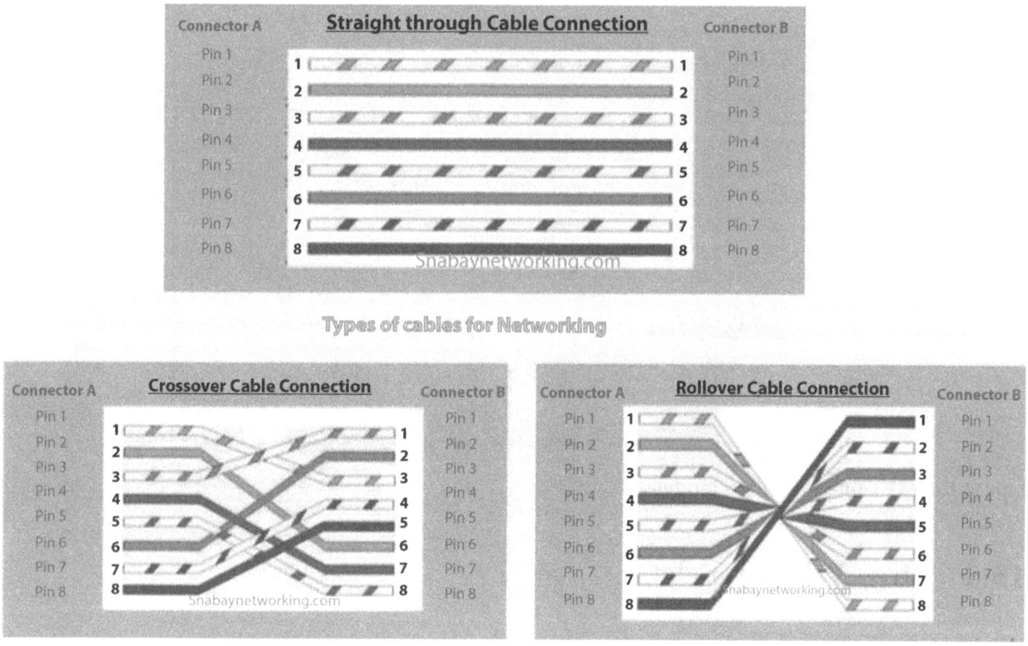

Figure 15.2 Types of cabling.

All connected devices are devised to connected through an ethernet cable.

15.5 Tools

(a) Firewall: this tool is for security purposes. All things related to security are managed here. Data is also saved here. We can create and delete all the users, and also provide them some limit like the number of devices they can use Wi-Fi on, with a monthly, weekly, and daily limit of data usage. Sophos firewall have been shown in Figure 15.3.

(b) Tower: it is used for signal purposes. This device is directly connected with a router, *i.e.*, firewall, core switch, etc. ISP Tower (BTS) Ground Image have been shown in Figure 15.4.

Figure 15.3 Sophos firewall.

Figure 15.4 ISP tower (BTS).

(c) Core Switch: it is not a kind of network switch. It refers to a data switch, which is used as a backbone or physical core of a network.

(d) Distribution switch: with this switch, we can create VLAN and MLT, which are necessary so devices can communicate with each other.
- VLAN: - refers to "virtual local area network." Any broadcast domain which is partitioned and isolated in the computer network at data link layer.
- MLT: it is the abbreviation for "multi-link trucking." Multi-link trucking allows the grouping of several physical ethernet links into one logical ethernet link to provide a high-speed link between the router, switches and servers.

(e) Access switch: it is a device which allows a user to connect to the network. It is not a high-level switch compared to a distribution switch.

It is a hierarchical internet working model defined by Cisco which includes the core layer, distribution layer, and access layer. Image of Core switch have been shown in Figure 15.5 and distribution switch, access switch, hierarchy level of networking have been shown in Figures 15.6-15.8.

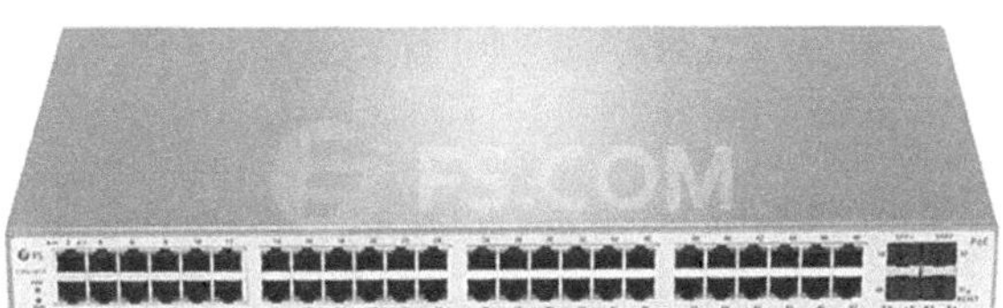

Figure 15.5 Core switch.

Figure 15.6 Distribution switch.

Figure 15.7 Access switch.

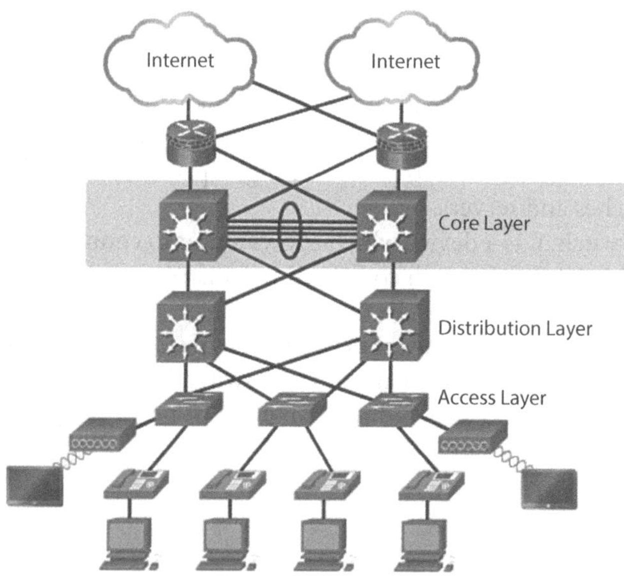

Figure 15.8 Hierarchy level of networking.

Its hierarchy is:

1. Core layer
2. Distribution layer
3. Access layer

All have their own usage which are describe earlier.

15.5.1 Terms, Uses, and Communication

1. Packet: it is a fundamental unit of data transmitted over the Internet. For example, your PC sends a request to open YouTube to open a video, it sends the request as pieces, which are known as packets. Each packet has its own IP address.
2. Hub: it is a simple device of network that sends packets to the all connected devices.
3. Bridge: it is a device which connects two network devices together and only allows packets through that are needed.
4. Switch: this device is connected through multiple devices together and also filter the packets based on each and individually allocated destination within the connected device.
5. Router: this device receives and analyzes packets and then routes them towards the destination via the ethernet cable.
6. IP address: each and every device that connects to the internet, whether it is a personal computer, a laptop, a mobile, a tablet, or anything else, has an assigned unique IP address. It is the abbreviation for "Internet protocol address".
   It has two types:
   (a) IPv4: abbreviation for "Internet protocol version 4," which has four numbers between 0 to 255, separated by period. As the use of IPv4 increases day-by-day so, therefore its size needed to be grown to the point where IPv4 exhausted.
   (b) IPv6: it is abbreviation for "Internet protocol version 6." IPv6 is standard formatted as eight groups of four hexadecimal digits. It has a limit of $3.4 * 10^{38}$ possible address.
7. Domain name: a domain name is a friendly name of any networking device, which is human-friendly. This name is followed by specific "top-level domain" or TLD. For example, the Google domain name is google.com. "Google" describes the organization and ".com" describes the top-level domain.
8. DNS: it is abbreviation for "domain name system." It directly acts on the Internet. When someone wants to access the device with a given domain name, a DNS is required.
9. Packet switching: when a packet is sent from one device, it does not follow a straight path or does not reach the destination directly. In fact, sometimes, some messages with the same destination does not arrive via the same path.

10. Protocol: a network protocol is a set of rules which determine how devices communicate with each other. Network protocols are the main reason why you can easily communicate with people all over the world, so it plays a critical or vital role in modern digital communications.

15.6 Effect on Society

Wi-Fi's effect on society:

like all the other things, everything has an effect on society. These effects are either positive or negative.

15.6.1 Positive Effect of Wi-Fi on Society

- Today, almost all things have become online. Almost all public data, their property details, and their assets can be monitored online. The government has taken almost all data and stored them on the cloud for easily acknowledgment.
- It provides effective communication through emailing and instant messaging services to any part of the world.
- It improves business interactions transactions and saves on vital time.
- Online Banking and online shopping have made life less complicated.
- We can access or get knowledge of the latest news from any part of the world without depending on the TV or newspaper.
- Education has received a huge boost as uncountable books and journals are available online from e-libraries across the world. This has made research easier. Students can now opt for online courses, PDF, and resources by using the Internet.
- Application for jobs has also become easier as most vacancies are advertised, their job vacancies are online with online applications.
- Professionals can now exchange information and materials online by using any messaging tools, thus enhancing research.
- Today, all social networking sites are the preferred platforms for all kinds of activities, both business and personal, and sociability has dramatically increased—but it is a different kind of sociability.
- After the COVID-19 pandemic, all jobs became online. All big companies are providing a special type of facility, *i.e.*, work from home (WFH). These companies are using the Internet to provide tasks for all the employees and collecting the tasks back from the employees. These Internets are of any kind, like broadband connection, mobile network, and Wi-Fi (WAN, LAN).
- Also, after the pandemic, all classes of students became online due to COVID-19. Such a type of action prevented them from social distancing.

15.6.2 Negative Effect of Wi-Fi on Society

- Addiction of Internet (Wi-Fi) and problematic internet use. A lack of control over one's internet consumption can lead to a decrease in physical and psychological wellbeing. This addiction also associated symptoms such as distress, anger, loss of control, social withdrawal, familial conflicts and others pushing people towards isolation.
- There is a lot of evidence to suggest that children's cognitive development can be damaged by prolonged internet use, including the development of memory skills, attention span, abilities for critical reasoning, language acquisition, reading, and learning. However, a lot of research is needed to draw conclusions. Information is overloaded. Having too much information can make it difficult to adequately understand an issue or to make effective decisions. Information overload is associated with loss of control, feelings of being overwhelmed, reduced intellectual performance, and diminished job satisfaction. The way in which the internet and smartphones blur the distinction between different spheres of life—such as work and home. It also harms the boundaries between people's public and private lives.
- Negative effects that can result from such interpenetration include lower quality of life, lack of privacy, and negative impacts on social relationships.
- Another negative effect can be when friends and family members feel that they are left behind by new technology.
- Also, impact on people's behavior, belief, and attributes.
- The World Health Organization (WHO) has defined disorder due to addictive behaviors to include gambling and gaming disorders, which may involve both online and offline behavior of the people.
- Disorders due to addictive behaviors including gambling and gaming disorders, which may involve both online and offline behavior.
- Gaming addiction or gaming disorder is the problematic online behavior that has seen the largest evidence with all age group base across all addictions related to internet use. It has been argued that online video games are one of the most widespread inappropriate activities irrespective of culture, age and gender.
- Extensive internet use, of social media, is correlated with loneliness and social isolation. Intimate relationships can be degraded by internet use, particularly due to most viewing of online pornography. Malicious online behavior, particularly cyber-bullying, cyber-stalking, an online predation, affects a significant percentage of internet users.
- There is evidence that children's cognitive development can be damaged by prolonged internet use, including the development of memory skills, attention span, abilities for critical reasoning, language acquisition, reading, and learning abilities. More research is, however, needed to draw more reliable conclusions.

Internet activity addiction:

- Online video games, virtual reality
- Sports wagering, online casino, etc.
- Instagram, YouTube, Facebook, Twitter, etc.
- Cybersexting, pornography

It will reduce by such activity:
(A) Prevention and health promotion - reducing risk and harm policy option

- Initiate information and prevention campaigns.
- Increase education regarding internet use and its consequences.
- Stimulate employers to develop policies that protect workers against harmful work-related internet use.

(B) Providing support services

- Strengthen the health and social services' support available for internet users that engage in harmful use.
- Support communities and networks affected by individual online users.

(C) Governmental actions at EU and national level

- Establish governmental units to address the problem of harmful internet use.

(D) Better protection offered by industry

- Promote technology that better protects against harmful internet use.
- Promote technology that better protects social institutions and social inclusion.

(E) Research policy

- Promote more research into the effects of internet use and effective interventions.

Mutation accumulation produced by cellular DNA damage is likely to be both cumulative and irreversible as well because later mutations are highly unlikely to reverse previously occurring mutations. Wi-Fi has been placed in most schools, hotels, restaurants, coffee shops, commercial aircrafts, and airports, as well as in many homes, and Wi-Fi hot spots are becoming increasingly common in cities around the world, thus, we should expect massive cumulative Wi-Fi effects in many people. A second tentative inference is that false assurances of safety on the part of the industry are likely to lead to much more severe effects on people exposed to Wi-Fi or other electromagnetic field (EMF); rather than leading them to protect themselves or their children by avoiding exposures or demanding that others stop non-voluntary exposures, they are likely to avoid protective changes or be prevented from doing such protective changes. A third inference is that these effects may be among the more difficult ones for us to attribute to EMF exposure. We are much more aware of effects that occur rapidly than those that take months or years before they become readily apparent.

15.6.3 Some Myths about Wi-Fi Technology

I. Does Wi-Fi cause cancer?

Currently, there is no definite answer to this question. That is because there is no solid evidence suggesting that Wi-Fi, or EMFs in general, directly causes cancer. In 2011, the World Health Organization's International Agency for Research on Cancer (IARC) stated that EMFs are "possibly carcinogenic to humans." The label was established by 30 scientists who evaluated studies on EMFs and cancer. Studies involving EMFs and cancer are conflicting. For example, according to a 2017 research review, EMFs from wireless devices increase the risk of glioma, a type of brain tumor. But a 2018 study states that there is no clear association between EMFs and brain tumors. So, without any solid evidence we cannot say anything, there is a lot of research required to establish any kind of relation between Wi-Fi and cancer.

II. Male infertility

According to a 2014 animal study, long-term exposure to Wi-Fi radiation reduces the reproductive function of male rats (Tok *et al.*, 2014). A 2016 animal study found that radiation emitted from Wi-Fi causes DNA damage in the testes of rats (Akdam *et al.*, 2016)

In a 2015 study, researchers examined the sperm motility of more than 1000 men (Yildirim *et al.*, 2015). The men answered questions about their Wi-Fi and cellphone use. Researchers determined that men who used wireless (mobile) internet had reduced sperm motility, compared with men who used wired Internet.

But the researchers did not state the possible mechanisms behind these results. They also noted that the questionnaire did not ask if participants were smokers. Smoking may affect sperm motility. These results show that there is not any proper evidence or mechanism which we can say that male infertility is also caused by Wi-Fi radiation.

III. Heart health

In a 2015 animal study, exposure to Wi-Fi altered the heart rhythm and blood pressure in rabbits. According to the researchers, this suggests that Wi-Fi affects the heart. But the mechanisms are unclear, and more human studies are needed (Yildirim *et al.*, 2015).

15.6.4 False Claims About Wi-Fi Health Risks

As concerns over Wi-Fi increase, there've been several myths about its effects on health.

According to these claims, Wi-Fi causes conditions like:

- Autism
- Skin issues
- Headaches
- Pain

- Dizziness
- Nausea
- Fatigue
- Heart palpitations

There is no hard evidence that Wi-Fi is linked to these conditions.

Wi-Fi uses electromagnetic radiation to connect electronic devices. Some people believe that it can contribute to the development of cancer. But there is no solid evidence that Wi-Fi causes health risks in humans. Most of the available research involves animals. Even then, the results have been inconclusive.

15.7 Results

We all already know that every coin has two faces. Like Wi-Fi technology also has both positive and negative effects. Education has received a huge boost as uncountable books and journals are available online from e-libraries across the world. This has made research easier. Students can now opt for online courses, PDF, and resources by using the Internet. It improves business interactions and transactions, and saves on vital time. After the COVID-19 pandemic, almost all jobs became online. All big companies provided a special type of facility, *i.e.*, WFH. These companies are using the Internet to provide tasks to all the employees and collecting the tasks back from the employees. These Internets are of any kind, like broadband connection, mobile network, and Wi-Fi (WAN, LAN). But it has some negative effects, the most common negative effect is that it almost disconnects the people in their own home. One creates one's own world and disconnect with one's family. They live in their home like a stranger. It also means it affects social relationships. The WHO has defined disorder due to addictive behaviors to include gambling and gaming disorders, which may involve both online and offline behavior of people. Related to Wi-Fi's health risks, there are a lot of misunderstandings, and these are: autism, skin issues, headaches, pain, dizziness, nausea, fatigue, and, male infertility. We cannot clearly say that these are symptoms of Wi-Fi radiation because there is no evidence.

15.8 Conclusion

We cannot say that Wi-Fi radiation has harmful effects on human beings because there is more research required. All the research conflicts with one another. Some researchers say that it has a harmful effect and some say it does not affect human beings. But sometimes, due to irregularities of light, back up equipment emits a harsh noise. But it portrays the effect on all age groups of society depending on their use. Today, all can learn new technology, knowledge related to subjects, competitive exams, and much more. On the other hand, there are a lot of websites which provide online gambling, online video games, pornography, etc. Some people are addicted to these things, and they feel demotivated, hesitant, and have low IQ levels. They used to keep and maintain social distance. This type of activities is seen mainly in all age groups and they face all the problems mentioned above.

Acknowledgment

This author is thankful to Dr. Sanjay Kumar, Associate Professor, Geography, PG Department Maharaja College Ara.

References

Akdag, M. Z., Dasdag, S., Canturk, F., Karabulut, D., Caner, Y., & Adalier, N. (2016). Does prolonged radiofrequency radiation emitted from Wi-Fi devices induce DNA damage in various tissues of rats?. *J. Chem. Neuroanatomy*, 75, 116-122.

Tök, L., Nazıroğlu, M., Doğan, S., Kahya, M. C., & Tök, Ö. (2014). Effects of melatonin on Wi-Fi-induced oxidative stress in lens of rats. *Indian J. Ophthalmol.*, 62(1), 12.

Yildirim, M. E., Kaynar, M., Badem, H., Cavis, M., Karatas, O. F., & Cimentepe, E. (2015). What is harmful for male fertility: cell phone or the wireless Internet?. *Kaohsiung J. Med. Sci.*, 31(9), 480-484.

16

Application of Remote Sensing and GIS in Disaster Management: An Applied Review

Sadaf

P.G. Department of Geography, Veer Kunwar Singh University, Ara, Bihar, India

Abstract

It is a very unfortunate and well-known fact that disasters, whether natural or man-made, are destructive events. It strikes in both developed and developing countries as well, resulting in the loss of human lives and damage to buildings, communication systems, agricultural land, forest, and the natural environment. There is an imminent need to reduce disaster impacts on economic, infrastructure, environmental, social, and the tragic loss of lives. Currently, remote sensing and geographic information system (GIS) play crucial roles in mapping, monitoring, and disaster risk assessment. The main objective of this chapter is to assess and review the application of remote sensing and GIS in disaster management. On the basis of the review of various research papers, the study concludes that remote sensing and GIS techniques are ideal and the latest tools for various phases of disaster management, such as prevention, preparedness, relief, reconstruction, early warning, and monitoring systems.

***Keywords*:** Disaster, remote sensing, GIS, risk management, mapping

16.1 Introduction

Disasters, whether natural or man-made, have become a major problem in developed as well as in developing countries, and a serious threat to sustainable development. Disasters are extreme events within the earth's system, resulting in loss of life, injury, and disease to humans, and damage or loss of valuable goods, such as buildings, communication systems, agricultural land, forest, and the natural environment (Debnath *et al.*, 2023). In the last few decades, the frequency and intensity of natural disasters have increased (Sud *et al.*, 2023). There are 4022 natural disasters that occurred between 2001 and 2010 worldwide, reportedly killing a total of 1,221,332 people. The historical data (Figure 16.1) reported that the world has witnessed a significant reduction in disaster deaths through earlier predictions due to more resilient infrastructure, emergency preparedness, and response systems.

It is evident that there is an inverse relationship between the level of development and loss of human lives in the case of a disaster. Developing countries witnesses about 95% of

Email: sadafmalik849@gmail.com

Sanjay Kumar, Suraj Kumar Singh, Shruti Kanga, Gowhar Meraj, Majid Farooq and M. S. Nathawat (eds.)
Disaster Management and Environmental Sustainability, (191–206) © 2024 Scrivener Publishing LLC

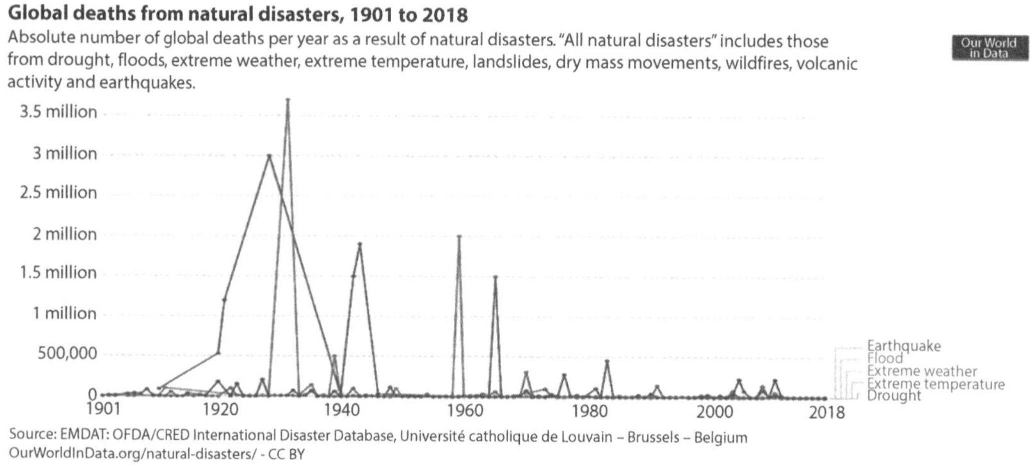

Figure 16.1 Source: EMDAT (2019): OFDA/CRED International Disaster Database, Université catholique de Louvain Brussels – Belgium.

the disaster-related casualties, where about more than 4.2 million people live. Economic losses due to natural hazards may represent as much as 10% of the gross national product in developing countries. It is more possible to predict the occurrence of certain natural phenomena and to succeed in massive evacuations due to the accuracy of warning systems in developed countries (Westen, 2000; Rafiq *et al.*, 2022).

High population growth, intense urbanization, and industrialization encourage the presence of a high-density population in high-risk areas, which may be responsible for more casualties by natural disasters (Kobiyama *et al.*, 2006).

Thus, low-income regions (Figure 16.2) are often the most vulnerable to disaster events. Improving infrastructure, the standard of living, and warning and mitigation systems will be the key to reduce the destruction from disasters in the coming days.

The extent, recurrence, and impacts of disaster events are increasing in developing countries (Nadi *et al.*, 2010). It is difficult to make preventive plans for some abrupt occurrences of natural disasters like tsunamis, tornados, and earthquakes. While mapping of some disasters such as floods and landslides is easier, and it can be easily predicted by the affected people because the vulnerable areas are generally known (Sorensen *et al.*, 2012).

India, the second most populated country in the world, is vulnerable to natural disasters due to its geographical location, seismic risk zone, and intertropical convergence point. This subcontinent is also hit by recurring atmospheric phenomena; heavy monsoon rains, floods, episodes of high heat, and drought (Goswami *et al.*, 2015). Population density and high population growth in vulnerable areas worsen the disaster event.

During the last 30 years, India has been exposed to more than 400 major disasters and it results in enormous loss of life and property (Kanga *et al.*, 2022). Flood disasters have caused a significant number of casualties and major destruction to infrastructure, property, livelihoods, and crops in the country during the last 20 years (1995 to 2014). Disasters such as cyclones (storms), abnormal weather, and sediment disasters have occurred more frequently and they are highly destructive. Major natural disasters like earthquakes and tsunamis have a huge impact once they occur, although they have a low frequency of occurrence (Table 16.1). On the basis of the occurrence of past disasters, it is evident that major

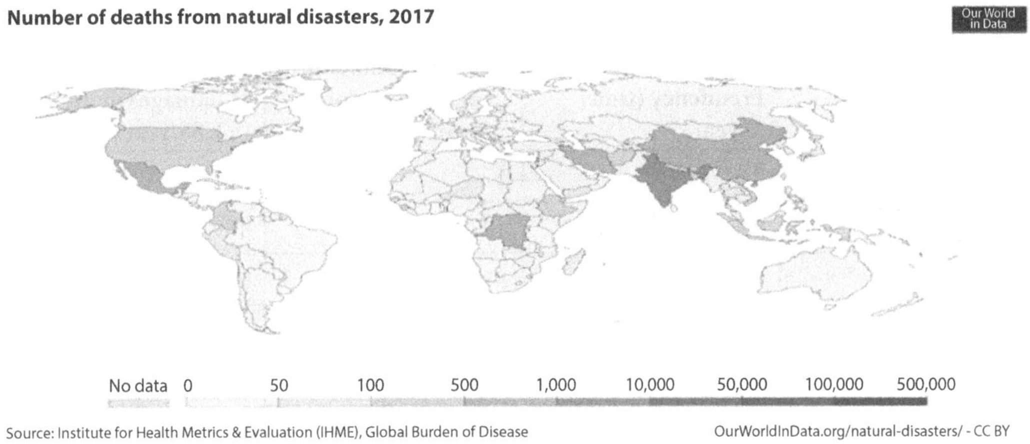

Figure 16.2 Source: Institute for Health Metrics & Evaluation (HME), Global Burden of Disease.

natural disasters are floods, earthquakes, tsunamis, cyclones sediment disasters, and abnormal weather in India. It is a well-known fact that proper understanding and management of natural disasters are very important for maintaining safer and sustainable communities (Zhou *et al.*, 2009; Chen *et al.*, 2006). Due to their broad applicability, remote-sensing and geographic information system (GIS) techniques are valuable and the latest tools for disaster management (Thomas *et al.*, 2007).

The use of remote sensing (RS) and GIS techniques are very important in natural disaster management (Luiz A. Manfré *et al.*, 2012). Multispectral satellite imagery has the capacity to provide valuable data for large and inaccessible locations. With the help of GIS software packages, users can apply multiple assessment techniques in the prevention of various disaster risk factors. Remote sensing and GIS techniques have been used in the identification of high-risk areas and the modeling of disaster events. Nowadays, RS data has been widely used for natural resource management. Remote sensing is the best source of data for large-scale applications and studies about disaster management because users can get the remotely sensed data from different sensors of various platforms with a wide range of spatiotemporal, radiometric and spectral resolutions (Melesse *et al.*, 2007).

16.2 Classification of Disasters

It is very important to know the difference between the term "hazard" and "disaster." A potentially damaging phenomenon is clearly known as a hazard. For example, any disaster such as an earthquake, volcanic eruption, tsunami, and cyclone is not considered a disaster when it occurs in uninhabited areas. However, it is called a disaster when it occurs in an inhabited area and brings damage to human society. There are various ways to categorize the different types of disasters. Westen (2000) classified disasters in three ways: 1) natural disasters are events that are caused by purely natural phenomena and affect human societies (such as earthquakes, volcanic eruptions, and hurricanes), 2) human-made disasters are events that are caused by human activities (such as atmospheric pollution, industrial chemical accidents, major armed conflicts, nuclear accidents, and oil spills), and

Table 16.1 India: breakdown of disasters.

Type of disaster	Frequency (time/year)	Average number of deaths (persons/year)	Average economic damage (thousand US$/year)
Drought	0.2	1	102,056
Earthquake	0.4	1079	189,100
Tsunami	0.1	819	51,140
Abnormal weather	1.5	484	20,000
Flood	8.2	1532	1,979,169
Sediment disaster	1.3	101	2,725
Storm	2.8	852	681,656
Wildfire	0.1	0	0

Source: Compiled by the survey team based on (Kanga *et al.*, 2022).

3) human-induced disasters are natural disasters that are accelerated by human influence. Krishnamoorthi (2016) classified disasters in two parts: 1) natural disasters are events that take place purely by natural phenomena and result in heavy loss to livelihoods and societies (for example, earthquakes, hurricanes, tornadoes, and volcanic eruptions) and 2) man-made disasters are the events that occur due to human activities such as pollution, nuclear accidents, industrial chemical accidents, oil spills, and major armed conflicts.

Natural disasters are being classified into five major and 30 sub-types (Figure 16.3 and Table 16.2). They tried to classify the major disasters' definitions as follows: geophysical events are the events originating from solid earth. Meteorological events are caused by short-lived/small to meso-scale atmospheric processes. Hydrological events are caused by deviations in the normal water cycle and/or overflow of bodies of water caused by wind setup. Climatological events caused by long-lived/meso- to macro-scale processes. Biological disasters are caused by the exposure of living organisms to germs and toxic substances.

16.3 Disaster Management Cycle

It is very important to understand the disaster management cycle for reducing the impacts of disasters effectively. The disaster management cycle comprises of five phases in which two phases take place before a disaster occurs, disaster prevention and disaster preparedness, and three phases that happen after the incidence of a disaster, disaster relief, rehabilitation, and reconstruction (Singh *et al.*, 2023).

Remotely sensed data can provide a valuable source of information in each phase and it also helps the scientists and authorities for decision making with the help of valuable data (Karen *et al.*, 2009). Figure 16.4 shows the disaster management cycle.

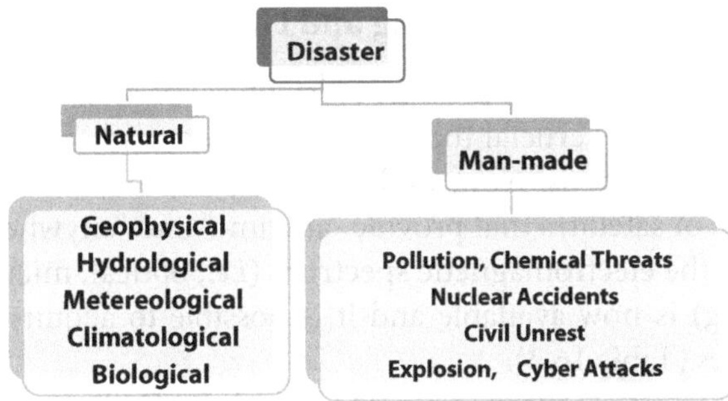

Figure 16.3 Types and sub-types of disasters.

Table 16.2 Classification of major and sub-categories of natural disaster.

Geophysical	Hydrological	Meteorological	Climatological	Biological
• Earthquake • Volcanic activity • Mass movement i. Rock fall ii. Landslide iii. Avalanche iv. Subsidence	• Flood • Landslide • Wave action	• Convective storm • Extra-tropical Cyclone • Fog • Tropical cyclone	• Drought • Glacial lake outburst • Wildfire • Extreme temperature i. Heat wave ii. Cold wave	• Animal stampede • Epidemic i. Viral infectious disease ii. Bacterial infectious disease iii. Parasitic infectious disease iv. Fungal infectious disease v. Prion infectious disease vi. Insect infection

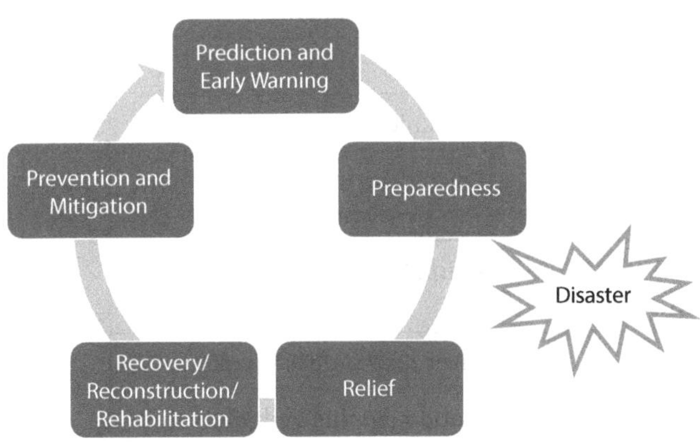

Figure 16.4 Disaster management cycle.

16.3.1 Application of Remote Sensing and GIS in Disaster Management

Remote sensing and GIS are used almost in every phase of a disaster management cycle. Remote sensing technology is crucial today because a very high-resolution dataset is now available for various applications. Now, we have an agile satellite system and low equatorial orbit constellation of satellites that provide ”anytime” and ”anywhere” satellite imagery. Complete coverage of the electromagnetic spectrum (*i.e.*, optical, microwave, thermal, and hyperspectral imaging) is now available and it is possible to acquire data from different electromagnetic ranges (Table 16.3).

Ground-based observation based on LIDAR and RADAR provides unprecedented information. Drone (UAV) and aerial sensors are playing a major role in disaster monitoring and rescue operation and they also provide critical information in a cost-effective

Table 16.3 Different sensors used in disaster management.

Sensor example	Types of information	Characteristics
AVHRR, MODIS, IKONOS	Building stock assessment	Visible rays (0.4–0.7 mm)
IKONOS, MODIS	Population density	
ASTER, PRISM	Digital elevation model	
MODIS, AVHRR, SPOT, LANDSAT TM	Vegetation mapping	Near Infrared 0.7–1.0 mm
MODIS	Flood mapping	
AIRS	Water vapor	Shortwave infrared 0.7–3.0 mm
MODIS	Active fire detection	Thermal infrared 3.0–14 mm
MODIS	Burn scar mapping	
MODIS, AVHRR	Hotspots	
Hyperion	Volcanic activity	
RADARSAT SAR, PALSAR	Earth deformation and ground movement	Microwave, radar 0.1–100 cm
METEOSAT, MICROWAVE IMAGER (aboard TRMM)	Rainfall	
AMSR-E	River discharge and volume	
AMSR-E	Flood mapping and forecasting	
QUICKSCAT RADAR	Surface winds	
Precipitation RADAR (aboard TRMM)	3D storm structure	

Source: Krishnamurthy, 2016.

and timely manner. They can be used to solve complex planning and management problems. Geographic information system (GIS) is a computer-based technology that captures, collects, and stores spatial data for decision-making, disaster loss estimation, and risk modeling.

16.3.2 Phases of Disaster Management

Disaster management consists of three important phases: the pre-disaster phase (planning, preparedness, and mitigation), the on-disaster phase or impact phase (response, recovery, and evacuation, etc.), and post-disaster phase (rehabilitation, damage assessment, providing food and medical facilities, etc.).

All phases of disaster management (planning, preparedness, mitigation, response, recovery, evacuation, rehabilitation, damage assessment, reduction, etc.) can benefit from RS and GIS techniques. The easy availability of different kinds of maps such as roads, railways, hospitals, schools, shelters, and other engineering structures simplify disaster management modeling (Figure 16.5). Some of the most frequent disasters that occur in India are included in this study. Apart from these disasters, COVID-19 was also included because India badly suffered from this biological disaster.

a) Application of Remote Sensing and GIS in Earthquake Management

A sudden motion or trembling of the earth caused by the abrupt release of slowly accumulated strain by faulting or volcanic activity is known as an earthquake (Kanga *et al.*, 2022). It is one of the most common natural disasters having the potential of a huge loss of life and property. It occurs when a plate ruptures along a geological fault with differences in tectonic settings in a particular geographic expanse. In the aspect of earthquake disasters, RS and GIS have important roles in identifying settlement infrastructure based on data such as fault location and seismic record (Sekac *et al.*, 2020). For the damage assessment, distance from active faults, geological structure, soil types, depth of the water table, topography, and construction types of buildings play an important role.

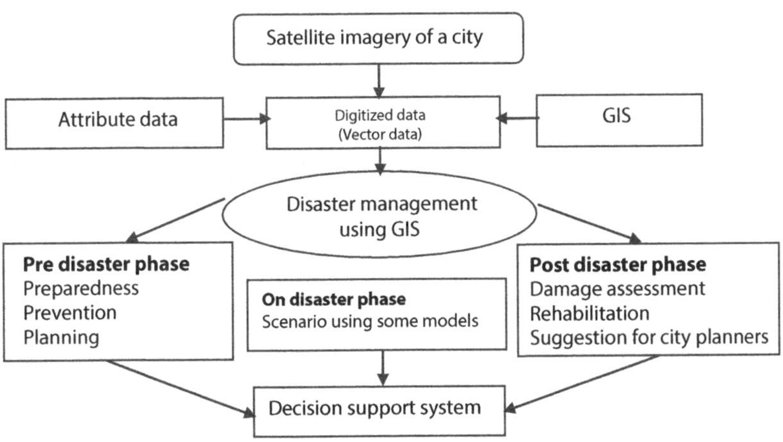

Figure 16.5 Flow chart for disaster management using GIS by Vijayaraghavan *et al.*, 2012.

Satellite RS can play a role in the mapping of lineaments, faults, and the study of the tectonic setting of an area, and it may be helpful for earthquake disaster prevention (Kanga *et al.*, 2022). Satellite laser ranging (SLR) and very long base baseline interferometry (VLBI) are very useful for the monitoring of crustal movement near active faults. Global positioning system (GPS) has become very important in the measurement of fault displacements. In the mapping of earthquake deformation fields, synthetic aperture radar (SAR) interferometry (InSAR) has been used. There are several methods involved in the detection of earthquakes, faulting, and damages such as Lid AR (light detection and ranging), which are useful for fault interpretation with the help of the highest resolution, **digital elevation model**. Very high-resolution optical data is providing the best imagery for damage assessment of buildings and infrastructure (Matsuoka and Yamazaki, 2005). Remote sensing and GIS have also played a valuable role in land use planning which is very helpful to identify vulnerable areas from earthquake events (Alparslan *et al.*, 2008, Sun *et al.*, 2008). Some sensors such as ½, ENVISAT ASAR, and ALOS PALSAR help in the assessment of fault displacement (Krishnamurthy, 2016).

b) Application of Remote Sensing and GIS in Flood Management

Flood is one of the most devastating natural hazards in the world and it affects the largest amount of lives and property damage (CEOS, 2003). On the basis of magnitude, frequency, duration, flow velocity, time of occurrence, and real extent, flood is classified into different types such as river flood, flash flood, coastal flood, and dam-break flood. Remote sensing and GIS techniques are playing crucial roles in flooding emergency response. In most phases of flood disaster management, satellite data has been successfully and operationally used (CEOS, 1999). For assessing hydrological and hydro-geological risks, for meteorological evaluation, interpretation, and validation, and for the development of numerical weather prediction models, multi-channel and multi-sensor data derived from GEOS satellites are used (Farooq *et al.*, 2022). Synthetic aperture radar (SAR) and RADARSAT have been very useful for flood mapping (Chkraborti, 1999). Quantitative precipitation estimates (QPE) and forecasts (QPF) use RS data as a very useful source of information to provide early warnings of flood hazards to communities. Integrated SAR and TM imagery are beneficial to monitor changes in coastal geomorphology and land cover, flood and erosional damage, and help in planning and maintenance of mitigations (Singh Roy, 1995; Farooq *et al.*, 2022). The NOAA's AVHRR images combined with radar data have been used to estimate precipitation intensity, amount, and coverage, measure moisture and winds, and determine ground effects such as surface soil wetness (Singh *et al.*, 2022). Therefore, RS data can be applied in flood management for the preparation of a hazard assessment map, generation of the hydrological model, quantitative assessment of soil, generation of a flood risk map, and early warning.

c) Application of Remote Sensing and GIS in Drought Management

These days, many parts of the world are facing a water crisis due to the growing human population, expansion of agricultural, energy, and industrial sectors, and also because of climate change and contamination of water supplies. The water crisis is being further intensified by droughts and it may reduce the water supply, worsen water quality, and crop failure. Therefore, understanding drought components and their modeling is very important for water resource planning and management (Belal *et al.*, 2012). Remote sensing and GIS techniques are useful drought detection techniques. Some indexes such as the normalized difference vegetation index (NDVI), and land surface temperature (LST), derived from the RS data, have been used to monitor agricultural droughts in relation to plant growth (Han *et al.* 2010). It is evident that the vegetation temperature condition index (VTCI index) has been used to monitor the drought of an area in real-time. The autoregressive integrated moving average (ARIMA) model is one of the most widely used in time series studies. It helps in assessing moving averages, exponential smoothing, neural networks, forecasting capability, and time-related changes. Remote sensing techniques make it possible to help in the detection, monitoring, and mitigation of disasters with the help of spectral bands that cover the whole range between visible and microwaves (Kotecha *et al.*, 2023).

The main advantage of using RS and GIS techniques for drought risk management is that they not only give a general picture of the drought situation but also give an idea about the probable damage due to drought hazard. Satellite images are the key source that can provide information about the drought in large areas at satisfactory costs (Wipulanusat *et al.* 2009). The main problem is to quantify the impact of drought on society and agriculture. So, various reliable indices have been used to detect the intensity and spatial as well as temporal dimensions of drought occurrences for decision-making and for alleviation of poverty (Singh *et al.*, 2022). Satellite data can be helpful for forecasting the drought one month earlier and delineating the potential zone more accurately also the impact of drought on the crop can be detected early which may be helpful in global food security and trade (Meraj *et al.*, 2021; Singh *et al.*, 2023). Remote sensing and GIS techniques provide suitable rapid and continuous data and information for disaster monitoring.

d) Application of Remote Sensing and GIS in Landslide Management

Landslides are generally frequently occurring disasters but their impact is in a small portion in hilly and mountain regions. Landslides can be classified in many ways; depending on the type of movement (slide, topple, flow, fall, spread), the material involved (rock, debris, soil), the speed of movement (mm/year - m/sec), and the triggering mechanism (earthquake, rainfall, human interaction). Satellite imagery can be used for landslides inventory mapping and mapping of factors (such as lithology, geomorphological setting, faults, land use, vegetation, and slope) that affect the occurrence of landslides. Imagery with adequate spatial resolution obtained from SPOT and IRS can be used to assess past landslides inventory. Aerial photographs are essential for landslide disaster monitoring. The best air photo scale for the interpretation of landslides is between 1:15.000 and 1:25.000 (Sajan *et al.*, 2022). Large-scale aerial photo interpretation is useful for landslide risk management. Very high resolution (VHR) imagery derived from IKONOS-2 can be used successfully for landslide inventory. Multispectral imagery derived from the SPOT has been used for geomorphological (such as classification of lithology, slope, soils, faults, vegetation, and land use) mapping (Singh *et al.*, 2022). Global positioning system (GPS), RADAR, and photogrammetry can be very useful for monitoring the landslide movements. For landslide disaster preparedness, DEM (digital elevation model) derived from SPOT and IRS imagery can be used. Meteosat and NOAA, combined with rain gauge data, have been used for predicting the landslide disaster. Although it is very difficult to predict landslide occurrence, warning systems for landslides are operational in some parts of the world (Meraj *et al.*, 2021).

e) Application of Remote Sensing and GIS in Cyclone Management

Tropical cyclones, hurricanes, or typhoons are among the most dangerous and destructive natural disasters affecting life, property, and the environment across the world (Poulos, 2010; Gallina *et al.*, 2016. The effect of tropical cyclones is comparatively much higher than other natural disasters (Li and Li, 2013). Due to climate change, the frequency and intensity of tropical cyclones will likely increase in the coming decades (Mendelsohn *et al.*, 2012; Upadhyay *et al.*, 2023; Krishna Mohan *et al.*, 2014; Deo and Ganer, 2014). The devastating impacts of tropical cyclone disasters can be reduced by using appropriate management methods. The use of RS and GIS techniques have significantly increased tropical cyclone risk management. Remote sensing data is very useful in every phase of cyclone disaster management (Lavane *et al.*, 2023; Meraj *et al.*, 2022; Wang *et al.*, 2010). Satellite RS can be used for tropical cyclone disaster risk management through the evaluation of cyclone hazard, vulnerability, and mitigation capacity assessment (Li and Li, 2013; Van Westen, 2013; Appelquist and Balstrøm, 2015; Fang *et al.*, 2016). The optical and SAR satellite images, DEM, and field data have been used in assessing the impacts and recovery of tropical cyclone disasters across the globe. Several multi-criterial integrated procedures such as equal weight, analytic hierarchy process (AHP), ranking, and rating validation of risk assessment have been used for tropical cyclone disaster management. Therefore, the RS technique is providing very useful data for cyclone disaster management (Joyce *et al.*, 2009; Martino *et al.*, 2009; Klemas, 2009).

f) Application of Remote Sensing and GIS in COVID-19 Management

During the latter part of 2019 and early 2020, the world witnessed the origin of one of the world pandemics, COVID-19. In some way, almost everybody has been affected by the deadly and infectious COVID-19 virus. Remote sensing and GIS technology play a greater role to understand, track, and monitor COVID-19 cases. With the advancement in GIS technology, it can provide dynamic maps to understand the geographical distribution of diseases, frequency of cases, fatality mapping, the spatial cluster of diseases, disease association with other factors and network analysis, etc.

One of the important applications of geospatial technology is digital contact tracing, which presents a solution to hinder the transmission of infectious diseases among people. Most countries have used dashboards to present information and to show the trends of COVID-19 all around the world. India also developed dashboards to show the trends of

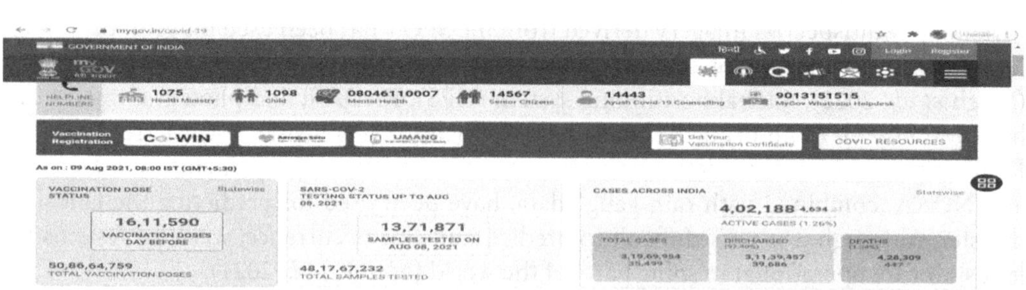

Figure 16.6 COVID-19: dashboard of India. Accessed on 07/08/2021 https://www.mygov.in/covid-19.

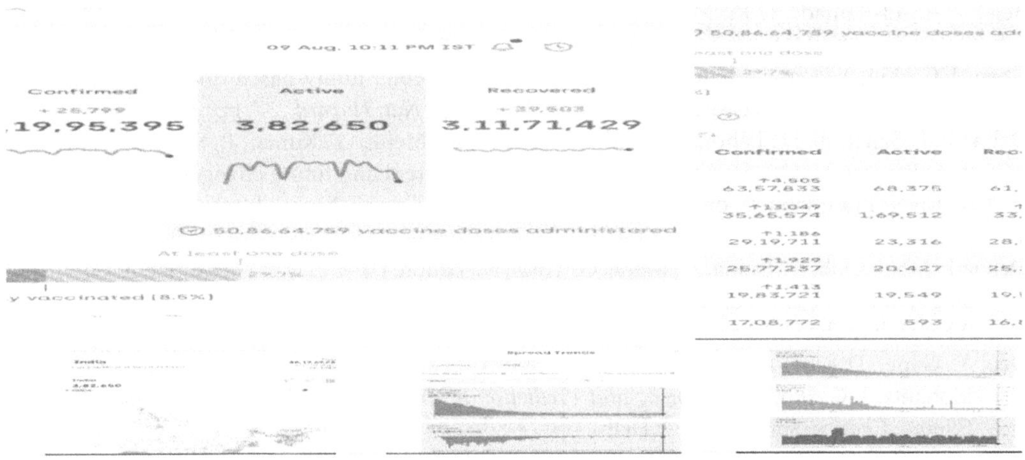

Figure 16.7 India COVID-19: dashboard showing COVID 19 information. Accessed 07/08/2021. https://www.covid19india.org/.

COVID-19 cases, active cases, recovery information, number of deaths, and vaccination (Figures 16.6 and 16.7).

16.4 Conclusion

This chapter presented a review of the application of remote sensing (RS) and GIS in the field of disaster management. On the basis of the reviews of several pieces of above said literature, it is evident that RS and GIS techniques have enormous potential to address various aspects of disaster management. The study highlights that RS data has increasingly been used in the three phases of disaster management: pre-disaster management, disaster management, and post-disaster management. Remote sensing and GIS techniques have proven most useful in assessing natural or man-made disasters such as the detection of earthquakes, faulting, volcanic activity, landslides, flooding, and wildfire. The need and importance of spatial data are increasing day by day due to the frequent occurrence of disasters all over the world. The application of RS and GIS technology in the field of disaster management is likely to grow in the future. It has been concluded that RS and GIS techniques are ideal tools in various phases of disaster management, such as prevention, preparedness, relief, reconstruction, early warning, and monitoring.

References

Alparslan, E., Ince, F., Erkan, B., Aydo¨Ner, C., O¨Zen, H., Do¨Nertas,, A., Ergintav, S., Yagsan, F.S., Zaterogullar, A., Eroglu, I., Deger, M., Elalm, H. And O¨Zkan, M., 2008, A GIS model for settlement suitability regarding disaster mitigation, a case study in Bolu Turkey, *Eng. Geol.*, 96, pp. 126–140.

Appelquist, L.R., Balstrøm, T., 2015. Application of a new methodology for coastal multi-hazard-assessment & management on the state of Karnataka, *India. J. Environ. Manag.*, 152, pp. 1-10.

Belal, A.A., El-Ramady, H.R., Mohamed, E.S. *et al.* (2014). Drought risk assessment using remote sensing and GIS techniques. *Arab. J. Geosci.*, 7, pp. 35–53.

Chen, L.C., Liu, Y.C. and Chan, K.C. (2006). Integrated community-based disaster management program in Taiwan: a case study of Shang-An village. *Nat. Hazard.*, 37, pp. 209–223.

Debnath, J., Sahariah, D., Lahon, D., Nath, N., Chand, K., Meraj, G., Kumar, P., Singh, S. K., Kanga, S., & Farooq, M. (2023). Assessing the impacts of current and future changes of the planforms of River Brahmaputra on its land use-land cover. *Geosci. Front.*, https://doi.org/10.1016/j.gsf.2023.101557.

Debnath, J., Sahariah, D., Saikia, A., Meraj, G., Nath, N., Lahon, D., Annayat, W., Kumar, P., Chand, K., Singh, S. K., & Kanga, S. (2023). Shifting Sands: Assessing Bankline Shift Using an Automated Approach in the Jia Bharali River, India. *Land*, 12(3), 703.

Deo, A., Ganer, D. (2014). Tropical Cyclone Activity over the Indian Ocean in the Warmer Climate, Mohanty, U.C. (Ed.), *Monitoring and Prediction of Tropical Cyclones in the Indian Ocean and Climate Change*. Springer, New Delhi, pp. 72-80.

EMDAT (2019): OFDA/CRED International Disaster Database, Université catholique de Louvain – Brussels – Belgium.

Fang, Y., Yin, J., Wu, B. (2016). Flooding risk assessment of coastal tourist attractions affected by sea level rise and storm surge: a case study in Zhejiang Province, China. *Nat. Hazard.*, 84 (1), 611e624.

Farooq, M., Meraj, G., Rishabh, Kanga, S., Nathawat, R., Singh, S. K., & Ranga, V. (2022a). Slum Categorization for Efficient Development Plan—A Case Study of Udhampur City, Jammu and Kashmir Using Remote Sensing and GIS. In *Geospatial Technology for Landscape and Environmental Management: Sustainable Assessment and Planning*, (pp. 283-299). Cham: Springer International Publishing.

Farooq, M., Mushtaq, F., Meraj, G., Singh, S. K., Kanga, S., Gupta, A., Kumar, P., Singh, D., & Avtar, R. (2022b). Strategic Slum Upgrading and Redevelopment Action Plan for Jammu City. *Resources*, 11(12), 120.

Gallina, V., Torresan, S., Critto, A., Sperotto, A., Glade, T., Marcomini, A. (2016). A review of multi-risk methodologies for natural hazards: consequences and challenges for a climate change impact assessment. *J. Environ. Manage.*, 168, pp. 123-132.

Goswami S *et al.* (2016). A review on application of data mining techniques to combat natural disasters, *Ain Shams Eng. J.*

Gulati, B., Sharma, R., Kanga, S., Singh, S. K., Sajan, B., Meraj, G., Kumar, P., & Ramanathan, A. L. (2023). Unraveling the Relationship Between Stubble Burning and Air Quality Degradation in Punjab: A Temporal and Spatial Analysis (2019-2022). *J. Clim. Change*, 9(2), 43-53.

Han P, Wang PX, Zhang SY, Zhu DH (2010). Drought forecasting based on the remote sensing data using ARIMA models. *Math. Comput. Model.*, 51:1398–1403.

Joyce, K.E., Belliss, S.E., Samsonov, S.V., McNeill, S.J., Glassey, P.J. (2009). A review of the status of satellite remote sensing and image processing techniques for mapping natural hazards and disasters. *Prog. Phys. Geogr.*, 33 (2), pp. 183-207.

Kanga, S., Meraj, G., Farooq, M., Singh, S. K., & Nathawat, M. S. (Eds.). (2022a). *Disaster Management in the Complex Himalayan Terrains*. Geography of the Physical Environment. Springer, Cham.

Kanga, S., Singh, S. K., Meraj, G., Kumar, A., Parveen, R., Kranjčić, N., & Đurin, B. (2022b). Assessment of the Impact of Urbanization on Geoenvironmental Settings Using Geospatial Techniques: A Study of Panchkula District, Haryana. *Geographies*, 2(1), 1-10. ISSN No. 2673-7086.

Karen E. Joyce *et al.*, (2009). A review of the status of satellite remote sensing and image processing techniques for mapping natural hazards and disasters, *Prog. Phys. Geogr.* 33(2), pp. 183–207.

Klemas, V.V. (2009). The role of remote sensing in predicting and determining coastal storm impacts. *J. Coast. Res.*, pp. 1264-1275.

Kobiyama, M.; *et al.* (2006). Prevenção de Desastres Naturais, 1st ed.; Organic Trading: Florianópolis, Brazil, pp. 20–25

Kotecha, M. J., Kanga, S., Pankhaniya, S. K., Agrawal, S., Meraj, G., & Singh, S. K. (2023). Urban Heat Island (UHI) Resilience Plan in Varying Climatic Conditions Using Geospatial Approach: A Case Study of Rajkot City. In *Climate Change, Agriculture and Society: Approaches Toward Sustainability* (pp. 307-338). Cham: Springer International Publishing.

Krishnamohan, K., Mohanakumar, K., Joseph, P. (2014). Climate change in tropical cyclones and monsoon depressions of North Indian Ocean. In: Mohanty, U., Mohapatra, M., Singh, O., Bandyopadhyay, B., Rathore, L. (Eds.), *Monitoring and Prediction of Tropical Cyclones in the Indian Ocean and Climate Change*. Springer, pp. 33-39.

Krishnamoorthi, N. (2016). Role of Remote Sensing and GIS in NaturalDisaster Management Cycle, *Imp. J. Interdiscip. Res. (IJIR)* Vol-2, Issue-3, pp. 144-146.

Lahon, D., Sahariah, D., Debnath, J., Nath, N., Meraj, G., Kumar, P., Hashimoto, S. and Farooq, M. (2023). Assessment of Ecosystem Service Value in Response to LULC Changes Using Geospatial Techniques: A Case Study in the Merbil Wetland of the Brahmaputra Valley, Assam, India. *ISPRS Int. J. Geo-Inf.*, 12(4), p.165.

Lavane, K., Kumar, P., Meraj, G., Han, T. G., Ngan, L. H. B., Lien, B. T. B., Ty, T. V., Thanh, N. T., Downes, N. K., Nam, N. D. G., Minh, H. V. T., Singh, S. K., & Kanga, S. (2023). Assessing the Effects of Drought on Rice Yields in the Mekong Delta. *Climate*, 11(1), 13.

Li, K., Li, G.S. (2013). Risk assessment on storm surges in the coastal area of Guangdong Province. *Nat. Hazard.*, 68, pp. 1129-1139.

Luiz A. Manfré *et al.* (2012). An Analysis of Geospatial Technologies for Risk and Natural Disaster Management, *ISPRS Int. J. Geo-Inf.*, 1, pp. 166-185.

Martino, L., Ulivieri, C., Jahjah, M., Loret, E. (2009). Remote sensing and GIS techniques for natural disaster monitoring. In: Olla, P. (Ed.), *Space Technologies for the Benefit of Human Society and Earth*. Springer, pp. 331-382.

Matsuoka, M. and Yamazaki, F. (2005). Building damage mapping of the 2003 Bam Iran Earthquake using Envisat/ASAR intensity imagery. *Earthquake Spectra*, 21, S285, DOI: 10.1193/1191.2101027.

Melesse, A., & Wang, X. (2007). Impervious Surface Area Dynamics and Storm Runoff Response. *Remote Sensing of Impervious Surfaces*; CRC Press/Taylor & Francis, 19, pp. 369-384.

Mendelsohn, R., Emanuel, K., Chonabayashi, S., Bakkensen, L. (2012). The impact of climate change on global tropical cyclone damage. *Nat. Clim. Change*, 2 (3), pp. 205-209.

Meraj, G., Farooq, M., Singh, S. K., Islam, M. N., & Kanga, S. (2022). Modeling the sediment retention and ecosystem provisioning services in the Kashmir valley, India, Western Himalayas. *Model. Earth Syst. Environ.*, 8(3), 3859-3884.

Meraj, G., Kanga, S., Kranjčić, N., Đurin, B., & Singh, S. K. (2021a). Role of natural capital economics for sustainable management of earth resources. *Earth*, 2(3), 622-634.

Meraj, G., Singh, S. K., Kanga, S., Islam, M. N., & Singh, S. K. (2021b). Modeling on comparison of ecosystem services concepts, tools, methods and their ecological-economic implications: A review. Modeling Earth Systems and Environment *Model. Earth Syst. Environ.*, 1-20.

Nadi, B. *et al.* (2010). Application Geospatial Technology in Disaster Management, *J. Environ. Sci. Eng.*, Volume 4, No.1 (Serial No.26).

Poulos, H., (2010). Spatially explicit mapping of hurricane risk in New England, USA using ArcGIS. *Nat. Hazard.*, 54 (3), pp. 1015-1023.

Rafiq, M., Meraj, G., Kesarkar, A. P., Farooq, M., Singh, S. K., Kanga, S., & Bhardwaj, V. (2022). Hazard Mitigation and Climate Change in the Himalayas–Policy and Decision Making. In S. Kanga, G. Meraj, M. Farooq, S. K. Singh, & M. S. Nathawat (Eds.), *Disaster Management in the Complex Himalayan Terrains*. Geography of the Physical Environment. Springer, Cham. https://doi.org/10.1007/978-3-030-89308-8_12.

Sajan, B., Mishra, V. N., Kanga, S., Meraj, G., Singh, S. K., & Kumar, P. (2022). Cellular automata-based artificial neural network model for spatial-temporal dynamics and land use/land cover change prediction. *Agronomy*, 12, 2772.

Saptarsi Goswami *et al.* (2016). A review on application of data mining techniques to combat natural disasters, *Ain Shams Eng. J.*

Sekac, T. *et al.* (2020). Application of Geospatial Technology in Earthquake Risk Assessment in Papua New Guinea. ed. Pal, I. *et al.*, *An Interdisciplinary Approach for Disaster Resilience and Sustainability*., pp. 185-195.

Singh, A., Singh, S. K., Meraj, G., Kanga, S., Farooq, M., Kranjcic, N., Đurin, B., & Sudhanshu. (2022b). Designing Geographic Information System Based Property Tax Assessment in India. *Smart Cities*, 5, 364–381.

Singh, H., Meraj, G., Singh, S., Shrivastava, V., Sharma, V., Farooq, M., Kanga, S., & Kumar, P. (2022a). Status of Air Pollution during COVID-19-Induced Lockdown in Delhi, India. *Atmosphere*, 13(12), 2090.

Singh, S. K., Gupta, S. K., Meraj, G., Kumar, P., & Kanga, S. (2023b). Uncovering the hydro-meteorological drivers responsible for forest fires utilizing geospatial techniques. *Theor. Appl. Climatol.*, https://doi.org/10.1007/s00704-023-04497-y

Singh, S. K., Meraj, G., Mondal, N., Bera, A., Verma, M. K., Tomar, J. S., & Kanga, S. (2022). Assessment of seasonal vegetation dynamics over parts of thar desert using geospatial techniques. *ANGRAU*, 105.

Singh, S., Singh, S. K., Prajapat, D. K., Pandey, V., Kanga, S., Kumar, P., & Meraj, G. (2023a). Assessing the Impact of the 2004 Indian Ocean Tsunami on South Andaman's Coastal Shoreline: A Geospatial Analysis of Erosion and Accretion Patterns. *J. Mar. Sci. Eng.*, 11(6), 1134.

Sorensen, M.B.; Spada, M.; Babeyko, A.; Wiemer, S.; Grünthal, G. (2012). Probabilistic tsunami hazard in the Meditarranean Sea. *J. Geophys. Res.*, 117, B01305.

Sud, A., Kanga, R., Singh, S. K., Meraj, G., Kanga, S., Kumar, P., Ramanathan, A. L., & Bhardwaj, V. (2023). Simulating Groundwater Potential Zones in Mountainous Indian Himalayas—A Case Study of Himachal Pradesh. *J. Geophys. Res.: Earth Surf.*, 10(3), 65.

Sun, C.G., Chun, S.H., Ha, T.G., Chung, C.K. and Kim, D.S. (2008). Development and application of a GIS-based tool for earthquake-induced hazard prediction. *Comput. Geotech.*, 35, pp. 436–449.

Thomas, D.S.K.; Eturĝay, K. & Kemeç, S. 2007). The role of Geographic Information System/Remote Sensing in Disaster Management. In *Handbook of Disaster Research*, 1st ed.; Rodríguez, H., Quarantelli, E.L., Dynes, R., Eds.; Springer: Newark, NJ, USA, 2007; pp. 83–96.

Upadhyay, R. K., Tripathi, G., Đurin, B., Šamanović, S., Cetl, V., Kishore, N., Sharma, M., Singh, S. K., Kanga, S., Wasim, M., Rai, P. K., & Bhardwaj, V. (2023). Groundwater Potential Zone Mapping in the Ghaggar River Basin, North-West India, Using Integrated Remote Sensing and GIS Techniques. *Water*, 15(5), 961.

Van Westen, C.J. (2013). Remote sensing and GIS for natural hazards assessment and disaster risk management. In: Schroder, J.F., Bishop, M.P. (Eds.), *Treatise on Geomorphology*. Academic Press, Elsevier, San Diego, pp. 259-298.

Vijayaraghavan, C, Thirumalaivasan D. & Venkatesan, R. (2012). Utilization of Remote sensing and GIS in Managing Disasters – A Review, *Int. J. Sci. Eng. Res.*, Volume 3, Issue 1, pp. 1-6.

Wang, F., Xu, Y.J. (2010). Comparison of remote sensing change detection techniques for assessing hurricane damage to forests. *Environ. Monit. Assess.*, 162 (1-4), 311e326.

Westen, C.V. (2000). International Archives of Photogrammetry and Remote Sensing. Vol. XXXIII, Part B7, pp. 1609-1617.

Wipulanusat W, Nakrod S, Prabnarong P (2009.) Multi-hazard risk assessment using GIS and RS applications: a case study of Pak Phanang Basin. *Walailak J. Sci. Tech.*, 6(1):109–125.

17

Population Change and Its Impact on the Shortfall of Groundwater: A Case Study of Berhampore in Murshidabad, West Bengal

Subodh Singha* **and Harsha Kumar Das Gupta**

Department of Geography, C.M.J. University Jorabat, Meghalaya, India

Abstract

Considering the landscape architecture, the British rulers selected the location along the levee of the Bhagirathi River as an army cantonment. The origin of Berhampore was established as an army cantonment. Prior to the establishment of this town, Cossimbazar, Monindranagar, Hatinagar, Babulbons, and Soidabad were the locations of traders, mainly British, other clans were Portuguese, Danish, Vatia, Marwari, and Hindu traders. To support the British rulers, mainly Hindu Jamiders took a vital role in trade and commerce with the foreigners. For the service of the aforesaid, the social group's concentration of the population gradually increased. After the establishment of railway links, influx of population rapidly increased. After the partition of Bengal, the concentration of refugees increased the population.

Thus, the requirement for quality water demand increased both for domestic and agricultural works, small cottage industry, and various commercial and other sectors rapidly increased, influx of refugees rapidly increased, as well as the rural population concentration. The lifting of groundwater for various purposes gradually increased the use of water, and naturally, the demand for water increased.

Keywords**:** Shortfall, groundwater, FAR, social upliftment, satellite imagery

17.1 Introduction

During the British rule, Berhampore was known as a cantonment. Cossimbazar, Bishnupur, Khagra, Babulbona, and Hatinagar were the big business centers. Due to an administrative center, the public concentration, growth of urbanization, trade, and commerce, the growth of economic activity rapidly increased at Berhampore as a service center. During the British rule, the service sector of Berhampore gradually increased, pond water and dug wells were the common source of drinking water for all levels of people (Debnath *et al.*, 2023). As per the geological review, the groundwater level was a good source for both domestic and other use (Debnath *et al.*, 2022). The growth of the population after 1937 rapidly increased. For agricultural use and industrial use, in all sectors of human lifestyle, demand for water increased. From the middle of 1988, the population concentration increased the demand.

**Corresponding author*: subodh12@gmail.com

Sanjay Kumar, Suraj Kumar Singh, Shruti Kanga, Gowhar Meraj, Majid Farooq and M. S. Nathawat (eds.)
Disaster Management and Environmental Sustainability, (207–218) © 2024 Scrivener Publishing LLC

Industry, population growth in the rural sector, demand for rural water supply both in villages and agriculture fields, and reckless use of groundwater, made the situation gradually grave. In the present situation, nearby hills in and around Berhampore, are suffering from shortfall of water (Debnath *et al.,* 2023). Moreover, solid, liquid, and other wastes reduced the percolation rate of groundwater. Land use map of Berhampore Municipality has been shown in Figure 17.1.

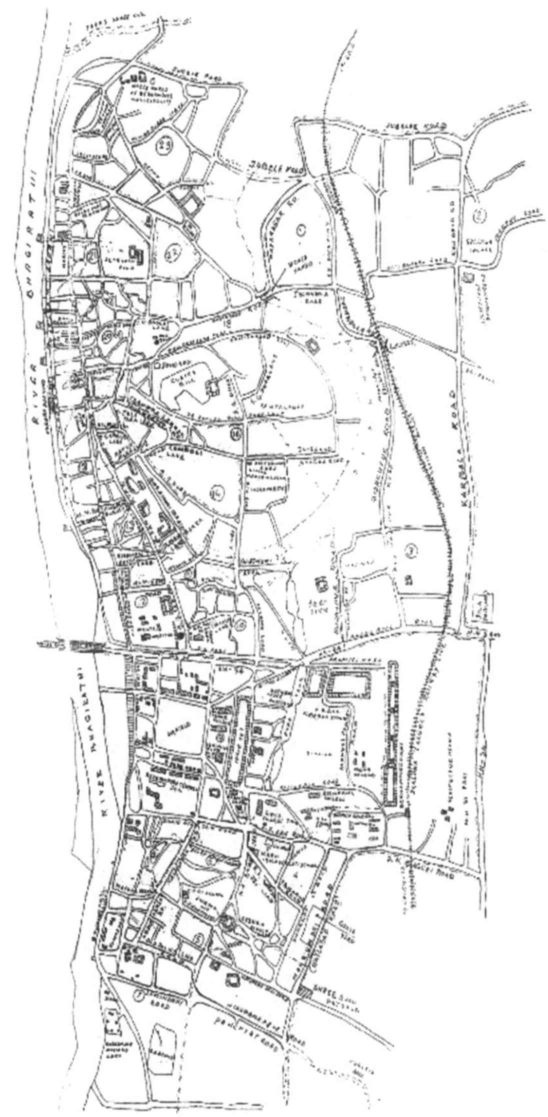

Figure 17.1 Land use map of Berhampore Municipality. *(Continued)*

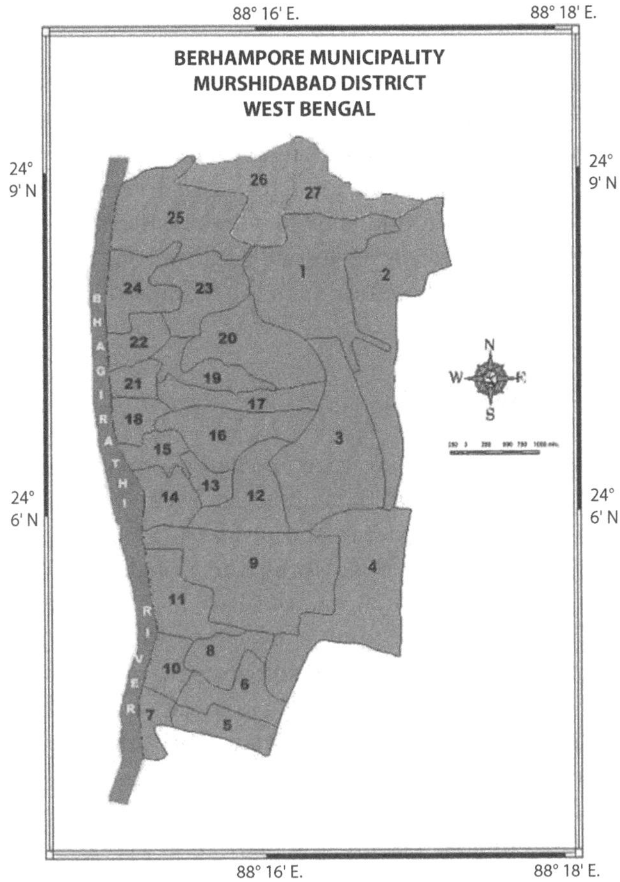

Figure 17.1 (Continued) Land use map of Berhampore Municipality.

17.2 Berhampore and Groundwater Condition

The upper surface of the groundwater is called the water table. The water table conforms roughly to the topography of the land, but its depth depends on a number of factors. The major factor is infiltration. It depends on the following: 1) intensity of rainfall, 2) duration of rainfall, 3) temperature, 4) soil characteristics, 5) vegetation cover, 6) urban area, 7) depth of surface, 8) detention and thickness of saturated area, 9) entrapped air in soil pores, 10) turbidity of the water, 11) compaction of soil, 12) geology, 13) slope of rock structure, 14) water holding capacity, etc. (Gulati *et al.*, 2023).

In the geological setup, hydrology reacts. The water table conspicuously fluctuates from season to season and in the situation. The drought-heavy human consumption leads to withdrawal of groundwater, and thus, the water table sinks to deeper levels. The rate of fluxion depends on the rate of groundwater movement, varying from a few centimeters to tens of meters per day (Farooq *et al.*, 2022; Meraj *et al.*, 2021). The movement of groundwater is governed by the permeability of the medium and its ability to allow the passage of water without impairment of the structure of the soil or rock fabric of the medium. The nature of land use in Berhampore is not methodical, the changing of slope created various problems in the runoff and flow direction; water stagnation is now a very common occurrence at Berhampore.

To consider the location of Berhampore prior to British rule in the early records of the district, prepared by "Walls" that the location of Berhampore was at the west of Bhagirathi attached to Gokarno Mouzas. Many times, Bhagirathi changed its entry in Murshidabad. The three major shifts were Chukka, Suti, and Chaurasia. To consider the geomorphology of Berhampore town, the base was founded by Mayurakhy, Bremhani, and The South West built by Ajoy. 18,000 years prior to the Holocene age, the zone gradually developed. The impact of the Pleistocene age took a major role in the foundation, 20,000 years ago.

17.3 Origin of Berhampore

In 1914, O'Malley published a compiled, gazette where he mentioned the name of Sayed Cossim, an unknown fakir, he is known in Jangipur as Sayed Cossim. Another Cossim was known as a Pir, a resident of Cossimbazar, and probably, his full name was Sayed Khaja Cossim Fakir Darbesi Ala. The name of Cossimbazar probably came out of the name of those persons (Farooq *et al.*, 2022). In 1556, the East India Company built a company Quarter, known as Khunti where the administrator was John Ken. Job Charnak at that time was the fourth employee of East India Company. Charnak was transferred to Patna and again he returned in 1680 to Cossimbazar. In 1681, the East India Company invested 140,000 pounds as business capital. When Murshidabad became the capital, the British resident took the responsibility of administration and their residence was known as Residency. During the time of the Charnak regime, the trade at Cossimbazar increased rapidly. And after the Plessey War, Cossimbazar became the main trade center, mainly for silk, spices, etc.

Not only the European traders, Marwari, Gujarati, and Bihari traders also came to Cossimbazar for their business. Their residential areas were known as Mahajan Tuli/ tola. In 1813, Bhagirathi suddenly changed her course, and the flow along Katiganga stopped. Due to the epidemic, three-fourths of the total population lost their lives and the best trade entry of Bengal was destroyed. The whole of Cossimbazar became a dilapidated town. The waste raw materials of Cossimbazar were then used as building materials for the construction of Berhampore. On the basis of the two Cossimbazar Rajbaries namely, Sripur and Baspur, the trade rejuvenated, but not up to the old status. Apart from Cossimbazar, adjacent areas like Bishnupur, Madhupur, Kantanagar, and Khagra became small trade and residential centers.

17.4 Social Upliftment of Berhampore

Due to the British rule in Bengal, for their service, they created a new middle-class Bengali people. The role of the business strategies of the East India Company gradually increased. They started controlling the economic activity and administrative modification, gradually increasing in the new middle-class people in Bengal. These populations originated mainly in Dhaka and Kolkata, but the development of Berhampore town in Murshidabad was the mark of multi-dimensional progress of the middle-class people where the concentration of business trend flourished. From 1917, due to the famine, the trade, policies, culture, and administration judiciary almost closed, or almost rapidly abolished. The British rulers shifted their administration to Calcutta and then to Delhi.

Gradually, the importance of Berhampore became the accommodation place of scattered Nawab family members. In a short spell, wholesale trading started improving for silk, and muslin trades. Bell metal and bullion gold traders as well but those trades gradually came under the control of British rulers. The settlement pattern of Berhampore shifted many times due to various reasons. From the end of the 19th century, the cantonment was removed from Berhampore and gradually, the market economy of Gorabazar was reduced, and old shops serving the soldiers and British people were closed. During the early days, Gorabazar was mainly the residential area of north Indian Muslims and Hindu shop owners. From 1862 to 1872, the Azimganj-Nalhati railway line was established and the land price of Berhampore started increasing rapidly. The affluent society of both Hindus and Muslims purchased land at Berhampore at a high price. In the settlement report of 1897–98, it was recorded that the house rent was high at Khagra and Berhampore. In comparison, the rent was low at Gorabazar.

From 1906 to 1907, the Lalgola–Sealdah railway line was established and Berhampore became the distribution center of goods. Naturally, the importance of Berhampore rapidly increased and the population also increased for the service sectors. During the Second World War, the population concentration in Berhampore suddenly increased as it became a safe shelter for the people.

Berhampore is the source of income for various sectors of trade and commerce, with an influx of refugees, and after the partition, the business on gold and bell metal increased all over the district and a new horizon of economic activity flourished (Lavane *et al.*, 2023).

For service, trade, commerce, and other economic sectors, a new class of citizens gradually increased in Berhampore. Those who were engaged in unorganized labor, mainly bus drivers, truck drivers, porters, rickshaw poolers, sweepers, servants, etc. gradually started concentrating in Berhampore.

Due to the establishment of British rule, the agricultural-based rural economy reduced their functional activity. The establishment of industry gradually forced to concentrate rural blacksmiths, carpenters, waivers, and earthen pot makers for their earnings and started concentrating at Berhampore. For the clearing of sewerage and night soil and for the supply of drinking water, they established the cantonment and civil line. At the beginning of the British rule, they shattered the local administration of urban and rural places. But in 1932, the British rulers established the Bengal Municipality Act. On April 1, 1869, an accumulation of local residents demarcated the municipal area and established a cantonment and town committee. The British rulers shifted the subsidiary treasury circuit court Sadar Dewani and Nijamat adulates to Calcutta but for administrative reasons, they again started functioning at Berhampore. On July 1, 1876, they established Berhampore Municipality. The total members were 19, out of which, five were nominated in 1876, and there were 6 wards in Berhampore, Gorabazar 3, cantonment 1, Berhampore 4, Khagra 3, Saidabad 2, and Cossimbazar 1. Population density, Berhampore Municipality, 2014 has been shown in Figure 17.2.

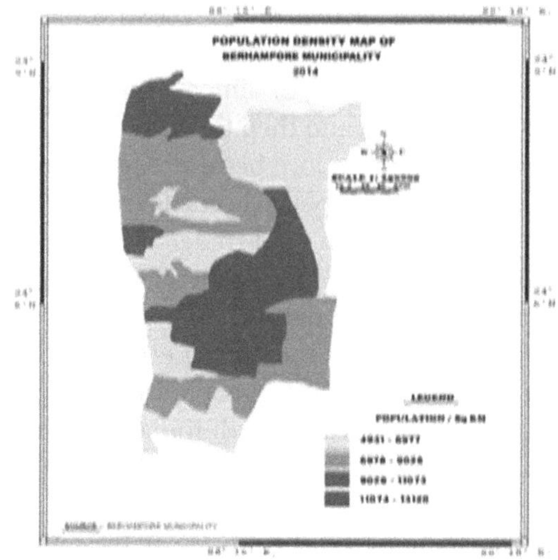

Figure 17.2 Population density, Berhampore Municipality, 2014.

Due to the rapid growth of urbanization, the requirement for water rapidly increased, deep tub wells became the only immediate solution for urban water supply. But the growth of population in the villages of Murshidabad increased the demand for water.

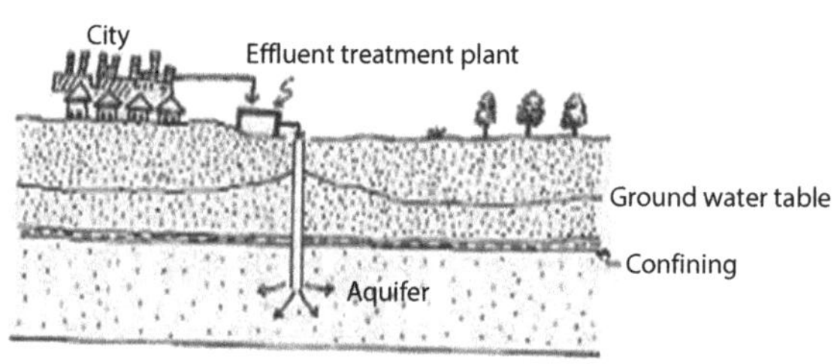

Figure 17.3 Shows the effluent treatment plant using the ground water.

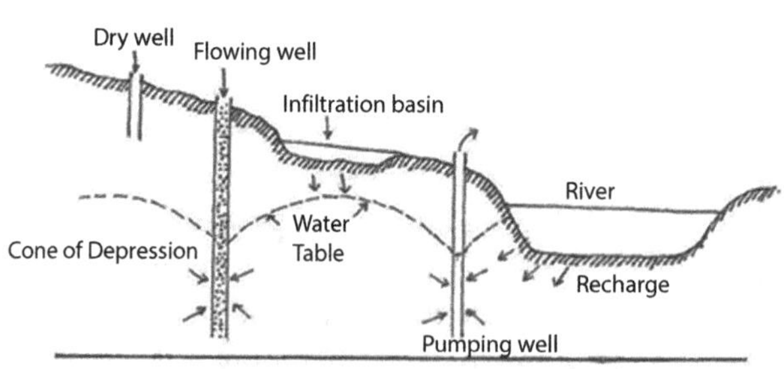

Source: Valdiya, 2006

Figure 17.4 Shows the underground structure of ground water pumping.

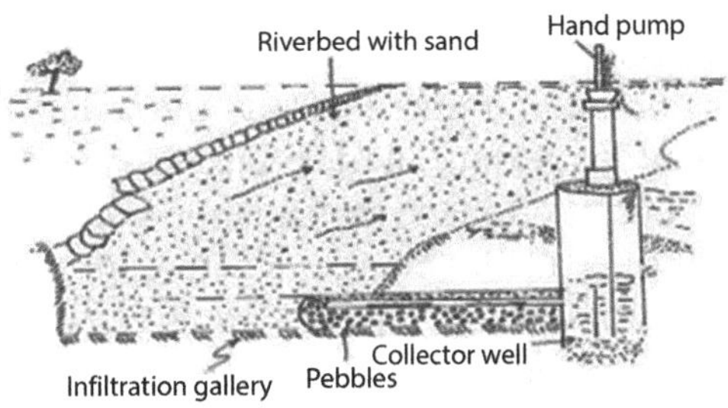

Figure 17.5 Shows the ground water well near the river bed.

Water for domestic, industrial, agricultural, and for other uses increased, like building construction, road, cleaning of various automobile servicing, administrative installations, and many other purposes. For the aforementioned reasons, the installation of deep tube wells rapidly increased due to social and political issues. Due to overuse, many problems have cropped up. The cones are formed over succession, thus, the wells became out of order in 2–3 y (Figures 17.3–17.5).

It has been recorded that in many situations, without having the opportunity of a river lift, deep tube wells were installed, and the use of shallow tube wells for all-purpose use rapidly increased without any planning. A lot of people were not aware of the misuse of water. All these happened already and is still happening today. In many situations, ignoring the future consequences, by the name of the NREGA project, people collect soil from the Vairab River bed from places, without any planning. Moreover, these are happening without any permission from the water resource department. It is happening for vested political interest. The local people collect that soil for filling the ponds and changing the slope for building construction. For those aforesaid reasons, the percolation runoffs are badly influenced. For those reasons, people do not have groundwater throughout the year. To meet the shortfall of water, big deep tube wells were installed (Figure 17.6).

In rural areas of Murshidabad and all over West Bengal, for political interest and votes in elections, many projects of both the central and state governments were introduced in

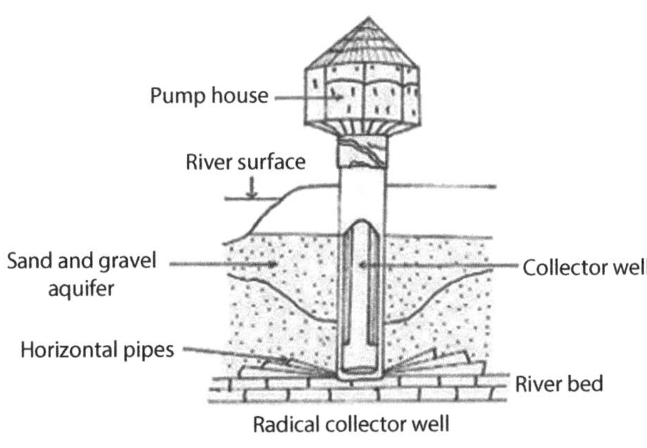

Figure 17.6 Shows the structure of the radical collector well.

the field, like agriculture, village road, and rural water supply. The administration ordered to complete those overnight. Moreover, for immediate orders by the State and Central Government, the administration had no time to assess the water resource availability in the state and/or in districts. For this reason, all the projects failed. Moreover, expansion of urban places and growth of population made the situation graver. For those reasons, the infiltration of water and also the percolation rate from natural resources were reduced. The cause is mainly urban wastes, both solid and liquid (Figures 17.7–17.8).

Water supply from feeder streams reduced for unplanned rural settlement, use of plastics, polythene, chemical toxic wastes, construction of roads in towns and villages, intercepted the normal runoff of rainwater all over Murshidabad. Construction of new railway lines, highways, and bridges disturbed the natural run off. Planning of recharge well is a dream now. Moreover, in urban places, FAR (floor area regulation) is not maintained. In some cases, it is deliberately violated by administration, with unhealthy relations with builders, promoters, political leaders, mafias, police, and all sectors. For those aforesaid reasons, the shortfall of ground water is now very common in Murshidabad.

To maintain the normal flow of Vandardaha Hill, during the British period, "the department of public works" constructed many sluice gates to feed Vandardaha Hill throughout the year, linking with Bhagirathi (Singh *et al.*, 2022). Due to strict administration of the government, those sluice gates were operated in a systematic manner. But because of some vested interests, those sluice gates were closed. This happened in between the year 1968 and 1972, with an unhealthy understanding among political leaders, police, administration, and promoters, all the sluice gates were gradually closed.

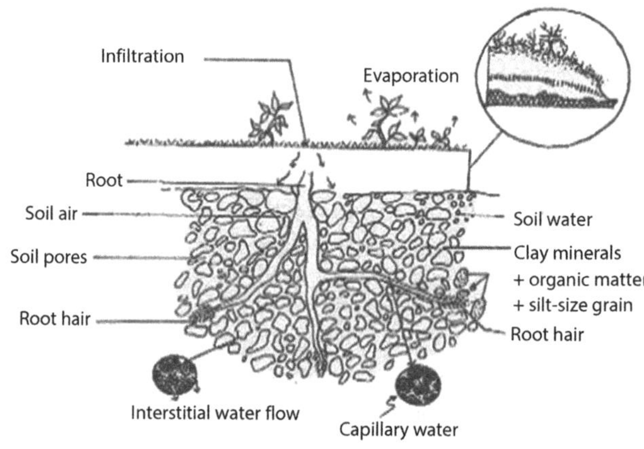

Figure 17.7 Shows the interstitial spaces with the soil.

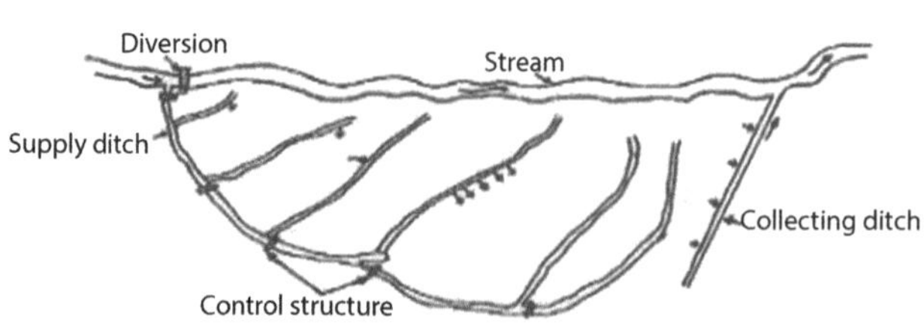

Figure 17.8 Shows the collecting ditches for the water to be used for supply.

The promoters started construction, violating all norms and ethics. Gradually, the public started suffering from shortfall of groundwater. Due to all links from Bhagirathi becoming closed, the biodiversity and ecology of the total area were gradually disturbed. All the linking hills gradually became dry due to shortfall of flow along the Vairab River and its old channels (Figure 17.9). Thus, the planning of improving the groundwater storage is only possible if we rejuvenate the flow of the Vairab River at Akhriganj (Singh *et al.*, 2023). Also, opening of the sluice gates is a most urgent matter to maintain the groundwater table of Berhampore town. Those old channels also will be able to maintain the health of Berhampore town.

Due to unsatisfactory planning, shortfall of groundwater is now a common scenario all over the district of Murshidabad. Watershed planning should be taken up first to improve the water resource management. Many small projects are taken up by the State Government for the conservation of rainwater, but public awareness is the most important factor for their own interest, and Murshidabad is far behind, even Kesiari block and West Midnapur. They are more active on groundwater conservation, and use ethics. They are trying to maintain water table equilibrium throughout the year (Figure 17.10).

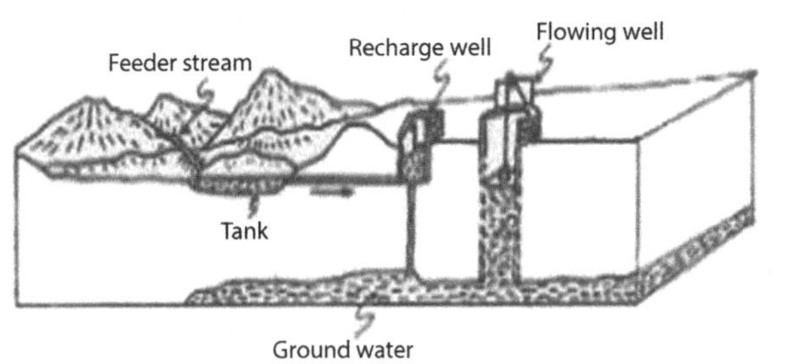

Figure 17.9 Shows the structure of the recharge well.

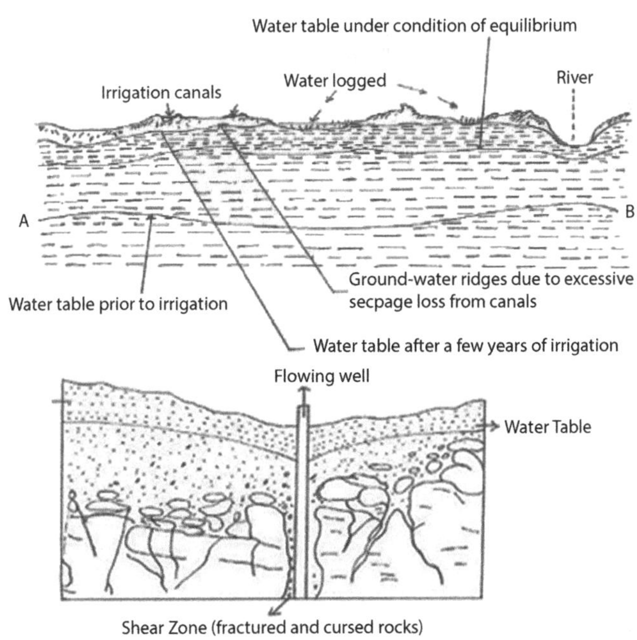

Figure 17.10 Shows the water table under conditions of equilibrium.

In many pleases of Murshidabad, River lift planning may be introduced to meet the scarcity of agricultural water supply. As the matter is not much lucrative for the vested groups, for the reason the matter is neglected by administration. In Rajasthan with the help of "Pane Panchayat", they are successful. Lifting of groundwater by deep tube well is injurious habit to meet the immediate shortfall. But recharging is difficult in 3rd world country.

17.5 Analysis

In every town, withdrawal of groundwater causes the depilation of groundwater reservoirs. In normal rules, the water table will be lowered. Pumping out of groundwater creates a cone of depression around the well. It happens when the rate of inflow is less than the quantity of withdrawal. The larger the number of well and the rate of withdrawal, the greater would be the lowering of the water table (Figure 17.11). This practice is called overdraft. At present, in Berhampore town, the shortfall of groundwater is a very common occurrence at almost every ward. The findings are as follows:

Ward Number	**Have water throughout the year**	**Does not have water from December to June**
	22, 21, 4, 1, lo 21, 20, 19. 17. 16 23. 18 7, 2, 4, 26	21,20,19,17,16,23,18,15,14,8, 6,5,3,1

The nature of water lifting and supply at Berhampore:

Dug well	**Municipal supply**		**Shallow tube well**	**Deep tube well (private)**
3%	13%		56%	28%
	6% River lift	8% Deep tube well		

Western part of West Bengal Alluvial layer.

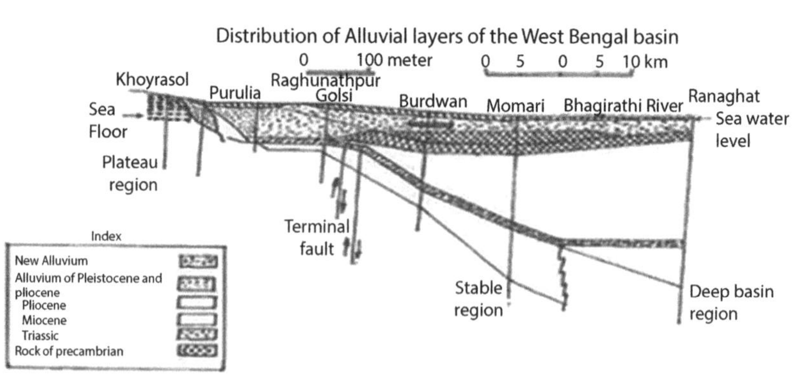

Source: Oil and Natural Gas Commission

Figure 17.11 Shows the distribution of alluvial layers of the West Bengal Basin.

17.6 Conclusion

In view of the groundwater problem at Berhampore, the growth of population concentration at the district head quarter is the basic one. Urban facilities are not adequate at subdivisions and in big villages. Therefore, people are concentrating at Berhampore for health, education, legal help, market, recreations, safe accommodation, business, banks, insurance, and for many more facilities and services.

Thus, the growth of urban population is quite obvious. The demand for water supply has increased accordingly. In Berhampore, the administration failed to maintain the FAR rules, thus, the gaps among the buildings are not adequate. In urban planning, some liberties are given to the state government. But liberty turns to rustic administrations, without knowing, recklessly, municipal rules are framed on the outlook of tax collection and other revenues. Thus, the management failed to understand that their revenue earnings from water supply will be gone (Figure 17.12).

Figure 17.12 Showing the Vandardaha Hill (Berhampore) simulated image and three inch rainfall watershed 2017.

References

Debnath, J., Sahariah, D., Lahon, D., Nath, N., Chand, K., Meraj, G., Kumar, P., Singh, S. K., Kanga, S., & Singh, S. K. (2022a). Geospatial modeling to assess the past and future land use-land cover changes in the Brahmaputra Valley, NE India, for sustainable land resource management. *Environ. Sci. Pollut. Res.*, 1-24.

Debnath, J., Sahariah, D., Lahon, D., Nath, N., Chand, K., Meraj, G., Kumar, P., Singh, S. K., Kanga, S., & Farooq, M. (2023b). Assessing the impacts of current and future changes of the planforms of River Brahmaputra on its land use-land cover. *Geosci. Front.* https://doi.org/10.1016/j.gsf.2023.101557.

Debnath, J., Sahariah, D., Saikia, A., Meraj, G., Nath, N., Lahon, D., Annayat, W., Kumar, P., Chand, K., Singh, S. K., & Kanga, S. (2023c). Shifting Sands: Assessing Bankline Shift Using an Automated Approach in the Jia Bharali River, India. *Land*, 12(3), 703.

Farooq, M., Meraj, G., Rishabh, Kanga, S., Nathawat, R., Singh, S. K., & Ranga, V. (2022a). Slum Categorization for Efficient Development Plan—A Case Study of Udhampur City, Jammu and Kashmir Using Remote Sensing and GIS. In *Geospatial Technology for Landscape and Environmental Management: Sustainable Assessment and Planning* (pp. 283-299). Cham: Springer International Publishing.

Farooq, M., Mushtaq, F., Meraj, G., Singh, S. K., Kanga, S., Gupta, A., Kumar, P., Singh, D., & Avtar, R. (2022b). Strategic Slum Upgrading and Redevelopment Action Plan for Jammu City. *Resources*, 11(12), 120.

Gulati, B., Sharma, R., Kanga, S., Singh, S. K., Sajan, B., Meraj, G., Kumar, P., & Ramanathan, A. L. (2023). Unraveling the Relationship Between Stubble Burning and Air Quality Degradation in Punjab: A Temporal and Spatial Analysis (2019-2022). *J. Clim. Change*, 9(2), 43-53.

Lavane, K., Kumar, P., Meraj, G., Han, T. G., Ngan, L. H. B., Lien, B. T. B., Ty, T. V., Thanh, N. T., Downes, N. K., Nam, N. D. G., Minh, H. V. T., Singh, S. K., & Kanga, S. (2023). Assessing the Effects of Drought on Rice Yields in the Mekong Delta. *Climate*, 11(1), 13.

Meraj, G., Kanga, S., Kranjčić, N., Đurin, B., & Singh, S. K. (2021a). Role of natural capital economics for sustainable management of earth resources. *Earth*, 2(3), 622-634.

Singh, A., Singh, S. K., Meraj, G., Kanga, S., Farooq, M., Kranjcic, N., Đurin, B., & Sudhanshu. (2022b). Designing Geographic Information System Based Property Tax Assessment in India. *Smart Cities*, 5, 364–381.

Singh, H., Meraj, G., Singh, S., Shrivastava, V., Sharma, V., Farooq, M., Kanga, S., & Kumar, P. (2022a). Status of Air Pollution during COVID-19-Induced Lockdown in Delhi, India. *Atmosphere*, 13(12), 2090.

Singh, S., Singh, S. K., Prajapat, D. K., Pandey, V., Kanga, S., Kumar, P., & Meraj, G. (2023a). Assessing the Impact of the 2004 Indian Ocean Tsunami on South Andaman's Coastal Shoreline: A Geospatial Analysis of Erosion and Accretion Patterns. *J. Mar. Sci. Eng.*, 11(6), 1134.

18

Development of Future Rule Curves for KLRS Pulichintala Reservoir Operation Using SWAT and GA Models

Jerripotu Gopala Rao[1], Saurabh Singh[2], Bishnu Prasad Panda[4], Deepak Kumar Prajapat[1]*, Sujeet Kumar[3] and Azmeera Balu[4]

[1]*Department of Civil Engineering, Suresh Gyan Vihar University, Rajasthan, Jaipur, India*
[2]*Centre for Climate Change and Water Research, Suresh Gyan Vihar University, Rajasthan, Jaipur, India*
[3]*Department of Mechanical Engineering, Suresh Gyan Vihar University, Rajasthan, Jaipur, India*
[4]*Central Institute of Petrochemicals Engineering and Technology, Jaipur, India*

Abstract
This study applied the genetic algorithm (GA) technique to connect with the reservoir simulation model to search optimal operation reservoir rule curves. The KLRS Pulichintala project, located in Guntur District, Andhra Pradesh State (India), was an illustrative application including monthly streamflow data from 2014-2019 and future inflow from 2021-2099 generated by the Soil Water Assessment Tool (SWAT) hydrological model. The SWAT uses 79 years of future climate data of precipitation and temperature from the driving model of CCCma-CanESM2. The model results showed better performance for monthly timestep of calibration and validation coefficients of R2 values are 0.86, 0.99 and NSE values are 0.84, 0.86. Using the RCP4.5 dataset, 60–70% of demand will be satisfied, and only 50–60% of demand would be satisfied using the RCP8.5 climate projection data.

Keywords: SWAT simulation, KLRS Pulichintala reservoir, Krishna River, GA model

18.1 Introduction

Nowadays, water system issues have become more complex and are linked to climate change and land use (LULC) as a result of population growth and economic change, which are skyrocketing. Water resource management often involves both demand management of demand positions and water supply sites to resolve issues. Improving the function of the reservoir to enhance efficiency is another way of managing the supply site, which does not require the physical development of the reservoir (Singh *et al.*, 2022; Sud *et al.*, 2023

**Corresponding author*: deepak.parjapat@mygyanvihar.com

Sanjay Kumar, Suraj Kumar Singh, Shruti Kanga, Gowhar Meraj, Majid Farooq and M. S. Nathawat (eds.)
Disaster Management and Environmental Sustainability, (219–228) © 2024 Scrivener Publishing LLC

Upadhyay *et al.*, 2023; Farooq *et al.*, 2022a; Kanga *et al.*, 2022a; Gulati *et al.*, 2023). The Intergovernmental Panel on Climate Change (IPCC) reported that the Earth's mean temperature increased by 0.6°C in the 20th century, and there is investigated change in temperature parameter from 1.4°C to 5.4°C, based on projections made by various climate prediction models (IPCC 2007). In the SWAT model, the LULC map and soil properties map play the leading role in updating the streamflow, including rainfall and temperature. The future LULC maps of 2025, 2055, and 2085 are derived from the cellular automata (CA), with the help of the Markov chain model of the Krishna River Basin (KRB), India (Chanapathi *et al.*, 2020; Singh *et al.*, 2022b).

Also, the assessment of the sustainability of the Krishna River in current and future climatic scenarios were investigated (Chanapathi *et al.*, 2020; Debnath *et al.*, 2022a). The applicability of the SWAT model in water flow prediction and the sediment burden in a semi-arid Mediterranean karst basin were checked. This is achieved through monthly and annual release and sediment loading data (Abbaspour *et al.*, 2013, 2015 and Shi *et al.*, 2011). The water utilization index (WUI) works under supply and demand (Neelima *et al.*, 2014). The most important crop cultivated in the ICA during the Kharif season was paddy (0.0234 MHA), 37.2% of which were under the right channel, and 62.8% were below the left channel by application of remote sensing data, and the CROPWAT model is also applied.

Taylor (2012) stated that representative concentration pathways (RCPs) scenarios estimate the radiative forcing (RF) that are based on the multi-model ensembles of Coupled Model Inter-Comparison Phase Five (CMIP5). Since general circulation models (GCM) have coarser resolutions, regional climate models (RCM) have been widely used due to their high spatial resolution for the depiction of future temperature and precipitation in a basin (Anjum *et al.*, 2019). Several studies reported that climate variability and its frequency is reflected by climate models such as GCMs and RCMs.

The study of GA originated in the mid-1970s (Singh *et al.*, 2023b). Optimal rule curves provide necessary guidelines for reservoir operation that have been used to assess the performance of any reservoir to satisfy downstream water supply, irrigation, industrial, hydropower, and environmental conservation requirements. Several recent researchers gave summaries of the essentials of the GA model (Singh *et al.*, 2023b) and it is necessary to use a science-based approach to plan and operate existing reservoir systems. The fitness function of the GA (Sajan *et al.*, 2022) minimizes the squared deviation (SD) of monthly irrigation demand and SD in the mass balance equation. The GA is based on a Darwinian natural selection process that combines the survival concept of the most robust test with that of unaffected genetic carriers. The function depends on the demand values and area-elevation-storage curves. It has to satisfy the irrigation demands downstream (Singh *et al.*, 2023a; Debnath *et al.*, 2023b).

The above literature and feedback show a better understanding and implementation for further studies of SWAT analysis and minimization of water supply with the GA model. The principal aim of this study is to determine simulations of water resources and features of hydrological processes at the KLRS Pulichintala basin (determine the optimal release of the dam for domestic, agricultural, and industrial purposes) (Singh *et al.*, 2023a).

18.2 Study Area and Data

The KLRS Pulichintala Project, situated near Pulichintala Village in Guntur District, serves as a vital balancing reservoir on the Krishna River. Positioned approximately 85 km upstream of the Prakasam barrage and 115 km downstream of the Nagarjuna Sagar Dam. It has a capacity to store 45.77 thousand million cubic meters of water. This reservoir plays a crucial role in stabilizing the existing ayacut. The reservoir covers an expansive area of 13.08 Lakh acres in the Krishna delta region and facilitates the early transplantation of paddy crops in June and July. The dam is situated at coordinates 16°45′15″N latitude and 80°03′24″E longitude. The location of the KLRS Pulichintala reservoir basin and the reservoir system details are shown in Figure 18.1.

For the Krishna River Basin, rainfall (0.25°x0.25°) and temperature (1°x1°) daily hydrometeorological data were used as meteorological data for the model, received from IMD, Pune, India. Additional data needed, such as solar radiation, wind speed, and relative humidity, were generated automatically by SWAT in the construction weather generator.

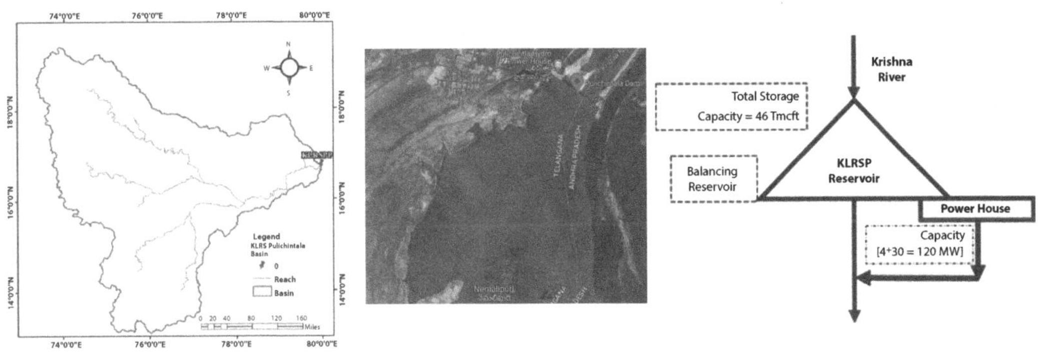

Figure 18.1 Location of the KLRS Pulichintala basin and system across the Krishna River.

Table 18.1 KLRS Pulichintala Reservoir details and purposes.

Project name	KLRS Pulichintala Project
Purpose	Irrigation, flood
Project type	Major
Districts benefited	Krishna, Guntur, West Godavari, Prakasham
Project sharing	None
Culturable command area (Th. Ha.)	520
Capacity (TMC)	45.77
Dead Storage (TMC)	3.61
FRL (Ft)	175
Catchment area (Lakh Ha.)	20.028

Table 18.2 Future climate projections data collection and time period.

Source	Other climate variable		Date period of each
ESGF/CORDEX Data, WAS44, IITM, CCCma-HadCM2	Future/RCP4.5	Precipitation	2021–2099
		Temperature	2021–2099
	Future/RCP8.5	Precipitation	2021–2099
		Temperature	2021–2099
Classification map	Land use land cover map		2019
Soil	FAO soil map data		-

KLRS Pulichintala Reservoir daily release data were collected from the Water Resources Department in Andhra Pradesh state. The reservoir details are shown in Table 18.1.

The future climate projections data and their time period are shown in Table 18.2. The hydrological model used future climate data from the driving model of CCCma-CanESM2 (Canadian Centre for Climate Modelling and Analysis - second generation of Earth System Model of Coordinated Regional Climate Downscaling Experiment (CORDEX) data from Earth System Grid Federation (ESGF), the downstream demand and hydrologic data.

18.3 Methodology

18.3.1 Soil Water Assessment Tool Model

$$SW_t = SW_o + \Sigma_{i-1}^{n}(R_{day} + Q_{surf} + E_a + W_{seep} + Q_{gw})$$

$$Q_{surf} = \frac{\left(R_{day} - 0.2S\right)^2}{\left(R_{day} - 0.2S\right)}, R > 0.2S$$

Where, Q_{surf} = the accumulated runoff on day i (mm H_2O), R_{day} = the amount of precipitation on day i (mm H_2O), S = the retention parameter on day i (mm H_2O), the retention parameter S = 25.4* ((1000/CN)-10), CN is the curve number for the day and A discharge will follow when $R_{day} > 0.2S$.

The statistical parameters used for calibration and validation, such as mean square error (MSE), mean absolute error (MAE), root mean square error (RMSE), correlation coefficient (R^2), Nash-Sutcliffe efficiency (NSE), percent bias (PBIAS), and RMSE-observations standard deviation ratio (RSR), compare the values of the generated synthetic streamflow data to the observed streamflow data.

18.3.2 Genetic Algorithm Model

In this study, the intended objectives comprise downstream supply, hydroelectric energy production, and flood control, especially in certain months of the years like March, April, May, June, July, and August, for 12 months. The function is defined by minimizing the sum of squared distances from each target amount of water in a year as described below.

$$\text{Minimize } Z = \left(\frac{R_t - D_t}{D_t}\right)^2 \ \forall\, t, t = 1,2\ldots$$

Where, R_t the variable is monthly discharge (quantity of water) from the reservoir within a given month, t. and D_t is the downstream water requirement for domestic and industrial use for that specific month, t. The above function can have a minimum value (zero) only when $R_t = D_t$. So, in this function our only variable R_t. Where, R_{min} is minimum monthly demand in 30 years and R_{max} is maximum monthly demand in over 30 years. The GA algorithm code will try to make Rt = Dt minimize the above objective function under the following constraints.

$$S_{(t+1)} = S_{(t)} + Q_{(t)} + P_{(t)} - E_{(t)} - R_{(t)}$$
$$R_{min} < R_{(t)} < R_{max}$$
$$S_{(t)} \geq S_{min} \leq S_{max} \text{ and } R_{(t)} \leq D_{(t)}$$

The constraints used for such optimization are the capacity of the reservoir or tank and the limits of the decision variables. Finally, the findings are discussed, followed by the conclusion, and the salient characteristics of the PJP reservoir system are given in Table 18.1.

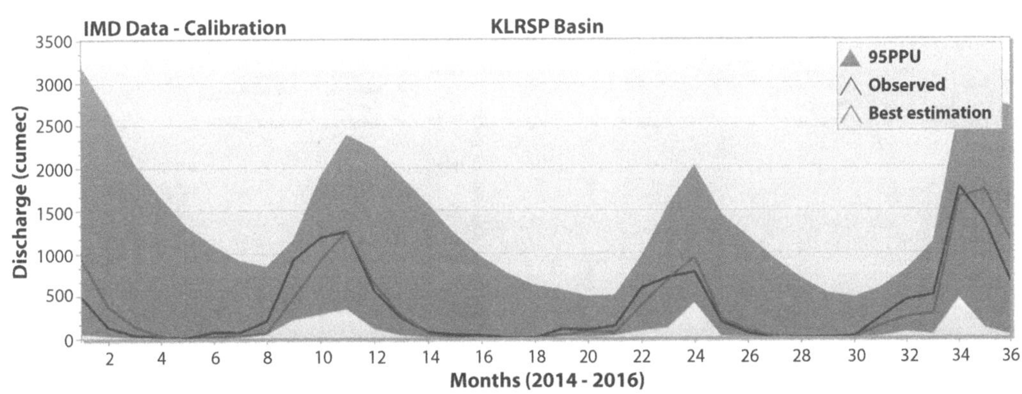

Figure 18.2 SWAT model calibration plot for basin.

18.4 Results and Discussion

18.4.1 KLRS Pulichintala Basin SWAT Results Discussion

Here, the calibration and validation coefficients of R^2 values are 0.86, 0.99, and NSE values are 0.84, 0.86, and finally, PBIAS values are 0.7 and 40.8 of IMD data set. The model results showed a better performance for monthly timestep.

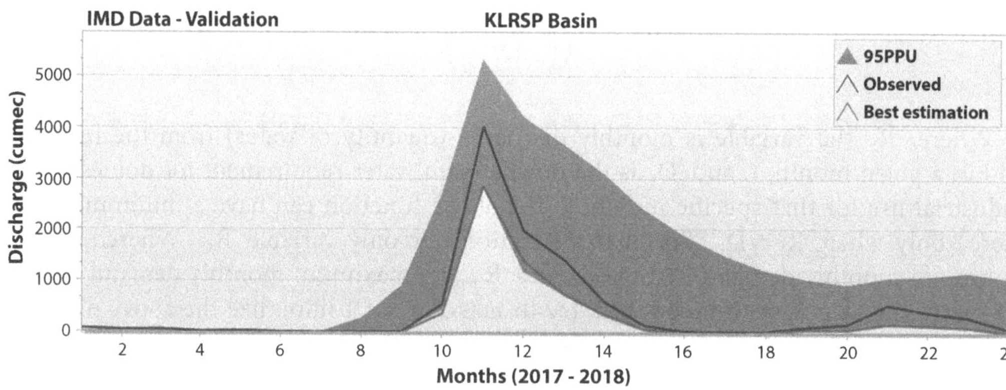

Figure 18.3 SWAT model validation plot for basin.

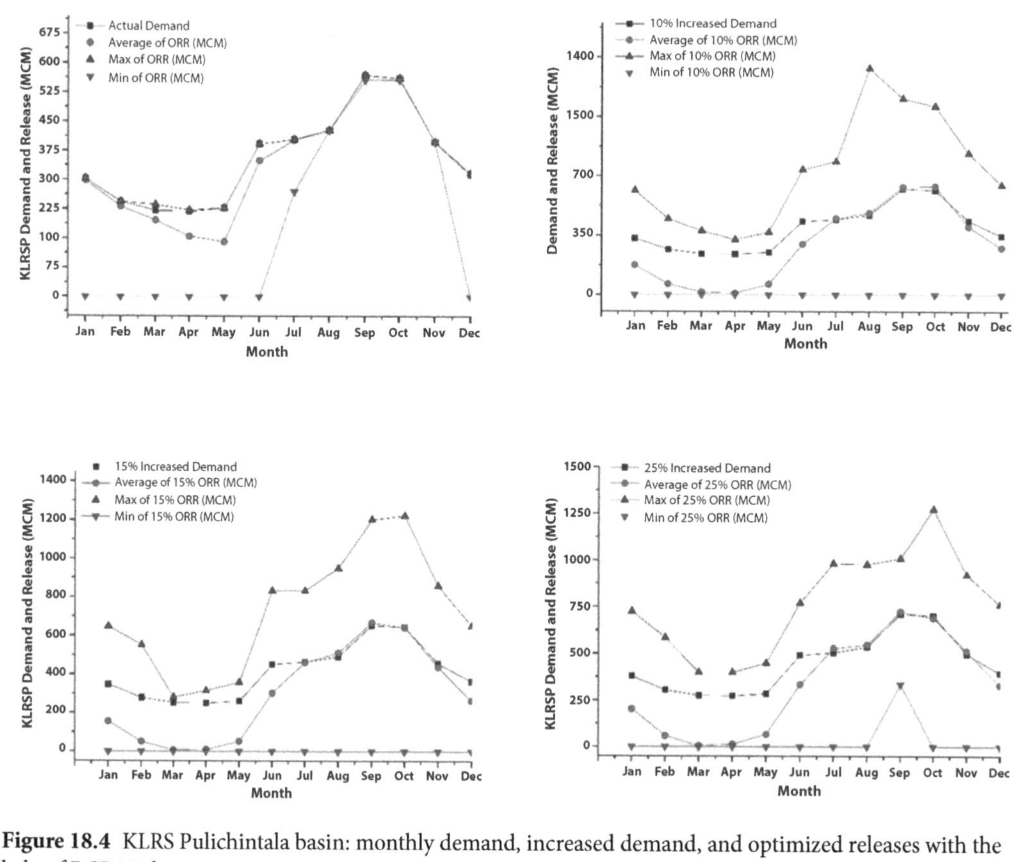

Figure 18.4 KLRS Pulichintala basin: monthly demand, increased demand, and optimized releases with the help of RCP4.5 dataset.

Based on this, the SWAT model was successfully calibrated and validated with the necessary statistics evaluated. The hydrographs shown in Figures 18.2 and 18.3 indicate that the calibrated flows have similar trends as that of the observed flows.

The model's r-factor value was much below 1, which shows that the model simulation outcomes (calibration and validation) are satisfactory. The p-factor value was higher in IMD data than in the calibration and validation periods, demonstrating that the SWAT model has a higher parameter uncertainty contribution to the simulation uncertainty.

18.4.2 KLRS Pulichintala Basin Optimization Results Discussion

Using the data from the RCP4.5, Figure 18.4 shows how reservoir releases affect different increased demands when the reservoir storage level is high. The monthly demand, increased demands (10%, 15%, and 25%) and ORR of a RCP4.5 (2021–2099) period were analyzed for KLRSP Reservoir. The KLRSP reservoir fulfilled 50% of demand during the monsoon season in all the four reservoir operating cases. The model-optimized results are not satisfactory because the storage is very low and demand is high with 25% increased demand. The same case with 10% increased demand SOP case.

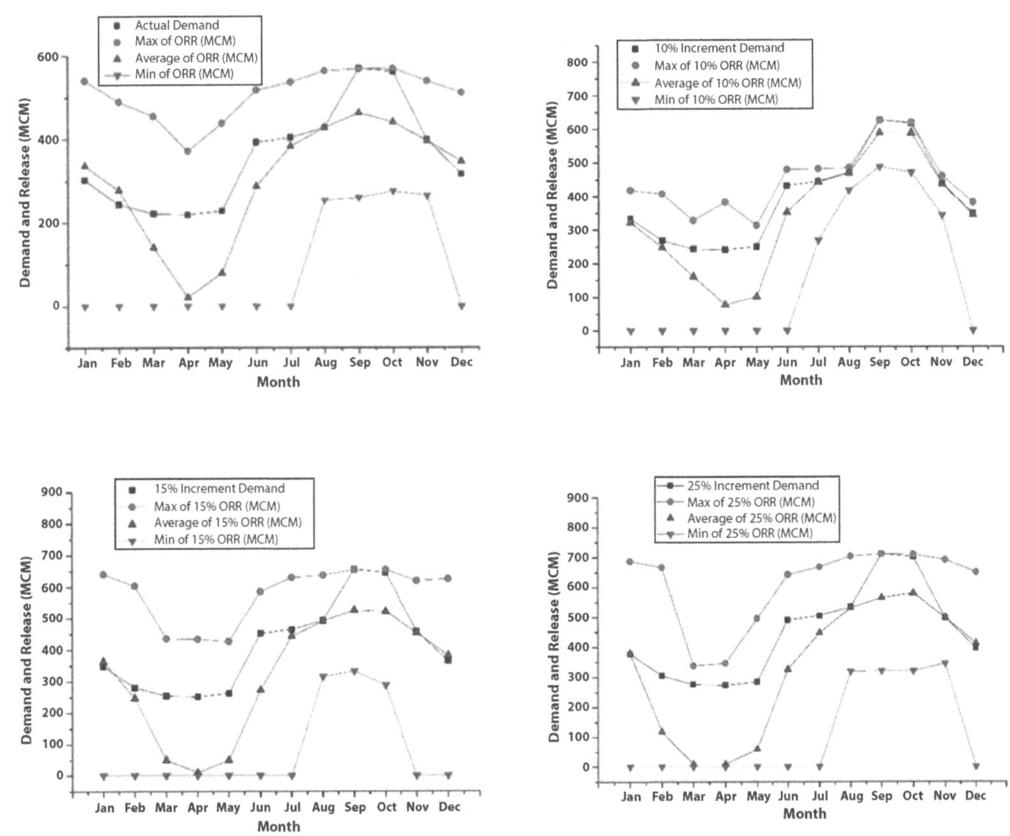

Figure 18.5 KLRS Pulichintala basin: monthly demand, increased demand, and optimized releases with the help of RCP8.5 dataset.

The KLRS Pulichintala basin optimal reservoir analysis outcomes is illustrated in Figure 18.5. In four rule plots, where the reservoir operation regulation was followed, the demand (downstream demand like as irrigation and domestic) was totally unmet in the months of March, April and May. The KLRSP reservoir is a balancing reservoir. In each of the four reservoir operating scenarios, the KLRSP reservoir met 50% of demand during the monsoon season, if agricultural demand was reduced under the RCP8.5 climate scenario and values peaked during the monsoon time.

The various increased demand and optimized releases or ORR of a reservoir, and observed base period of a reservoir, are shown in Figure 18.5. The KLRS Pulichintala reservoir was not able to fulfill the demand during the non-monsoon season in all four cases (Increased demand values of 10%, 15% and 25%), if the agricultural demand was reduced under the historical values had the highest values in on-soon period. The demand and maximum ORR values exhibit the same correlation. The optimized results from the GA model are quite good for the reservoir system.

18.5 Conclusion

At the location of the KLRS Pulichintala reservoir, the mean annual precipitation is 339.5 mm and the annual precipitation is 126.4 mm. The future climate projection (RCP4.5 and RCP8.5) shows the low values as compared to the base period. Therefore, the hydrologic model needs some efficient and effective algorithms. Furthermore, comprehensive understanding of the hydrological processes that take place within the basin and competence in a wide variety of hydrologic sensitive parameters are required. Also, using the RCP4.5 dataset, 60–70% of demand will be satisfied. Only 50–60% of demand would be satisfied using the RCP8.5 climate projection data. Finally, a proper hydrological and topography study is mandatory for the reservoir site before designing. Safety measures should be taken into consideration during the design or implementation of the reservoir or dam for extreme situations like floods and droughts.

Acknowledgments

The authors would like to acknowledge the India Meteorological Department (IMD), Pune, India, for their support in sharing the hydrometeorological data and thankful to the Irrigation & CAD Department of Andhra Pradesh State, India, for providing KLRS Pulichintala Reservoir data and Krishna River Management Board, India. The authors would like to acknowledge the Earth System Grid Federation (ESGF) for downscaled climate data.

References

Sud, A., Kanga, R., Singh, S. K., Meraj, G., Kanga, S., Kumar, P., Ramanathan, A. L., & Bhardwaj, V. (2023). Simulating Groundwater Potential Zones in Mountainous Indian Himalayas—A Case Study of Himachal Pradesh. *J. Geophys. Res.: Earth Surf.*, 10(3), 65.

Upadhyay, R. K., Tripathi, G., Đurin, B., Šamanović, S., Cetl, V., Kishore, N., Sharma, M., Singh, S. K., Kanga, S., Wasim, M., Rai, P. K., & Bhardwaj, V. (2023). Groundwater Potential Zone Mapping in the Ghaggar River Basin, North-West India, Using Integrated Remote Sensing and GIS Techniques. *Water*, 15(5), 961.

Singh, A., Singh, S. K., Meraj, G., Kanga, S., Farooq, M., Kranjcic, N., Đurin, B., & Sudhanshu. (2022b). Designing Geographic Information System Based Property Tax Assessment in India. *Smart Cities*, 5, 364–381.

Chanapathi T., Thatikonda S. (2020) Evaluation of sustainability of river Krishna under present and future climate scenarios. *Sci. Total Environ.*, 738:140322.

Chanapathi T., Thatikonda S. (2020) Investigating the impact of climate and land-use land cover changes on hydrological predictions over the Krishna River basin under present and future scenarios. *Sci. Total Environ.*, 721:137736.

IPCC. Climate Change 2007-Impacts, adaptation and Vulnerability. Contribution of working group II to the fourth assessment report of the IPCC. Cambridge: Cambridge University Press; 2007.

Sajan, B., Mishra, V. N., Kanga, S., Meraj, G., Singh, S. K., & Kumar, P. (2022). Cellular automata-based artificial neural network model for spatial-temporal dynamics and land use/land cover change prediction. *Agronomy*, 12, 2772.

Neelima T.L., Ramana K.V., Reddy M.D., Devi M.U., Ramulu V., Ramanjaneyulu A.V. (2014) Supply demand ratio and water utilization index in jurala project command area, Andhra Pradesh, India. *Int. sssJ. Bioresour. Stress Manag.*, 5(1):068

Shi P., Chen C., Srinivasan R., Zhang X., Cai T., Fang X., Qu S., Chen X., Li Q. (2011) Evaluating the SWAT Model for Hydrological Modeling in the Xixian Watershed and a Comparison with the XAJ Model. *Water Resour. Manage.*, 25(10):2595–2612.

Taylor, K. E., Stouffer, R. J., & Meehl, G. A. (2012). An overview of CMIP5 and the experiment design. *Bull. Am. Meteorol. Soc.*, 93(4), 485–498.

Singh, S. K., Meraj, G., Mondal, N., Bera, A., Verma, M. K., Tomar, J. S., & Kanga, S. (2022). Assessment of seasonal vegetation dynamics over parts of thar desert using geospatial techniques. *ANGRAU*, 105.

Singh, S., Singh, S. K., Prajapat, D. K., Pandey, V., Kanga, S., Kumar, P., & Meraj, G. (2023a). Assessing the Impact of the 2004 Indian Ocean Tsunami on South Andaman's Coastal Shoreline: A Geospatial Analysis of Erosion and Accretion Patterns. *J. Mar. Sci. Eng.*, 11(6), 1134.

Singh, H., Meraj, G., Singh, S., Shrivastava, V., Sharma, V., Farooq, M., Kanga, S., & Kumar, P. (2022a). Status of Air Pollution during COVID-19-Induced Lockdown in Delhi, India. *Atmosphere*, 13(12), 2090.

Singh, S. K., Gupta, S. K., Meraj, G., Kumar, P., & Kanga, S. (2023b). Uncovering the hydro-meteorological drivers responsible for forest fires utilizing geospatial techniques. *Theor. Appl. Climatol.*, https://doi.org/10.1007/s00704-023-04497-y

Debnath, J., Sahariah, D., Lahon, D., Nath, N., Chand, K., Meraj, G., Kumar, P., Singh, S. K., Kanga, S., & Singh, S. K. (2022a). Geospatial modeling to assess the past and future land use-land cover changes in the Brahmaputra Valley, NE India, for sustainable land resource management. *Environ. Sci. Pollut. Res.*, 1-24.

Debnath, J., Sahariah, D., Lahon, D., Nath, N., Chand, K., Meraj, G., Kumar, P., Singh, S. K., Kanga, S., & Farooq, M. (2023b). Assessing the impacts of current and future changes of the plan-forms of River Brahmaputra on its land use-land cover. *Geosci. Front.*, https://doi.org/10.1016/j.gsf.2023.101557.

Debnath, J., Sahariah, D., Saikia, A., Meraj, G., Nath, N., Lahon, D., Annayat, W., Kumar, P., Chand, K., Singh, S. K., & Kanga, S. (2023c). Shifting Sands: Assessing Bankline Shift Using an Automated Approach in the Jia Bharali River, India. *Land*, 12(3), 703.

Farooq, M., Meraj, G., Rishabh, Kanga, S., Nathawat, R., Singh, S. K., & Ranga, V. (2022a). Slum Categorization for Efficient Development Plan—A Case Study of Udhampur City, Jammu and Kashmir Using Remote Sensing and GIS. In *Geospatial Technology for Landscape and Environmental Management: Sustainable Assessment and Planning,* (pp. 283-299). Cham: Springer International Publishing.

Gulati, B., Sharma, R., Kanga, S., Singh, S. K., Sajan, B., Meraj, G., Kumar, P., & Ramanathan, A. L. (2023). Unraveling the Relationship Between Stubble Burning and Air Quality Degradation in Punjab: A Temporal and Spatial Analysis (2019-2022). *J. Clim. Change*, 9(2), 43-53.

Kanga, S., Meraj, G., Farooq, M., Singh, S. K., & Nathawat, M. S. (Eds.). (2022a). *Disaster Management in the Complex Himalayan Terrains.* Geography of the Physical Environment. Springer, Cham.

19

Understanding the Relationship Between River Health and Society for River Restoration: A Review

Pradeep kr Shrivastava[1], Deepak Kumar Prajapat[1] and Saurabh Singh[2]*

[1]Department of Civil Engineering, Suresh Gyan Vihar University, Rajasthan, Jaipur, India
[2]Centre for Climate Change and Water Research, Suresh Gyan Vihar University, Rajasthan, Jaipur, India

Abstract
River flows connect people, places, and other forms of life, inspiring and sustaining diverse cultural beliefs, values, and ways of life. Rivers have the potential to provide a wide range of benefits to society but are often exploited to deliver a narrow range of objectives, to the detriment of river health and other human needs (*i.e.*, sub-optimal investment decisions). Rivers are essential to human well-being. However, many rivers around the world are severely degraded or at risk, which undermines their ability to provide critical ecosystem services and related benefits. In an effort to better engage decision-makers in river conservation and restoration, this working chapter provides a synthesis of existing evidence regarding the relationship between river health and the benefits for society, and presents a conceptual framework to explain the links between rivers and social, economic, and strategic benefits. We find that sustainable management of river ecosystems requires a stronger inter-disciplinary approach, and reclaiming the 'water sector' from the margins to the center of policy-making. The costs of river development need to be better accounted for in planning processes, as well as an explicit consideration of who wins and who loses, and how to compensate the latter. Finally, a widespread shift in thinking is needed so that ecosystems are not viewed as consumers of water, but rather an essential component of water security.

Keywords: Water quality, river pollution, DO, BOD, coliform, India

19.1 Introduction

Rivers are essential to human well-being. The river flows connect the peoples, places and various form of life. These are the main channels for material circulation, energy flow, and information exchange between terrestrial and aquatic ecosystems and perform ecological functions (Silva *et al.*, 2021; Lahon *et al.*, 2023b; Lavane *et al.*, 2023). However, many rivers around the world are severely degraded or at risk due to rapid population growth or intensifying human activities (Luo *et al.*, 2018), which undermines their ability to provide critical ecosystem services and related benefits. In order to better engage decision-makers in conservation efforts, the (WWF) World Wide Fund for Nature believes there is an urgent

**Corresponding author*: saurabh.singh@mygyanvihar.com

Sanjay Kumar, Suraj Kumar Singh, Shruti Kanga, Gowhar Meraj, Majid Farooq and M. S. Nathawat (eds.)
Disaster Management and Environmental Sustainability, (229–240) © 2024 Scrivener Publishing LLC

need to synthesize and strengthen the evidence base regarding the relationship between improved river health and the benefits human societies derive from rivers–social, economic, or strategic (Sajan *et al.*, 2022; Singh *et al.*, 2022b).

This has included the development of hypothetical "causal chains" for fisheries, irrigated crop production, and hydropower. Water is essential for life. Agriculture, farming, and trade was very important for the development of a community. People began to build permanent communities in fertile river valleys. Settlers learned to use the water supply to irrigate the land. Rivers allowed people to bathe, to trade and travel. Rivers supplied drinking water for people and their animals and also supplied fishes for food (Meraj *et al.,* 2021b; Nath *et al.,* 2023; Pandey *et al.,* 2010; Rafiq *et al.,* 2022).

Historically, the first great civilizations all grew up in river valleys. The oldest civilization was developed during 3500 to 2000 B.C. along the Tigris and Euphrates rivers in the Middle East. The name given to that civilization was Mesopotamia, which means "between the rivers." The Nile valley in Egypt had been home to agricultural settlements as early as 5500 B.C., but the growth of Egypt as a civilization began around 3100 B.C. A third civilization grew up along the Indus River around 2600 B.C., in parts of what are now India and Pakistan. The fourth great river civilization emerged around 1700 B.C. along the Yellow River in China (Singh *et al.,* 2023a).

Civilizations tended to grow in river valleys for a number of reasons. The most obvious is access to a usually reliable source of water for agriculture and human needs. Plentiful water, and the enrichment of the soil due to annual flooding, made it possible to grow excess crops beyond what was needed to sustain an agricultural village (Singh *et al.,* 2022). Various forms of agriculture, practiced on about half of the Earth's land surface, provide the majority of food to the people. Agriculture also provides much of the fiber for cotton, wool, and linen clothing. This allowed for some members of the community to engage in non-agricultural activities such as construction of buildings and cities (the root of the word "civilization"), metal working, trade, and social organization. Additional advantages of locating near a river included easy transportation by water as well as good hunting and fishing (Singh *et al.,* 2023b).

The assessment of the health status of these rivers is of global concern. River health assessment studies are caried out by various researchers on different river systems of the world (Norris and Thoms 1999; Shan *et al.,* 2021; Anderson *et al.,* 2019). A river health evaluation index system is necessary to evaluate and maintain river health (Xia *et al.,* 2014). So various physical, chemical, biological and socioeconomic indexes have been developed to assess the health of a river (Upadhyay *et al.,* 2023; Sud *et al.,* 2023).

Song *et al.,* (2015) evaluated the health of the Weihe River in China and indicated the health of the river parts as moderate, poor, and sick. Factors affecting river health can be divided into natural (precipitation decrease, soil erosion, and less vegetation) and human factors (high water demand, population growth, and rapid increase in industries all affecting the river health).

Freshwater is arguably the most critical substance for life on Earth. It is necessary to understand the importance of the relationship between rivers and human activities. The aim of this study is to critically review evidence from the literature and to propose a framework for a more detailed exploration of specific causal linkages between river health and benefits to society, including indicators that might be used for assessing benefits.

19.2 Rivers, Society, and Health

Healthy rivers are a must for a healthy society. River health is the assessment of river condition, is often seen as being analogous with human health, giving many a sense of understanding. It is the integration of the physical, chemical, and biological characteristics of the river. River health is an indicator of the harmony between human and water resources, therefore, river health assessment is an important tool for humans to develop, utilize, and manage rivers in a sustainable way (Singh *et al.*, 2022a; Meraj *et al.*, 2021a). No doubt, rivers have an impact on the health of the civilization but the civilization, too, has impact on the health of the river ecosystem. The relationship of living organisms with each other and with their environment is extremely complex. Impacts on a species or a non-living element may have long-term consequences for a river ecosystem. Several key areas of human impact on river ecosystems are pollution, flow modifications, exotic species, and to some extent, on irrigation water requirements.

Pollution is the result of human infrastructure around a river. Agriculture uses vast quantities of water and is also a source of pollution, primarily by introducing nonpoint-source contaminants. Runoff from agricultural fields often contains eroded soil, fertilizers, animal manure, or pesticides that together, form a major source of water pollution. Pollution enters the river, sometimes in small amounts, at many different locations along the length of the river. Common sources of pollution come from rural and urban areas. Health risks are no longer merely a result of localized exposures to traditional forms of pollution. They are also a result of broader pressures on ecosystems, from depletion and degradation of freshwater resources, to the impacts of global climate change on natural disasters and agricultural production (Mateo-Sagasta *et al.*, 2017).

19.3 Impact of Changes in Water Ecosystem

Many aspects of the world's hydrological (water) cycle are regulated by the natural functions of ecosystems and associated geophysical processes (such as evaporation and the functioning of the climate system). Human interventions in watersheds, lakes, and river systems take many forms—deforestation, farming, irrigation, river damming, and extractions from subterranean aquifers. Wetlands play a crucial role in the filtering of fresh water, including the removal of various chemicals and potentially toxic elements (*e.g.*, heavy metals such as cadmium and lead).

The pattern and extent of change in incidence of a particular infectious disease depends on the particular ecosystems affected, type of land-use change, disease-specific transmission dynamics, socio-cultural changes, and the susceptibility of human populations. Infectious disease risks are affected by changes in the distribution and availability of surface waters and their quality (Keesing *et al.*, 2010).

Human societies also achieve benefits for well-being and healthy by the changes made to food- producing ecosystems in recent decades by constructing water resources projects which have provided substantial gains in food production. Such human-induced changes to ecosystems have been essential to meet growing needs for food and water (Meraj *et al.*, 2022). These changes have helped to reduce the poverty which, in turn, reduce malnourished people and improve human health. Poverty is a consistent determinant of undernutrition, lack of access to safe water and sanitation, and lack of access to other public services such as health services, garbage disposal, etc.

However, these gains have been achieved at increasing cost: degradation of ecosystem, and growing inequities and disparities across groups of people. The intensification of food production methods expanded the use of irrigation, forest-clearing, and the intensive exploitation of capture fisheries (*e.g.*, fishing in open marine or inland waters), all have entailed losses in natural resources and changes in ecosystems. The loss from nature of potential medicinal compounds is one consequence.

19.4 What Are The Challenges?

- The population is growing rapidly, putting more pressure on our water supply (demand is increasing)
- The amount of water is effectively reduced by pollution and contamination (supply is decreasing)
- Water sustainability, water security

Fresh water is essential for human health. It is used for growing food, drinking, personal hygiene, washing, cooking, and the dilution and recycling of wastes. Water scarcity jeopardizes food production, human health, economic development, and geopolitical stability (FAO 2011). Billions of people lack basic water services, and millions die each year from water-related diseases. Water is a basis of state/national/international conflict.

19.5 What Are The Solutions?

Basic human needs for water should be fully acknowledged as a top priority. Improvements in the efficiency of water use (ex: irrigation systems often perform poorly). Efficient management and modern technology can stretch even scarce water supplies much further. Israel, for example, supports its population, its growing industrial base, and intensive irrigation having water availability less than 500 m^3 per person per year. Water is often not used efficiently because it is underpriced. Direct and indirect subsidies (especially for agricultural use) are still common in both developed and developing countries. Rise in water prices can provide incentive for conservation and for the investments needed to spread more efficient technologies. Water-related diseases, including Guinea worm, diarrhea, malaria, and typhoid should be brought under control. Basic ecosystem water needs should be identified and met. Serious water-related conflicts should be resolved through formal negotiations.

19.5.1 Health Assessment Indicators

River health assessment (RHA) is a robust methodology for ascertaining the health of riverine ecosystems and offering solutions for river conservation and management (Zhang *et al.*, 2023). River health assessment involves the comparison of the present condition of a river

Table 19.1 River health assessment indicators.

S.N.	Indicator	Measure
1	Sediment sequence and composition	Rate of accumulation Minerology Geochemistry
2	Soil and sediment erosion	Rate of erosion Source of sediment Mode of transport
3	Stream flow	Total Annual Flow Variability
4	Stream channel morphology	Slope pattern Cross-sectional dimensions
5	Stream sediment storage and load	Sediment flux Mode of transport
6	Surface water quality	Turbidity Total suspended solids
7	Floodplains/wetlands structure and hydrology	Wetting and drying regimes Connectivity with the river area

as compared to the reference conditions. The reference condition is defined as really one of the best conditions of the river in historical period. Various river indicators/indexes have been developed to study or examine river health. Indicators of river health can be represented at spatial scales ranging from local to catchment, temporal scales from instantaneous to long-term, and may be direct or explanatory measures.

The International Union of Geological Sciences (IUGS) gives seven indicators that represent the condition of river systems (Table 19.1). Natural changes and human induced should both be considered during the analysis. The time scale for changes assessment is very important because some indicators.

19.6 River Restoration

For centuries, society has interacted with rivers. They are an integral part of our local environment on which we historically depended for food, shelter, and basic survival. Most towns and cities developed near rivers illustrates their importance to humans. But over time, urbanization has impacted rivers, and modern society has become less connected to our local environment and people's perception of rivers has changed.

River restoration is the process of managing rivers to reinstate natural processes to restore biodiversity, providing benefits to both people and wildlife (LoSchiavo *et al.*, 2013). The restoration and maintenance of "healthy" river ecosystems have become important objectives of river management. A healthy environment is known to improve quality of life. Restoring rivers helps to provide quality environments and puts people in closer contact with nature.

In urban environments, this is particularly relevant, as restoring rivers can provide people with much needed green space, to offer people a place away from stresses and strains often associated with cities—with a higher density of people, poor air quality due to vehicle emissions, poor housing, and lack of green space. Long-term monitoring of river health is necessary for a compressive evaluation of innovative restoration.

19.6.1 River Restoration Counters These Pressures Through a Wide Range of Social Benefits

a. Physical Well-Being

Many river restoration schemes in public parks have resulted in increased park usage, for longer periods, and an increase in sports and games. Research shows that regular physical activity helps reduce the risk of developing a range of serious diseases and improves overall quality of life. Many people are failing to meet recommended levels of activity which has led to an increase in obesity and coronary heart disease. River restoration in public open spaces offers free, accessible, and safe recreation for children and adults. Post-project research has highlighted renewed appreciation for and interaction with the natural environment, particularly where parks were previously regarded as unsafe due to antisocial behaviour.

b. Mental Well-Being

Society has recognized the benefits of the environment on mental health for centuries. More recently, there is evidence that contact with nature and living things makes people feel good and positively enhances their mental well-being while contributing to stress reduction. "Green exercise" refers to the synergistic benefit of participating in physical activities, while being directly exposed to nature. Reports have also demonstrated the value of outdoor activities in helping to combat depression and increasing happiness and well-being and confidence. "Less green nature means reduced mental well-being, or at least less opportunity to recover from mental stress. As natural green environments have increasingly come under pressure from economic development our own well-being has suffered as a consequence. Today, stress and mental ill-health are becoming more common, and the associated public health costs are growing."

c. Recreation

River corridors can provide opportunities for recreation as part of daily life allowing stress relief and enjoyment as well as activities such as fishing, cycling, and watching wildlife. In built up areas, river corridors enable safe, cost effective, and healthy commuting, which further contributes to physical health. River restoration schemes can facilitate recreation and learning to care more about their environment.

d. Social Cohesion

Rivers can play a vital role in the community by encouraging social interaction and bonding. Restoring a river between two estates can enable two separated communities to mix and foster a community spirit, pride in the environment, and a sense of belonging.

e. Aesthetic

River restoration improves the aesthetics of the environment by restoring natural landscape features and benefiting the associated flora and fauna, creating nicer surroundings. People wish to experience wildness yet feel safe in urban areas. Post-restoration improvements in social behavior such as reduced graffiti and littering further improve aesthetics and foster a sense of pride in the local environment.

f. Education

Restoration schemes provide huge learning potential about our natural world, helping to demonstrate how humankind depends on water and the need to reduce our collective impact on the environment. Key lessons include the water cycle, climate change adaptation, the threats of invasive species, and the ecology of our native species and habitats. Studies have shown schoolchildren learning about wildlife and natural habitats from their site visits were motivated to do something differently in their everyday life to help the environment.

g. Cultural

Rivers have always been part of our culture with human settlements located adjacent to rivers for transport, fishing, farming, and recreation. Restoring our cultural links to rivers is a long-term goal and an important part of all drivers raised above. Many religions connect spiritual and religious values to rivers and the earth, and landscapes have long provided inspiration for art, folklore, and architecture.

19.6.2 Other Benefits of River Restoration

River restoration can significantly increase the services provided by a healthy ecosystem, too often neglected by decision makers, such as flood control, groundwater recharge, pollution removal, recreational opportunities, and increased property values due to the increasing demand for more natural surroundings.

Conversely, river restoration schemes may imply significant changes in water management and land use and hence, negative economic impacts on certain economic activities; it may hamper navigation or agriculture for instance. In order to support river basin planning and decisions in river related policies, it is therefore very important to assess these impacts with appropriate and transparent methods, and weigh them against the benefits foreseen.

River restoration contributes to flood risk management by supporting the natural capacity of rivers to retain water. As flood risk consists of damage times occurrence, flood risk management needs to reduce either the damage, or the likelihood of floods to occur, or both. River restoration reduces the likelihood of high-water levels, and improves the natural functions of the river at same time.

19.6.2.1 Classic Flood Risk Management

In a natural river system, a river spreads water beyond its banks and over extended areas of a floodplain during periods of high water. In order to protect property and contain waters, the classic flood risk management approach is to constrain watercourses. This is commonly done by building dykes and straightening rivers to increase discharge capacity, dredging to deepen channels, and building reservoirs and artificial retention areas to store excess waters.

While this generally reduces the likelihood of flooding, in the event of extremely high waters, it increases the amount of damage if the engineered system is overwhelmed or fails. Without natural features such as wetlands and meanders, excess waters cannot be absorbed. Any breach will release an enormous amount of water, with potentially catastrophic consequences. Continuously reinforcing and building higher dykes cannot overcome this weakness, and is a very expensive option. Historically, engineering solutions upstream have created peak flows downstream, leading to more engineering. Moreover, climate change scenarios predict more extreme weather events and higher sea levels. A new approach is needed.

19.6.2.2 Modern Flood Risk Management

By reconnecting brooks, streams, and rivers to floodplains, former meanders and other natural storage areas, and enhancing the quality and capacity of wetlands, river restoration increases natural storage capacity and reduces flood risk. Excess water is stored in a timely and natural manner in areas where values such as attractive landscape and biodiversity are improved and opportunities for recreation can be enhanced. In these ways, river restoration directly contributes to climate change strategies aimed at mitigating the effects of increased and erratic peak flows and droughts.

19.7 River Pollution

As citizens of India, we have a constitutional duty to protect our environment. Similarly, the government also has a duty to protect the environment for the welfare of its citizens. There are many ways we can protect the river from pollution. Some immediate ways to control pollution are:

- Industries should install machineries to remove contaminants from their effluents and wastewater. One way to do so is the installation of an effluent treatment plant (ETP). This way, we can control pollution at the source itself.
- The towns and cities should also have facilities to clean the sewage effluent. All towns and cities must have sewage treatment plants (STPs) that clean up the sewage.
- Farmers should give up chemicals and pesticides in farming and should instead adopt organic methods of farming thus reducing chemical pollution of rivers.
- We should stop our religious practices that pollute river water.

The Central Water Commission is also monitoring water quality at 396 key locations covering all the major river basins of India. The CWC is maintaining a three-tier laboratory system for the analysis of the parameters. The level-I laboratories are located at 370 field water quality monitoring stations on major rivers of India where physical parameters such as temperature, colour, odor-specific conductivity, total dissolved solids, pH, and dissolved oxygen of river water are observed. There are 18 level-II laboratories located at selected division offices to analyze 25 nos. physico-chemical characteristics and bacteriological parameters of river water. Five level-III/II+ laboratories are functioning at Varanasi, Delhi, Guwahati, Hyderabad, and Coimbatore, where 41 parameters, including heavy metals/toxic parameters, and pesticides, are proposed to be analyzed.

19.8 Environmental Flow Requirements

The National Water Policy (2012) duly recognizes the ecological needs of the river, setting aside a portion of river flows on the pattern of natural flow regime and environmental management plan is an integral part of the planning of any water resources development project. The detailed environmental studies are required to be carried out including requirement of e-flows while planning any water resources project. However, the estimation of environmental flows is still a challenge primarily due to lack of both the understanding of and quantitative data on relationships between river flows and ecological characteristics of river. The assessment of environmental flows, rationally keeping in view of limited water resources and dependence of livelihood of large population, is still at infancy stage. Various studies show wide variations in the assessment of e-flows. In the case of the upper Ganga basin, various studies show requirements of e-flow varying from 5% to 100% at different locations and in different months. There is call from various groups to reserve the majority of river flows for e-flows during lean periods. This will hurt other sectors and genuine development needs of the society. Therefore, the allocation of water for environmental needs should not be decided in isolation. The water requirements of other demands of food, energy, livelihoods, etc., in the basin should also be taken into account while allocating scarce water resources to various competing users, including environmental needs.

19.9 Conclusions

After focusing on the major problems in the environment, we have done a small review for preserving the river system. By implementing this process and awakening the society, we can bring various solutions for rivers and society. Developing more green areas or other natural sources can bring changes in our environment. The restoration of rivers can bring huge changes to the environment. It will help to reduce water scarcity as well as improve the agricultural need. It will ultimately be very helpful to society. Thus, there is a strong relationship between river systems and society. Saving it will save us.

References

Anderson, E. P., Jackson, S., Tharme, R. E., Douglas, M., Flotemersch, J. E., Zwarteveen, M.& Arthington, A. H. (2019). Understanding rivers and their social relations: A critical step to advance environmental water management. Wiley Interdisciplinary Reviews: *Water*, 6(6), e1381.

dos Santos Silva, D., Gonçalves, B., Rodrigues, C. C., Dias, F. C., de Souza Trigueiro, N. S., Moreira, I. S. & Rocha, T. L. (2021). A multibiomarker approach in the caged neotropical fish to assess the environment health in a river of central Brazilian Cerrado. *Sci. Total Environ.*, 751, 141632.

FAO. *The State of the World's Land and Water Resources: Managing Systems at Risk.*; 2011. http://www.fao.org/3/i1688e/i1688e.pdf

Keesing F., Belden L.K., Daszak P., *et al.* Impacts of biodiversity on the emergence and transmission of infectious diseases. *Nature,* 2010;468(7324):647-652. doi:10.1038/nature09575

Limburg K.E., Swaney D.P., Strayer D.L. River Ecosystems. *Encycl. Biodivers.* Second Ed. 2013;6:469-484. doi:10.1016/B978-0-12-384719-5.00222-7

LoSchiavo, A. J., Best, R. G., Burns, R. E., Gray, S., Harwell, M. C., Hines, E. B., ... & Vearil, J. W. (2013). Lessons learned from the first decade of adaptive management in comprehensive Everglades restoration. *Ecol. Soc.*, 18(4).

Luo, Z., Zuo, Q., Shao, Q., 2018. A new framework for assessing river ecosystem health with consideration of human service demand. *Sci. Total Environ.*, 640–641, 442–453. https://doi.org/10.1016/j.scitotenv. 2018.05.361

Mateo-Sagasta J., Marjani S., Turral H., Burke J. *Water pollution from agriculture: a global review.* FAO y IWMI. Published online 2017:35. http://www.fao.org/3/a-i7754e.pdf

Norris, R. H., & Thoms, M. C. (1999). What is river health?. *Freshwater Biol.*, 41(2), 197-209.

Shan, C., Dong, Z., Lu, D., Xu, C., Wang, H., Ling, Z., & Liu, Q. (2021). Study on river health assessment based on a fuzzy matter-element extension model. *Ecol. Indic.*, 127, 107742.

Song, J., Cheng, D., Li, Q., He, X., Long, Y., & Zhang, B. (2015). An evaluation of river health for the Weihe River in Shaanxi Province, China. *Adv. Meteorol.*, 2015.

Thomas, K. V., Bijlsma, L., Castiglioni, S., Covaci, A., Emke, E., Grabic, R. & de Voogt, P. (2012). Comparing illicit drug use in 19 European cities through sewage analysis. *Sci. Total Environ.*, 432, 432-439.

Xia, J., Lin, L., Lin, J., & Nehal, L. (2014). Development of a GIS-based decision support system for diagnosis of river system health and restoration. *Water*, 6(10), 3136-3151.

Zhang, Z., Li, Y., Wang, X., Liu, Y., Tang, W., Ding, W., ... & Wu, W. (2023). Investigating river health across mountain to urban transitions using Pythagorean fuzzy cloud technique under uncertain environment. *J. Hydrol.*, 620, 129426.

Sud, A., Kanga, R., Singh, S. K., Meraj, G., Kanga, S., Kumar, P., Ramanathan, A. L., & Bhardwaj, V. (2023). Simulating Groundwater Potential Zones in Mountainous Indian Himalayas—A Case Study of Himachal Pradesh. *J. Geophys. Res.: Earth Surf.*, 10(3), 65.

Upadhyay, R. K., Tripathi, G., Đurin, B., Šamanović, S., Cetl, V., Kishore, N., Sharma, M., Singh, S. K., Kanga, S., Wasim, M., Rai, P. K., & Bhardwaj, V. (2023). Groundwater Potential Zone Mapping in the Ghaggar River Basin, North-West India, Using Integrated Remote Sensing and GIS Techniques. *Water*, 15(5), 961.

Lavane, K., Kumar, P., Meraj, G., Han, T. G., Ngan, L. H. B., Lien, B. T. B., Ty, T. V., Thanh, N. T., Downes, N. K., Nam, N. D. G., Minh, H. V. T., Singh, S. K., & Kanga, S. (2023a). Assessing the Effects of Drought on Rice Yields in the Mekong Delta. *Climate*, 11(1), 13.

Meraj, G., Farooq, M., Singh, S. K., Islam, M. N., & Kanga, S. (2022). Modeling the sediment retention and ecosystem provisioning services in the Kashmir valley, India, Western Himalayas. *Model. Earth Syst. Environ.*, 8(3), 3859-3884.

Meraj, G., Kanga, S., Kranjčić, N., Đurin, B., & Singh, S. K. (2021a). Role of natural capital economics for sustainable management of earth resources. *Earth*, 2(3), 622-634.

Meraj, G., Singh, S. K., Kanga, S., Islam, M. N., & Singh, S. K. (2021b). Modeling on comparison of ecosystem services concepts, tools, methods and their ecological-economic implications: A review. *Model. Earth Syst. Environ.*, 1-20.

Nath, N., Sahariah, D., Meraj, G., Debnath, J., Kumar, P., Lahon, D., Chand, K., Farooq, M., Chandan, P., Singh, S. K., & Kanga, S. (2023). Land Use and Land Cover Change Monitoring and Prediction of a UNESCO World Heritage Site: Kaziranga Eco-Sensitive Zone Using Cellular Automata-Markov Model. *Land*, 12(1), 151.

Rafiq, M., Meraj, G., Kesarkar, A. P., Farooq, M., Singh, S. K., Kanga, S., & Bhardwaj, V. (2022). Hazard Mitigation and Climate Change in the Himalayas–Policy and Decision Making. In S. Kanga, G. Meraj, M. Farooq, S. K. Singh, & M. S. Nathawat (Eds.), *Disaster Management in the Complex Himalayan Terrains*. Geography of the Physical Environment. Springer, Cham. https://doi.org/10.1007/978-3-030-89308-8_12.

Sajan, B., Mishra, V. N., Kanga, S., Meraj, G., Singh, S. K., & Kumar, P. (2022). Cellular automata-based artificial neural network model for spatial-temporal dynamics and land use/land cover change prediction. *Agronomy*, 12, 2772.

Singh, A., Singh, S. K., Meraj, G., Kanga, S., Farooq, M., Kranjcic, N., Đurin, B., & Sudhanshu. (2022b). Designing Geographic Information System Based Property Tax Assessment in India. *Smart Cities*, 5, 364–381.

Singh, H., Meraj, G., Singh, S., Shrivastava, V., Sharma, V., Farooq, M., Kanga, S., & Kumar, P. (2022a). Status of Air Pollution during COVID-19-Induced Lockdown in Delhi, India. *Atmosphere*, 13(12), 2090.

Singh, S. K., Gupta, S. K., Meraj, G., Kumar, P., & Kanga, S. (2023b). Uncovering the hydrometeorological drivers responsible for forest fires utilizing geospatial techniques. *Theor. Appl. Climatol.*, https://doi.org/10.1007/s00704-023-04497-y

Singh, S. K., Meraj, G., Mondal, N., Bera, A., Verma, M. K., Tomar, J. S., & Kanga, S. (2022). Assessment of seasonal vegetation dynamics over parts of thar desert using geospatial techniques. *ANGRAU*, 105.

Singh, S., Singh, S. K., Prajapat, D. K., Pandey, V., Kanga, S., Kumar, P., & Meraj, G. (2023a). Assessing the Impact of the 2004 Indian Ocean Tsunami on South Andaman's Coastal Shoreline: A Geospatial Analysis of Erosion and Accretion Patterns. *J. Mar. Sci. Eng.*, 11(6), 1134.

Lahon, D., Sahariah, D., Debnath, J., Nath, N., Meraj, G., Kumar, P., Hashimoto, S. and Farooq, M. (2023b). Assessment of Ecosystem Service Value in Response to LULC Changes Using Geospatial Techniques: A Case Study in the Merbil Wetland of the Brahmaputra Valley, Assam, India. *ISPRS Int. J. Geo-Inf.*, 12(4), p. 165.

20

Geospatial Mapping of Groundwater Potential Zones Using Multi-Criteria Decision Making AHP Approach in a Pisangan Watershed, Ajmer District (RAJ.)

Shilpi Yadav[1] and S.S. Khinchi[2]*

[1]*Department of Geography, Sophia Girls' College (Autonomous), Ajmer, India*
[2]*Department of P.G. Studies in Geography, Dr. BR Ambedkar Government P.G. College, Sri Ganganagar, India*

Abstract

Water shortage is a worldwide issue, owing to the link of water-food-energy and its influence on livelihood and worldwide economics. Integration of remote sensing data and the geographical information system (GIS) for the study of groundwater resources has become a discovery in the field of groundwater research, which assists in evaluating, monitoring, and conserving groundwater resources. The objective of this study is to explore the groundwater availability for agriculture in the Pisangan Watershed. Remote sensing data and GIS were used to locate potential zones for groundwater in the Pisangan Watershed. Maps of surficial-lithology, lineament-density, drainage-density, rainfall-distribution, base map, hydro geomorphological, geological, structural, drainage, slope, land use/land cover and groundwater prospect zones were prepared using the remote sensing data along with the present maps. The impact of each theme and sub units/classes to groundwater recharge based on previous studies was estimated using the analytical hierarchical process (AHP). The groundwater potential map was produced by the weighted index overlay model (WIOM) in GIS. The groundwater prospects were mapped in four categories: excellent, good, fair, and poor zone. The excellent zones cover 50.20 km^2 (3.60%) of the total size of the study area. While the good, fair, and poor zone covers 965.65 km^2 (69.28%), 377.02 km^2 (27.04%), and 0.92 km^2 (0.06%), respectively. Excellent and good groundwater promising zones occur in the Pisangan Watershed, which is very good for irrigation and agricultural activities.

***Keywords*:** Groundwater potential, GIS, remote sensing, analytic hierarchy process, overlay analysis

**Corresponding author*: shyamsunderkhinchi@yahoo.co.in

Sanjay Kumar, Suraj Kumar Singh, Shruti Kanga, Gowhar Meraj, Majid Farooq and M. S. Nathawat (eds.)
Disaster Management and Environmental Sustainability, (241–260) © 2024 Scrivener Publishing LLC

20.1 Introduction

Groundwater is one of the utmost natural resources that support human health and environmental diversity. Protecting it from pollution and carefully managing its use will ensure its future as an important part of ecologies and anthropological activity. The rate of groundwater flow is controlled by two main properties of the rock: porosity and permeability. The main sources of groundwater recharge are rainfall and surface flow, and for discharge, sources include run-off seepage into the streams and lakes, springs, evaporation, and pumping (Gulati *et al.*, 2023; Kanga *et al.*, 2022a).

It is said that approximately one third of the global population use groundwater for drinking. Groundwater is the main source for irrigation and domestic purposes. Eighty percent of the rural population use groundwater for domestic purposes and 50% of the urban population use groundwater for domestic purposes. Reliance on the usage of groundwater for domestic purposes, irrigation, and for other sectors, may result in misuse of groundwater resources (Kanga *et al.*, 2022b; Kotecha *et al.*, 2023). Groundwater is a dynamic and replenishable resource. The exploitation and examination of groundwater resources are needed to develop an understanding of the geology and geomorphology of that area. The data and thematic layers such as remote sensing satellite images, soil texture data, geology or lithology data, drainage density data, and rainfall data, are helpful for mapping of groundwater potential zones (Lahon *et al.*, 2023b; Lavane *et al.*, 2023; Meraj *et al.*, 2022).

Remote sensing data, combined with geographical information system (GIS) technique, is very capable in identifying the groundwater potential of the area. The study results that the integration of thematic layers prepared from traditional methods and remote sensing techniques using GIS give more accurate and precise results (Meraj *et al.*, 2021a; Lahon *et al.*, 2023a). Groundwater is available when water penetrates below the earth's surface, and soil underneath the earth's surface is porous. Groundwater table decreases when the pumping rate is higher than the rate of usage. Hence, it can be concluded that areas of high extraction rates may lead to the reduction of groundwater zones (Meraj *et al.*, 2021b; Nath *et al.*, 2023; Pandey *et al.*, 2010). Remote sensing is one of the most important sources for earth surface feature information of groundwater such as land use/land cover, land forms, and drainage pattern and density. With the help of remote sensing, surface information data can be easily integrated with GIS techniques to identify the ground water potential zones (Rafiq *et al.*, 2022; Sajan *et al.*, 2022). According to the World Bank report, India may be in ground water stress zone by 2025 and water scarce zone by 2050. This is due to the lack of education in groundwater misuse, improper maintenance and management of water, and the failure of government schemes and policies for rural areas may lead to groundwater and drinking purpose water problems in India (Singh *et al.*, 2022b). The benefits of the integration of remote sensing and GIS is, with the help of various layers and techniques, we can easily identify the sub-surface feature information of the potential of ground water zone in a short period of time (Sud *et al.*, 2023; Upadhyay *et al.*, 2023).

20.2 Remote Sensing and GIS Techniques

Remote sensing and GIS play a vital role in the effective management of water and land resources. The benefit of using remote sensing is to develop knowledge on spatial technology which is useful for analysis, interpretation, and evaluation. The National Remote Sensing Agency (NRSA) was first to identify the geospatial mapping of groundwater potential zones (Singh *et al.*, 2023a). Geographic information system (GIS) technique is used to categorize the results of remote sensing and assign the suitable weights to the related maps. These maps or thematic layers are used to identify the groundwater runoff flow and recharge zones. Remote Sensing and GIS play an important role in the delineation of groundwater potential zones. With the help of satellite data, we can easily identify the water holding capacity for different geomorphological, lithological, and structural units. From the land use land cover, slope, DEM, and rainfall data, we can identify the groundwater efficiency of the study area. Combined remote sensing and GIS techniques has shown that it is a time saving process for monitoring ground water level zone and low cost for obtaining slope, drainage density, geology, lineament, soil structure geomorphology maps, etc. (Singh *et al.*, 2022a).

20.3 Analytical Hierarchical Process

Analytical hierarchical process (AHP) is a multi-criteria decision-making method propounded by Prof Thomas L Satty in 1980. It is an approach to obtain proportion scales from paired difference. The information is taken from actual measurements such as weights, price, and from subjective conclusions. In this study, a total of seven parameters were used to delineate the ground water potential zones such as drainage density, digital elevation model (DEM), geology, geomorphology, land use and land cover, lineament, rainfall pattern, slope and soil texture. DEM data has been used to extract slope map, fill, flow direction and flow accumulation. The LANDSAT ETM images were used to classify the land use/land cover image (Singh *et al.*, 2023b). The drainage density map is created using QGIS software and weights are calculated. These parameters are prepared in geospatial GIS environment and weights are assigned for each class/object using AHP. The ground water potential zones are classified into four categories: poor, fine, good, and excellent. Analytical hierarchical process analysis diverse datasets into a pairwise matrix (group wise), which is used to calculate geometric mean and normalized weight of parameters (Singh *et al.*, 2022).

20.4 Weighted Overlay Method

In this method, initially, the geospatial data base was developed using the Survey of India (SOI) toposheet on a 1:50000 scale and LANDSAT 8 ETM satellite data. Various thematic layers such as drainage density, stream orders, and stream length are prepared by using GIS. Then the DEM data is used to acquire slope, contour, and flow accumulation map. The DEM data may be used to prepare the land use/land cover classification map (supervised classification) and lineament map. Therefore, all the thematic maps are used to examine the overlay and weights are assigned for each thematic layer and given ranks are assigned to evaluate the groundwater potential zone. For identifying ground water potential zones for the study area, the following equation is used:

$$\mathbf{Pr = RFwRFr + LGwLGr + GGwGGr + SGwSGr + LDwLDr + DDwDDr + LCwLCr + SCwSCr}$$

where Pr is groundwater potential index, RF is rainfall index, LG is lithology index, GG is geomorphology index, SG is slope gradient index, LD is lineament density index, DD is drainage density index, LC is land use and land cover index, SC is coil cover index, W is weight, and r is rank.

20.5 Study Region

The Watershed area of Pisangan Watershed is 1201.61 km^2 and located between 23°44'45" N to 23°58'30" N latitude and 78°35'45" E to 78°46'15" E longitudes. The Pisangan Watershed has good accessibility by road. It is well connected by National Highway (NH-8) and State Highways, making all the important places of the district easily accessible. The nearest railway station is Ajmer, which is spread over an area of 8481 km^2 and is bounded by Nagaur district to the north, Jaipur and Tonk district to the east, Bhilwara district to the south, and Pali district to the east. Apart from the National highway and State highway, the study area has a good road and rail network of district and village roads connecting the different areas of watershed with each other.

20.6 Flowchart of Methodology

Several thematic layers that favor the groundwater occurrence were combined and a groundwater potential map was prepared in GIS. The thematic maps (factor maps) used for the groundwater prospective mapping in this study are briefly described below.

20.7 Methodology

The base map of Pisangan Watershed was prepared based on the Ajmer satellite image and SOI toposheet on a 1:50,000 scale. The location map of the study area is shown in Figure 20.1. The drainage network for the study area was scanned from the Survey of India (SOI) toposheets and digitized in ArcGIS 10.2 platform. The slope map was prepared from ASTER DEM data in ArcGIS using spatial analyst module. The rainfall map was prepared using the data obtained from the Indian Meteorological Department (IMD) gauge stations. These data were then spatially interpolated using inverse distance weighted (IDW) method to obtain the rainfall distribution map. This interpolation method combines the concepts of proximity to follow Thiessen polygons with gradual change of the trend surface. The drainage density and lineament density maps were prepared using the line density analysis tool in ArcGIS. Satellite images from LANDSAT 8, on a scale of 1:50,000 (geo-coded, with UTM projection, spheroid, and datum WGS 84, Zone 43 North zone) were used for delineation

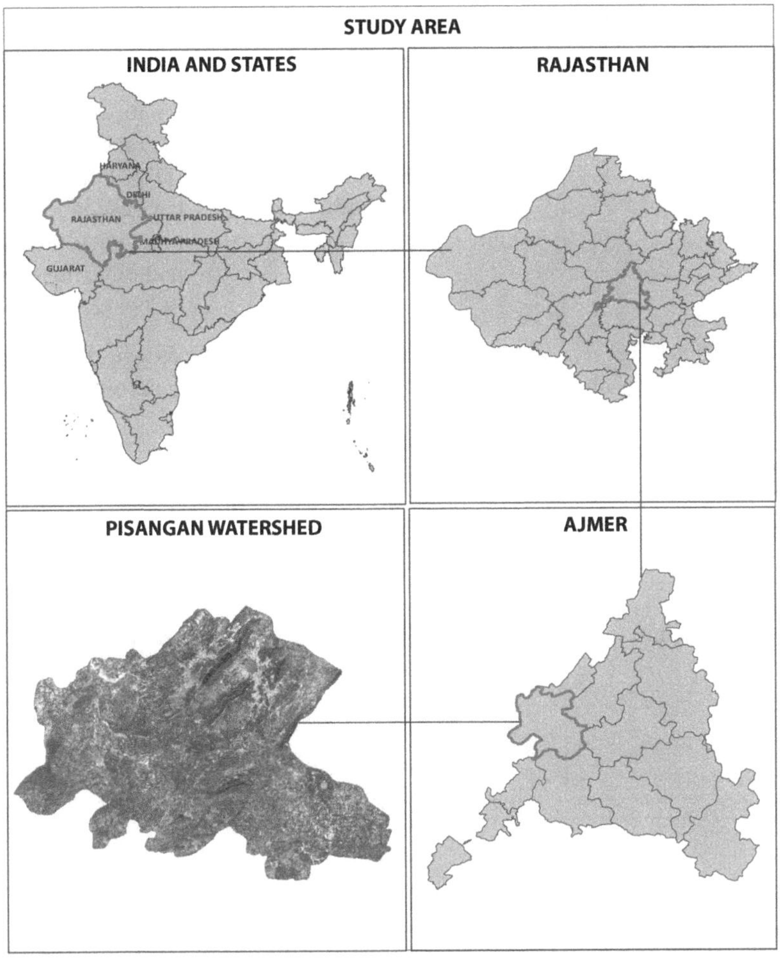

Figure 20.1 Location map of the study area of Pisangan Watershed.

of the thematic layers such as land use, lithology, lineament, and soil texture types. These thematic layers were converted into a raster format (30-m resolution) before they were brought into the GIS environment. The groundwater potential zones were obtained by overlaying all the thematic maps in terms of weighted overlay methods using the spatial analysis tool in ArcGIS 10.2. During weighted overlay analysis, the ranking was given for each individual parameter of each thematic map, and weights were assigned according to the multi-influencing factor (MIF) of that feature on the hydrogeological environment of the study area.

20.8 Multi-Influencing Factors of Groundwater Potential Zones

Seven influencing factors, such as lithology, geomorphology, slope, land use, lineament, drainage density, soil, and rainfall were recognized to delineate the groundwater potential zones. Regarding the interrelationship between these factors and their effect, each

Table 20.1 Data sources of collection for parameters related to groundwater potential.

S. no	Data type	Source of collection	Utility
1.	Remote sensing data	Landsat 8 and Aster DEM (Advanced Spaceborne Thermal Emission and Reflection Radiometer) Satellite images	Thematic mapping includes: digital elevation model (DEM), slope mapping, land use/land cover mapping, drainage density mapping, lineament density mapping, etc.
2.	Geology/lithology/ geomorphological data	RRSC, Jodhpur (Rajasthan)	Geology/lithology/ geomorphological mapping
4.	Rainfall data	From the website of IMD	For average rainfall
5.	Soil data	From the National Bureau of Soil Survey and Land Use planning	For estimating the soil texture map
6.	Watershed data	Collected from the Department of Watershed Development and Soil Conservation (Government of Rajasthan, Jaipur)	For boundary of watershed

relationship is weighted according to its strength. The representative weight of a factor of the potential zone is the sum of all weights from each factor. A factor with a higher weight value shows a larger impact, and a factor with a lower weight value shows a smaller impact on groundwater potential zones. Integration of these factors with their potential weights is computed through the weighted overlay analysis in ArcGIS. Data sources of collection for parameters related to groundwater potential have been shown in Table 20.1.

20.9 Results and Discussion

20.9.1 Weightage Calculation

The multi-influencing factors for groundwater potential zones, namely lineaments, drainage, lithology, slope, land use, rainfall, and soil were examined and assigned an appropriate weight. The effect of each influencing factor may contribute to delineate the groundwater potential zones. Moreover, these factors are interdependent. The effect of each major and minor factor is assigned a weightage of 1.0 and 0.5, respectively. The cumulative weightage of both major and minor effects is considered for calculating the relative rates (Table 20.2) and

Table 20.2 Weightage of both major and minor effects is considered for calculating the relative rates.

S. no.	Factor	Influence	Field/value	Value rank	Groundwater prospect
1.	LULC	6	Forest	5	Excellent
			Agriculture	5	Excellent
			Wasteland	1	Very poor
			Fallow land	2	Poor
			Water body	5	Excellent
			Settlement	1	Very poor
2.	**Lineament density**	5	1	1	Very poor
			2	2	Poor
			3	3	Moderate
			4	4	Good
			5	5	Excellent
3.	**Drainage density**	9	1	5	Excellent
			2	4	Good
			3	3	Moderate
			4	2	Poor
			5	1	Very poor
4.	**Soil type**	4	Clay loam	2	Poor
			Loam	3	Moderate
			Loamy sand	3	Moderate
			Sand	5	Excellent
			Sandy loam	4	Good
			Rock outcrop	1	Very poor

(*Continued*)

Table 20.2 Weightage of both major and minor effects is considered for calculating the relative rates. (*Continued*)

S. no.	Factor	Influence	Field/value	Value rank	Groundwater prospect
5	**Slope**	**13**	1	5	Excellent
			2	4	Good
			3	3	Moderate
			4	2	Poor
			5	1	Very poor
6.	**Geology/ geomorphology**	**25**	Alluvial plain	3	Moderate
			Structural hills	2	Poor
			Denudation hills	1	Very poor
			Pediplain	3	Moderate
			Eolian plain	3	Moderate
7.	**Rainfall**	**38**	1	1	Very poor
			2	2	Poor
			3	3	Moderate
			4	4	Good
			5	5	Excellent

methodology of groundwater potential zone using AHP technique shown in Figure 20.2. This rate is further used to calculate the score of each influencing factor. The proposed score for each influencing factor is calculated by using the formula.

$$\left[\frac{(A+B)}{\sum(A+B)}\right]\times 100$$

where, A is the major interrelationship between two factors and B is the minor interrelationship between two factors. The concerned score for each influencing factor was divided equally and assigned to each reclassified factor.

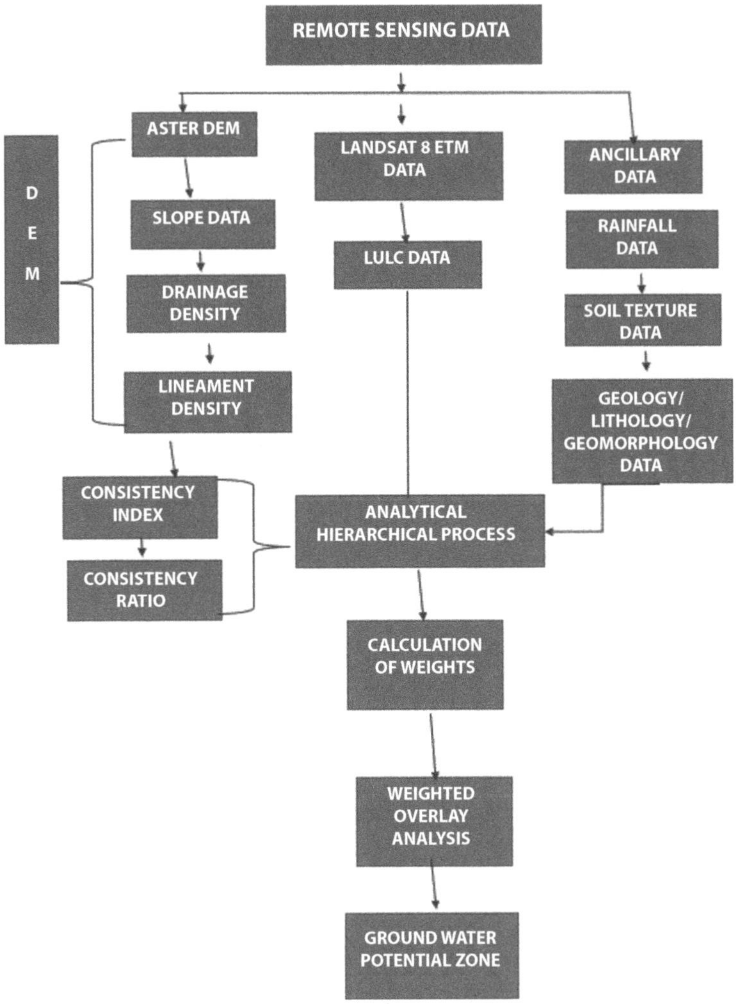

Figure 20.2 Methodology of groundwater potential zone using AHP technique.

20.9.2 Geomorphology

Some of the lithological units found in the study area of Pisangan watershed are phylite, shist, slate, granite, and gneissic. "Geomorphology is the study of landscapes. It entails the systematic description of landforms and the analysis of the processes that create them." The map as shown in Figure 20.3 describes the geomorphological units available in the study area of Pisangan Watershed shown in Table 20.3. Denudational hills, structural hills, eolian plain, alluvial plain, pediments, and pediplains are some of the geomorphological units found in the study area.

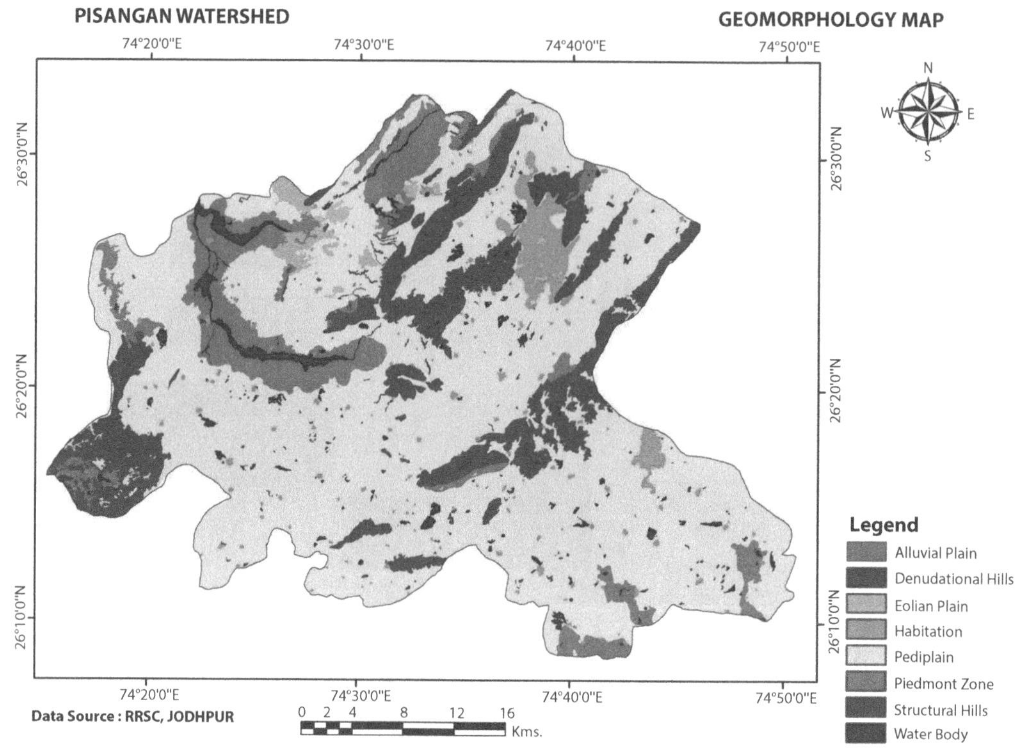

Figure 20.3 Geomorphology of the study area.

Table 20.3 Geomorphological units available in the study area of Pisangan Watershed.

S. no	Name	Area (km^2)
1	Alluvial plain	102.25
2	Structural hills	132.05
3	Denudation hills	90.08
4	Pediplain	1024.65
5	Eolian plain	20.48
6	Pediment zone	10.52
	Total	1379.38

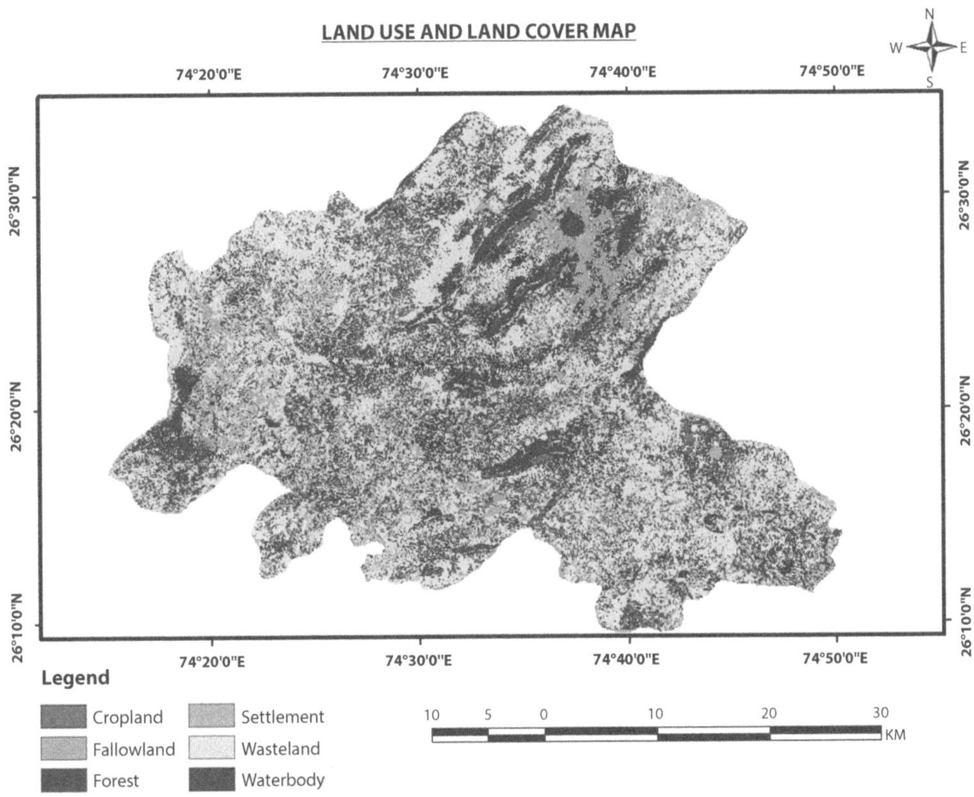

Figure 20.4 LULC map of the study area.

20.9.3 Land Use/Land Cover

The Figure 20.4 shows the comprehensive information on the spatial distribution of land use/land cover categories and the pattern of their change is a prerequisite for management and utilization of the land resources of the study area. There are six classes identified in the area: settlement, cropland, fallow land, forest, wasteland, and water bodies/rivers (Figure 20.4).

20.9.4 Lineament Density

Lineaments are structurally controlled linear or curvilinear features, which are recognized from the satellite imagery by their relatively linear alignments. These features show the surface topography of the underlying structural features. Lineaments represent the zones of faulting and fracturing resulting in increased secondary porosity and permeability. These factors are hydrogeologically very important as they provide the path ways for groundwater movement.

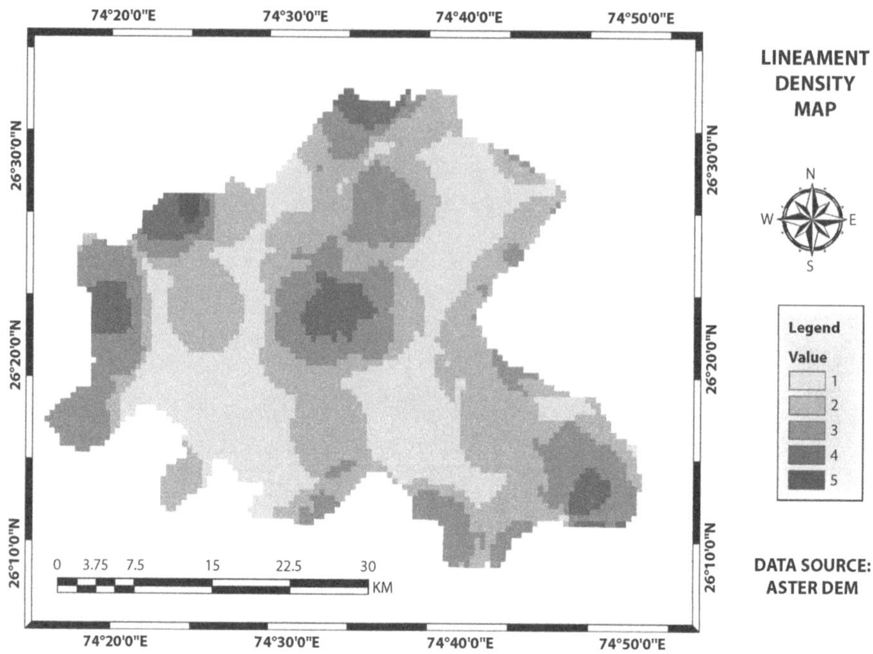

Figure 20.5 Lineament map of the study area.

Lineament density of an area can indirectly reveal the groundwater potential since the presence of lineaments usually denotes a permeable zone. The lineament density map of the study area reveals that high lineament density is observed in the center of the study area with a value ranging from 0 to 2.42 km/km^2 (Figure 20.5).

20.9.5 Drainage Density

Drainage density specifies the closeness of the spacing of stream channels and can be calculated as the total length of all the streams and rivers in the watershed divided by the total area of drainage watershed. The drainage density has an inverse relationship with groundwater prospect. The higher the drainage density is, the lower the probability of groundwater potential zone. Hence, the kernel density method in Arc GIS was carried out to calculate drainage density. The drainage network within the area was delineated from the DEM and updated from the satellite image. Hence, the kernel density method in Arc GIS was carried out to calculate drainage density using the formula as given below and the map shows that most of the study area is covered by moderate to low drainage density that refers to more infiltration and recharge to the groundwater (Figure 20.6).

$$DD = \blacktriangleleft \sum \blacktriangleright \sum \frac{LWS}{AWS}$$

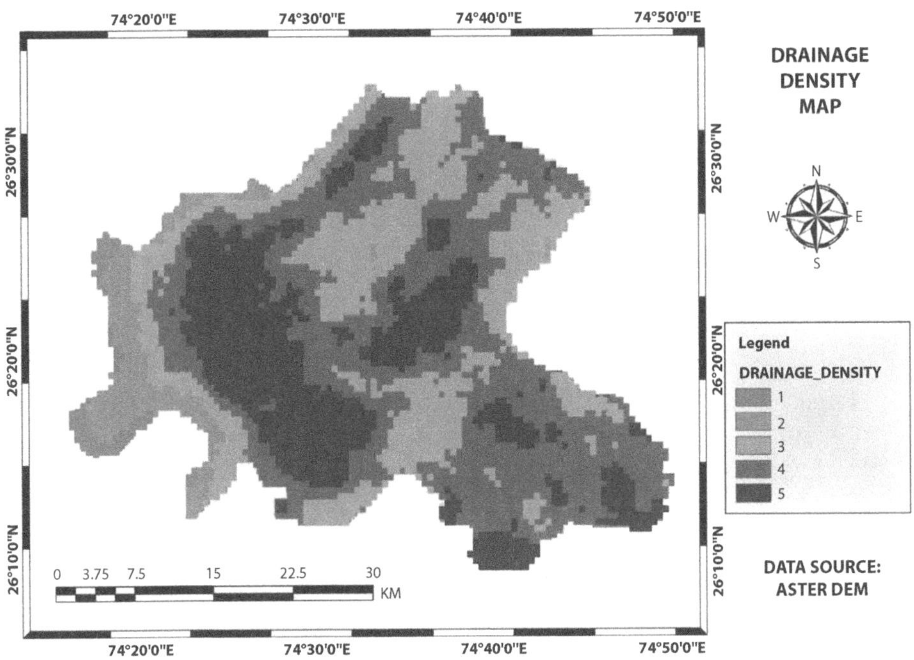

Figure 20.6 Drainage density map of the study area.

20.9.6 Slope

Slope is an important factor for the identification of groundwater potential zones. A higher degree of slope results in rapid runoff and increased erosion rate with feeble recharge potential. The slope map of the study area was prepared based on ASTER data using the spatial analysis tool in ArcMap 10.2. Slope grid is identified as "the maximum rate of change in value from each cell to its neighbors" (Burrough, 1986). Based on the slope, the study area can be divided into four slope classes. The areas having 0°–10° slope fall into the 'excellent' category because of the nearly flat terrain and relatively high infiltration rate. The areas with 10°–17° slope is considered as 'good' for groundwater storage due to slightly undulating topography with some runoff. The areas having a slope of 17°–55° cause relatively high runoff and low infiltration, and hence, are categorized as 'fine,' and the areas having a slope >190° are considered as 'poor' due to higher slope and runoff (Figure 20.7).

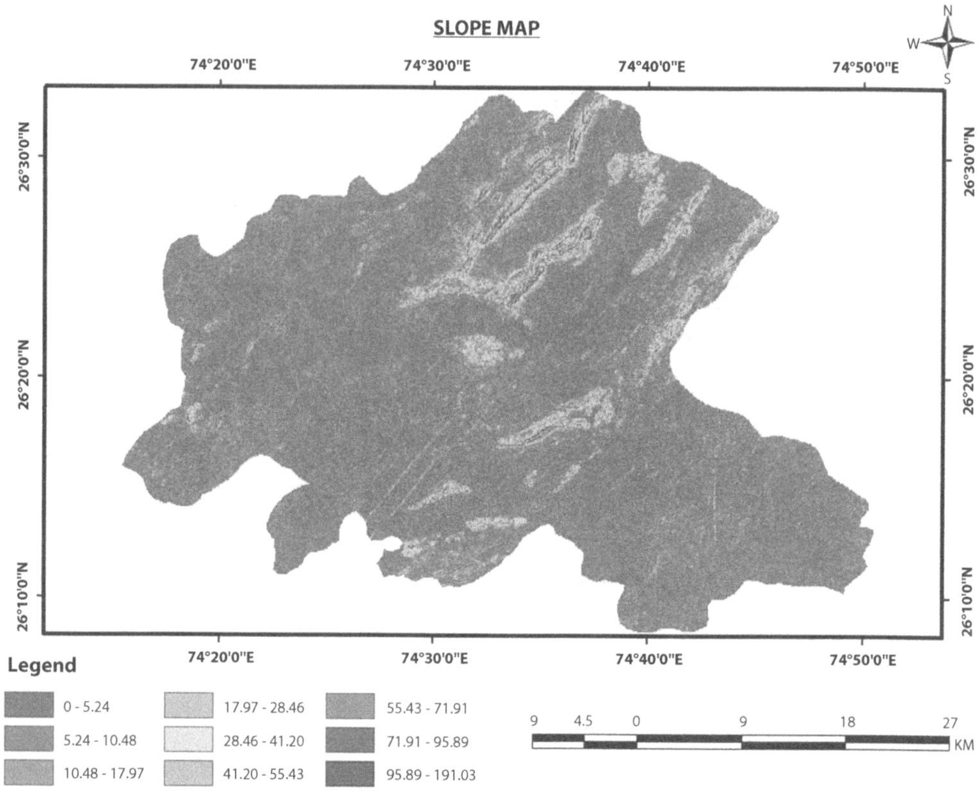

Figure 20.7 Drainage density map of the study area.

20.9.7 Soil

Soil is an important factor for delineating the groundwater potential zones. The analysis of the soil type reveals that the study area is predominantly covered by various soils. The watershed under study has 6 types of soil, *viz.*, sand, clay loam, loam, sandy loam, loamy sand, and rock outcrops (Figure 20.8).

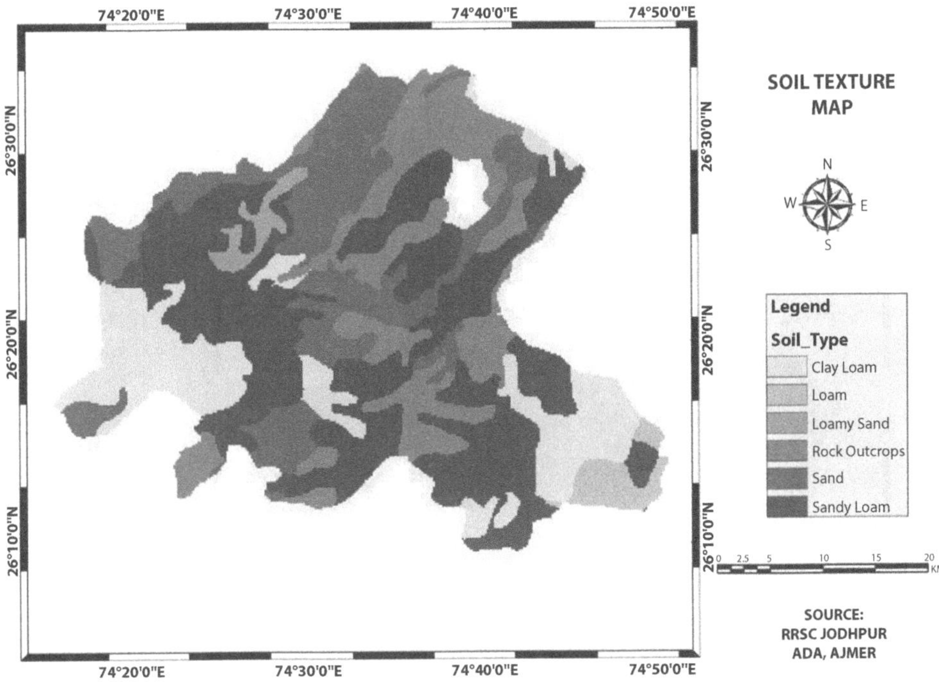

Figure 20.8 Soil texture map of the study area.

20.9.8 Rainfall

The study area depends mainly on southwest monsoon rains, which are brought by the troughs of low pressure establishing in the south Bay of Bengal between June to August. Rainfall distribution along with the slope gradient directly affects the infiltration rate of runoff water, hence, increases the possibility of groundwater potential zones. To determine the ground water potential zone, the average of annual historic rainfall event (10 y) was collected from five meteorological stations located within and near the study area. Using spatial analyst extension in ArcGIS 10.2, IDW interpolation was done to generate an estimated surface from these scattered set of point data (Figure 20.9).

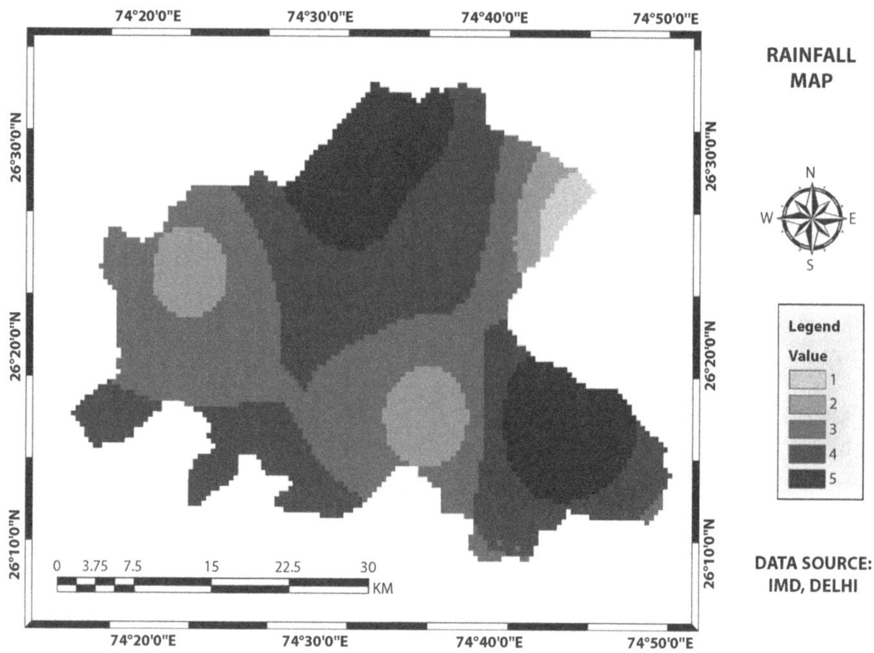

Figure 20.9 Rainfall map of the study area in mm.

20.9.9 Digital Elevation Model (DEM)

Any digital representation of the continuous variation of relief over space is known as a digital elevation model (DEM). The DEM is derived from ASTER Data. The DEM of Pisangan Watershed shows the area's elevation (height) at a class of 352–428 m, 428–475 m, 475–545 m, 545–657 m, and 657–888 m, respectively. The accuracy in environmental modeling (*e.g.*, watershed delineation) depends very much on how "truthful" the DEM is to the actual terrain surface (Figure 20.10).

20.9.10 Delineating the Groundwater Potential Zone

The groundwater potential zones for the study area were generated with the help of the integration of various thematic maps, *viz.*, drainage, slope, lithology, soil, lineament, rainfall, and land use, using remote sensing and GIS techniques. The demarcation of groundwater potential zones for the study area was made by pairing the interpreted layers through weighted multi-influencing factor and finally, assigned different potential zones. The groundwater potential zone of this study area can be divided into four grades, *viz.*, excellent, good, fine, and poor. The groundwater potential map determines that the excellent groundwater potential zone is concentrated in the southeastern and north-central region of the study area due to the distribution of alluvial plains and agricultural land with high infiltration ability. This indicates that soil type and slope play an important role in groundwater growth.

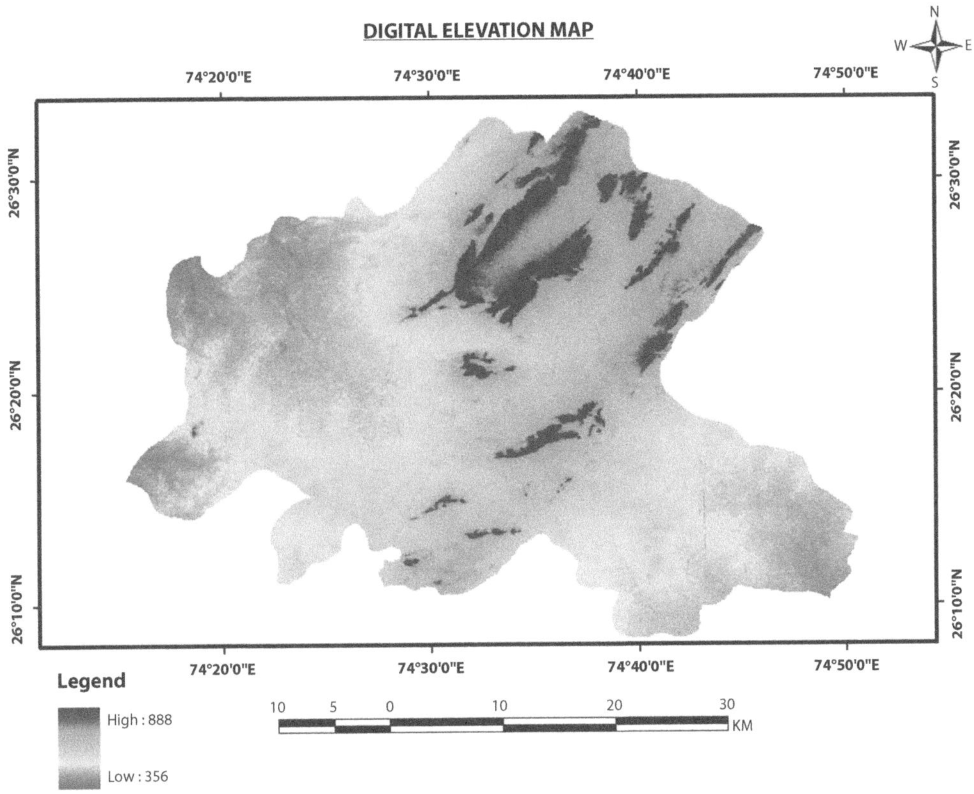

Figure 20.10 Digital elevation map of the study area in metres.

Moreover, the concentration of drainage density and lineament density also helps the infiltration ability of the groundwater system. About 0.92% of the total area falls under the "poor" zone, 27.04% falls under the "fair" zone, 69.28% falls under the "good" groundwater potential zone, and 03.60% of the study area falls under "very good" zone. Finally, the cumulative effect of the weighted multi-influencing factors through overlay analysis in GIS platform discovered the mapping of groundwater potential zones in the study area.

20.9.11 Overlay Analysis for the Delineation of Groundwater Potential Zone

The overlay analysis of the seven thematic layers was carried out in ArcGIS 10.2 as a summation of the real influence of the factors based on their criterion weight to generate the final potential zone. To arrive at these, all the layers were converted into an ultimate raster while their units were ranked. The groundwater potential zone was calculated based on Equation 20.1:

$$\mathrm{GWPI} = \sum_{i=1}^{n=7} W_i(R_i)$$

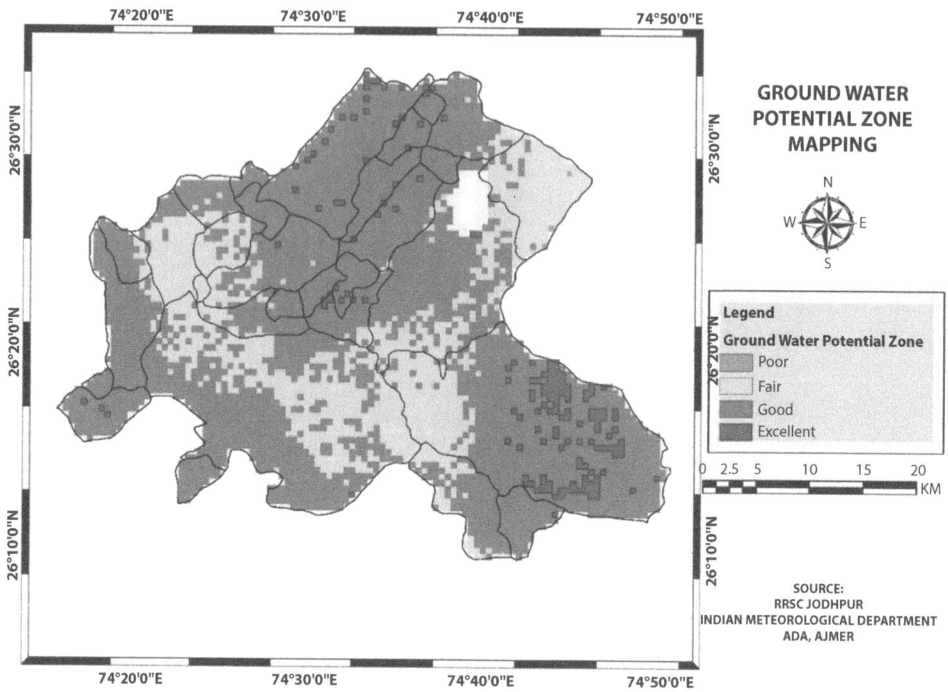

Figure 20.11 Ground water potential map of the study area.

where GWPI is the groundwater potential index, Wi is the criteria weight, Ri is the ranking of parameter factors, and i denotes each of the seven influencing factors with serial number from 1 to 7 (Figure 20.11).

20.10 Conclusion

Delineating the groundwater potential zones in the Theni district of Tamil Nadu using remote sensing, GIS, and MIF techniques is found to be efficient in minimizing the time, labor, and money, and thereby enables quick decision-making for sustainable water resources management. Satellite imageries, topographic maps, and conventional data were used to prepare the thematic layers of lithology, lineament density, drainage density, slope, soil, land use, and rainfall. The various thematic layers are assigned proper weightage through MIF technique and then integrated in the GIS environment to prepare the groundwater potential zone map of the study area. The groundwater potential map was produced by the weighted index overlay model (WIOM) in GIS. The groundwater prospects were mapped in four categories: excellent, good, fair, and poor zone. The excellent zones cover 50.20 km^2 (03.60%) of the total size of the study area. While the good, fair, and poor zones cover 965.65 km^2 (69.28%), 377.02 km^2 (27.04%), and 0.92 km^2 (0.06%), respectively. Excellent and good groundwater promising zones occur in the Pisangan Watershed which is very good for irrigation and agricultural activities.

The groundwater potential zones mapping using remote sensing and GIS techniques is a quick, reasonable, accurate/precise, and has large areal coverage. Thus, it provides a range by which the most possible zones within an area where groundwater occurrence is predictable.

References

Gulati, B., Sharma, R., Kanga, S., Singh, S. K., Sajan, B., Meraj, G., Kumar, P., & Ramanathan, A. L. (2023). Unraveling the Relationship Between Stubble Burning and Air Quality Degradation in Punjab: A Temporal and Spatial Analysis (2019-2022). *J. Clim. Change*, 9(2), 43-53.

Kanga, S., Meraj, G., Farooq, M., Singh, S. K., & Nathawat, M. S. (Eds.). (2022a). Disaster Management in the Complex Himalayan Terrains. *Geography of the Physical Environment*. Springer, Cham.

Kanga, S., Singh, S. K., Meraj, G., Kumar, A., Parveen, R., Kranjčić, N., & Đurin, B. (2022b). Assessment of the Impact of Urbanization on Geoenvironmental Settings Using Geospatial Techniques: A Study of Panchkula District, Haryana. *Geographies*, 2(1), 1-10. ISSN No. 2673-7086.

Kotecha, M. J., Kanga, S., Pankhaniya, S. K., Agrawal, S., Meraj, G., & Singh, S. K. (2023). Urban Heat Island (UHI) Resilience Plan in Varying Climatic Conditions Using Geospatial Approach: A Case Study of Rajkot City. In *Climate Change, Agriculture and Society: Approaches Toward Sustainability* (pp. 307-338). Cham: Springer International Publishing.

Lahon, D., Sahariah, D., Debnath, J., Nath, N., Meraj, G., Farooq, M., Kanga, S., Singh, S. K., & Chand, K. (2023a). Growth of water hyacinth biomass and its impact on the floristic composition of aquatic plants in a wetland ecosystem of the Brahmaputra floodplain of Assam, India. *PeerJ*, 1-32. http://dx.doi.org/10.7717/peerj.14811

Lahon, D., Sahariah, D., Debnath, J., Nath, N., Meraj, G., Kumar, P., Hashimoto, S. and Farooq, M. (2023b). Assessment of Ecosystem Service Value in Response to LULC Changes Using Geospatial Techniques: A Case Study in the Merbil Wetland of the Brahmaputra Valley, Assam, India. *ISPRS Int. J. Geo-Inf.*, 12(4), p.165.

Lavane, K., Kumar, P., Meraj, G., Han, T. G., Ngan, L. H. B., Lien, B. T. B., Ty, T. V., Thanh, N. T., Downes, N. K., Nam, N. D. G., Minh, H. V. T., Singh, S. K., & Kanga, S. (2023a). Assessing the Effects of Drought on Rice Yields in the Mekong Delta. *Climate*, 11(1), 13.

Meraj, G., Farooq, M., Singh, S. K., Islam, M. N., & Kanga, S. (2022). Modeling the sediment retention and ecosystem provisioning services in the Kashmir valley, India, Western Himalayas. *Model. Earth Syst. Environ.*, 8(3), 3859-3884.

Meraj, G., Kanga, S., Kranjčić, N., Đurin, B., & Singh, S. K. (2021a). Role of natural capital economics for sustainable management of earth resources. *Earth*, 2(3), 622-634.

Meraj, G., Singh, S. K., Kanga, S., Islam, M. N., & Singh, S. K. (2021b). Modeling on comparison of ecosystem services concepts, tools, methods and their ecological-economic implications: A review. *Model. Earth Syst. Environ.*, 1-20.

Nath, N., Sahariah, D., Meraj, G., Debnath, J., Kumar, P., Lahon, D., Chand, K., Farooq, M., Chandan, P., Singh, S. K., & Kanga, S. (2023). Land Use and Land Cover Change Monitoring and Prediction of a UNESCO World Heritage Site: Kaziranga Eco-Sensitive Zone Using Cellular Automata-Markov Model. *Land*, 12(1), 151.

Rafiq, M., Meraj, G., Kesarkar, A. P., Farooq, M., Singh, S. K., Kanga, S., & Bhardwaj, V. (2022). Hazard Mitigation and Climate Change in the Himalayas–Policy and Decision Making. In S. Kanga, G. Meraj, M. Farooq, S. K. Singh, & M. S. Nathawat (Eds.), Disaster Management in the Complex Himalayan Terrains. *Geography of the Physical Environment*. Springer, Cham. https://doi.org/10.1007/978-3-030-89308-8_12.

Sajan, B., Mishra, V. N., Kanga, S., Meraj, G., Singh, S. K., & Kumar, P. (2022). Cellular automata-based artificial neural network model for spatial-temporal dynamics and land use/land cover change prediction. *Agronomy*, 12, 2772.

Singh, A., Singh, S. K., Meraj, G., Kanga, S., Farooq, M., Kranjcic, N., Đurin, B., & Sudhanshu. (2022b). Designing Geographic Information System Based Property Tax Assessment in India. *Smart Cities*, 5, 364–381.

Singh, H., Meraj, G., Singh, S., Shrivastava, V., Sharma, V., Farooq, M., Kanga, S., & Kumar, P. (2022a). Status of Air Pollution during COVID-19-Induced Lockdown in Delhi, India. *Atmosphere*, 13(12), 2090.

Singh, S. K., Gupta, S. K., Meraj, G., Kumar, P., & Kanga, S. (2023b). Uncovering the hydro-meteorological drivers responsible for forest fires utilizing geospatial techniques. *Theor. Appl. Climatol.*, https://doi.org/10.1007/s00704-023-04497-y

Singh, S. K., Meraj, G., Mondal, N., Bera, A., Verma, M. K., Tomar, J. S., & Kanga, S. (2022). Assessment of seasonal vegetation dynamics over parts of thar desert using geospatial techniques. *ANGRAU*, 105.

Singh, S., Singh, S. K., Prajapat, D. K., Pandey, V., Kanga, S., Kumar, P., & Meraj, G. (2023a). Assessing the Impact of the 2004 Indian Ocean Tsunami on South Andaman's Coastal Shoreline: A Geospatial Analysis of Erosion and Accretion Patterns. *J. Mar. Sci. Eng.*, 11(6), 1134.

Sud, A., Kanga, R., Singh, S. K., Meraj, G., Kanga, S., Kumar, P., Ramanathan, A. L., & Bhardwaj, V. (2023). Simulating Groundwater Potential Zones in Mountainous Indian Himalayas—A Case Study of Himachal Pradesh. *J. Geophys. Res.: Earth Surf.*, 10(3), 65.

Upadhyay, R. K., Tripathi, G., Đurin, B., Šamanović, S., Cetl, V., Kishore, N., Sharma, M., Singh, S. K., Kanga, S., Wasim, M., Rai, P. K., & Bhardwaj, V. (2023). Groundwater Potential Zone Mapping in the Ghaggar River Basin, North-West India, Using Integrated Remote Sensing and GIS Techniques. *Water*, 15(5), 961.

Index

Page numbers followed by f and t indicate figures and tables, respectively.

Also of Interest

Check out these related titles from Scrivener Publishing

LAND AND ENVIRONMENTAL MANAGEMENT THROUGH FORESTRY, Edited by Abhishek Raj, Manoj Kumar Jhariya, Arnab Banerjee, Sharad Nema, and Kiran Bargali, ISBN 9781119910404. Written and edited by a group of experts in the field, this groundbreaking reference work sets the standard for engineers, students, and professionals working in forestry, agriculture, ecology and environmental science, offering the scientific community a way towards combating climate change.

GREEN CHEMISTRY FOR ENVIRONMENTAL REMEDIATION, Edited by Rashmi Sanghi and Vandana Singh, ISBN: 9780470943083. Presents high quality research papers as well as in depth review articles on the new emerging green face of multidimensional environmental chemistry.

BIOREMEDIATION AND SUSTAINABILITY: Research and Applications, Edited by Romeela Mohee and Ackmez Mudhoo, ISBN: 9781118062845. Bioremediation and Sustainability is an up-to-date and comprehensive treatment of research and applications for some of the most important low-cost, "green," emerging technologies in chemical and environmental engineering.

EMERGENCY RESPONSE MANAGEMENT OF OFFSHORE OIL SPILLS: Guidelines for Emergency Responders, By Nicholas P. Cheremisinoff, PhD, and Anton Davletshin, ISBN: 9780470927120. The first book to examine the Deepwater Horizon disaster and offer processes for safety and environmental protection.

COMPLEX BIOLOGICAL SYSTEMS: Adaptation and Tolerance to Extreme Environments, By Irina R. Fomina, Karl Y. Biel, and Vladislav G. Soukhovolsky, ISBN: 9781119510406. Written and edited by some of the most well-respected authors in the area of the adaptation of plants and animals to climate change, this groundbreaking new work is an extremely important scientific contribution to the study of global warming.

GLOBAL CLIMATE CHANGE DEMYSTIFIED, By James G. Speight, ISBN: 9781119653851. Tackling one of the most controversial subjects of our time, one of the world's foremost environmental and petroleum engineers explores the potential causes and ramifications of global climate change.

DEATH OF THE CHESAPEAKE: A History of the Military Role in Polluting the Bay, By Richard D. Albright, ISBN: 9781118686270. This unique book focuses attention on the failure of current efforts to cleanup the Chesapeake Bay and suggests an approach often used in cleaning up environmentally damaged sites.